BASIC
CONCEPTS
IN ALGORITHMS

BASIC
CONCEPTS
IN ALGORITHMS

Shmuel Tomi Klein

Bar Ilan University, Israel

World Scientific

NEW JERSEY · LONDON · SINGAPORE · BEIJING · SHANGHAI · HONG KONG · TAIPEI · CHENNAI · TOKYO

Published by

World Scientific Publishing Co. Pte. Ltd.

5 Toh Tuck Link, Singapore 596224

USA office: 27 Warren Street, Suite 401-402, Hackensack, NJ 07601

UK office: 57 Shelton Street, Covent Garden, London WC2H 9HE

Library of Congress Cataloging-in-Publication Data
Names: Klein, Shmuel T., author.
Title: Basic concepts in algorithms / Shmuel Tomi Klein, Bar Ilan University, Israel.
Description: Hoboken : World Scientific, [2021] | Includes bibliographical references and index.
Identifiers: LCCN 2021012980 | ISBN 9789811237584 (hardcover) |
    ISBN 9789811238529 (paperback) | ISBN 9789811237591 (ebook other) |
    ISBN 9789811237607 (ebook other)
Subjects: LCSH: Algorithms.
Classification: LCC QA9.58 .K54 2021 | DDC 518/.1--dc23
LC record available at https://lccn.loc.gov/2021012980

British Library Cataloguing-in-Publication Data
A catalogue record for this book is available from the British Library.

Cover image and design concept by Shoshanit Katz

For any available supplementary material, please visit
https://www.worldscientific.com/worldscibooks/10.1142/12298#t=suppl

Desk Editor: Amanda Yun

dedicated to

my spouse　　Rina

and our grandchildren

Ela, Itamar, Amichai and Boaz

Daniel, Lia, Elinadav, Michael, Ye'ela, Akiva and Emmanuelle

Bnaya, Ronny, Be'eri and Achinoam

Noga and Noam

Contents

Numerical Algorithms   217

List of Background Concepts

List of Algorithms

Preface

Every programmer is faced with a double challenge. There is, on the one hand, a need to master some programming language to overcome the technical hurdle of conveying our intentions to a machine that should be able to execute a well-defined sequence of commands. This aspect of *how* to write programs is taught in programming courses, generally at the very beginning of a Computer Science curriculum. Another facet of the problem is, however, the crucial issue of *what* actually should be done to accomplish a specific task. The necessary expertise to answer this question is acquired in the study of the theoretical sides of the art of writing computer programs. In most universities, the pertinent study fields include data structures, algorithms, complexity and computability, which should be learned in parallel to improving practical programming skills.

The present work is a direct extension of the topics dealt with in my first book, *Basic Concepts in Data Structures*, which appeared in 2016 and focused on the fundamental programming building blocks like queues, stacks, trees, hash tables and the like. I now turn to more advanced paradigms and methods combining these blocks and amplifying their usefulness in the derivation of *algorithms*, including the main steps of their design and an analysis of their performance.

The book is the result of several decades of teaching experience in data structures and algorithms. In particular, I have taught introductory and more advanced courses on algorithms more than fifty times, at Bar Ilan University and at several colleges, sometimes with up to five sessions given in the same semester. The book is self-contained but does assume some prior knowledge of data structures, and a grasp of basic programming and mathematics tools, which are generally earned at the very beginning of computer science or other related studies. In my university, a first course of algorithms is given in the first semester of the second year of the BSc program, with a prerequisite of *Discrete Mathematics*, *Introduction to Programming* and *Data Structures*, which are first-year courses. The format is three hours of lectures plus two hours of exercises, given by a teaching assistant, per week.

I have tried to reproduce my oral teaching style in writing. I believe in the didactic benefits of associative learning, in which one topic leads to another related one. Though the attention may be diverted from the central subject currently treated, it is the cumulative impact of an entire section or chapter that matters. There was no intention to compile a comprehensive compendium of all there is to know about algorithms; many relevant subfields like parallel algorithms, linear programming, computational geometry or quantum algorithms have not been mentioned. The choice of topics reflects, as usual, the personal taste of the author but consists of what many could agree to be the basic ingredients and major cornerstones of a body of knowledge on algorithms. A first one-semester course on algorithms, as given at Bar Ilan University, includes the contents of Chapters 1–5, 7 and 8, whereas a second course, given in the fourth semester, is based on Chapters 6 and 9–11. More advanced courses focusing on data compression and text algorithms are based on Chapters 6 and 7.

Each chapter comes with its own set of exercises, many of which have appeared in written exams. Solutions to most exercises appear in the appended section at the end of the book, on pages 315–332. To enhance comprehension, the reader is urged to try solving the problems before checking the suggested answers, some of which include additional insights. There are short inserts treating some background concepts: they are slightly indented, set in another font and separated from the main text by rules. Their list appears on page xi. Even though each chapter may be read and understood on its own, in spite of occasional pointers to earlier material, the book has been written with the intent of being read sequentially.

Most of the algorithms are also given formally in pseudo-code. I tried to restrict formal language to the bare minimum and did not adhere to any standard programming style. Assignments appear as left arrows ← rather than the ambiguous equal signs =, and begin–end pairs or brackets { } often used for grouping have been replaced by indenting. An exhaustive list of these algorithms appears on pages xiii–xv, just preceding this preface.

It will not be possible to mention all those I am indebted to for this project. Foremost, I owe all I know to the continuous efforts of my late parents to offer me, from childhood on, the best possible education in every domain. This also included private lessons, and I am grateful to my teacher R. Gedalya Stein, who interspersed his Talmud lessons with short flashes to notions of grammar, history, and more, and thereby set the seeds of the associative teaching techniques I adopted later. Most of what I learned as a child and teenager should be credited to my siblings Robi and Shirel Klein, in as diverse disciplines as Math, French, Hebrew, Talmud, soccer and guitar playing, as well as music and stamp collecting. My brother was the first to introduce me to computers, and he convinced me to share his enthusiasm in the early days of data processing. Another dominant tutor was Eli Reichental, whose informal lessons spanned a myriad of interests, from chess and history to the preparation for my bar mitzvah.

Many of my teachers at the Hebrew University of Jerusalem and at the Weizmann Institute of Science in Rehovot, as well as my colleagues at Bar Ilan University and elsewhere, had an impact on my academic career. I wish to especially thank my PhD advisor Aviezri Fraenkel, who even resigned from his partial job at Bar Ilan, freeing up a position that enabled them to hire me after my return from a postdoc in the US. I worked in 1981 at the Responsa Project, then headed by the late Yaacov Choueka, whose advice and guidance were invaluable. For the present work, I am much obliged to Dana Shapira for her thorough proofreading and constructive criticism.

Last but not least, I wish to thank my spouse and children for their ongoing encouragement and illuminating comments during the whole writing period. My grandchildren are still too young to understand what all this is good for, but their love and empathy fill our days with continuous enjoyment, so this book is dedicated to them.

Most of the work on this book was done during the Coronavirus lockdowns of 2020, and the last chapter was finished the day I received my second vaccine shot. May this coincidence symbolize that the conclusion of this writing project also announces the end of the pandemic.

Shmuel Tomi Klein Rehovot, January 2021

PART 1
Recursion

Chapter 1

Divide and Conquer

1.1 Minimum and maximum

In order to build up our knowledge incrementally, we shall start with some
very simple examples that will lead us gradually to the introduction of
the concepts of this chapter. Consider an array A of n elements. These
elements can be of any nature, as long as they are well defined and can
be compared to each other, but it will be convenient to think of them as
integers, unless otherwise specified. The task is to find a minimal element
in A, that is, an element $A[m]$ such that

$$A[m] \leq A[i] \qquad \text{for all} \quad 1 \leq i \leq n.$$

The obvious way to deal with this problem is to use a variable min, initialized by, say, the first element of the array, and then compare the value of
min with the other elements, storing in each iteration the minimal value
seen so far. Formally, this is shown in ALGORITHM 1.1.

Find-minimum-1(A, n)

> $min \leftarrow A[1]$
> for $i \leftarrow 2$ to n
>   if $min > A[i]$ then $min \leftarrow A[i]$

ALGORITHM 1.1: *Finding a minimum in an array A by linear search.*

The number of comparisons is $n - 1$, and we shall ask ourselves if one
can do better. One could, for example, try to process the elements in a
hierarchic structure, repeatedly comparing pairs. This could be used to

3

find first the minima of pairs, then of quadruples, then of 8-tuples, etc. We shall assume that the array A may be modified so that the minima of these pairs, quadruples, etc., may be stored in their respective leftmost elements. For the ease of description, let the size of the array be a power of 2, $n = 2^m$, which yields ALGORITHM 1.2.

Find-minimum-2(A, n)

---

  for $i \leftarrow 0$ to $m - 1$
    for $j \leftarrow 1$ to $n - 2^{i+1} + 1$ step 2^{i+1}
      $A[j] \leftarrow \min(A[j], A[j + 2^i])$

ALGORITHM 1.2: *Finding a minimum in an array A by hierarchical search.*

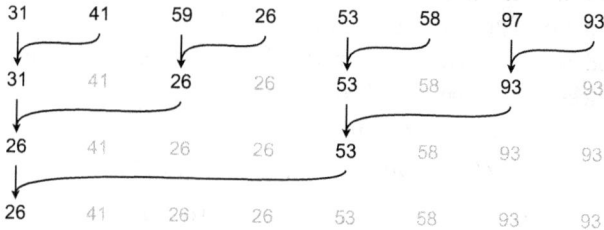

FIGURE 1.1: *Example of hierarchical search for a minimum.*

Figure 1.1 is an example of the hierarchical search and shows how the minimum value percolates to the leftmost position of the array in a logarithmic number of layers. To see if there is an improvement over the simple first algorithm, let us evaluate the number of comparisons in this case. It is

$$\frac{n}{2} + \frac{n}{4} + \cdots + 2 + 1 = n - 1,$$

that is, exactly the same as for ALGORITHM 1.1. Is this a coincidence?

In fact, any comparison in any of the algorithms to find a minimum will be between two elements, each of which can be either

- one of the original elements of the array or
- one of the results of a previous comparison.

Each of these algorithms can therefore be associated with a complete binary tree[1] with n leaves corresponding to the n elements; an internal node corresponds to the comparison of its two children nodes and contains the smaller of their two values. The minimum value is then stored in the root of the tree. The linear algorithm corresponds to a degenerate complete binary tree, as in Figure 1.2(a); the hierarchical algorithm corresponds to a full complete tree, as in Figure 1.2(b), which is equivalent to the tree in Figure 1.1; whereas Figure 1.2(c) depicts an arbitrary complete binary tree. However, for any complete binary tree with n leaves, the number of internal nodes, which is the number of comparisons, is equal to $n - 1$. We have to conclude that, no matter in which order the comparisons are performed, their number will always be $n - 1$, so the first, and simplest, algorithm mentioned above is just as good as any of the others.

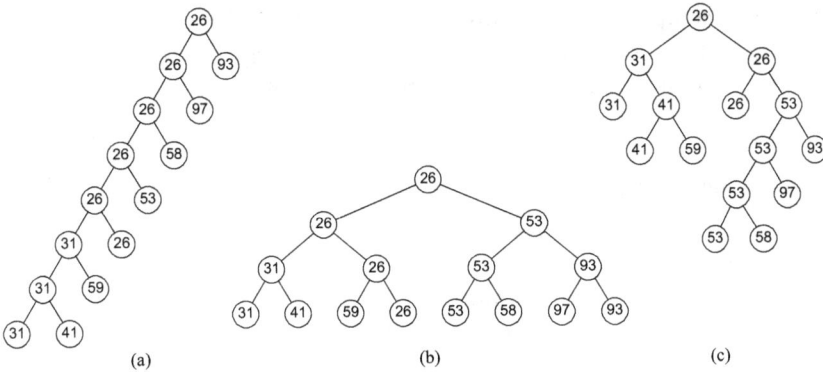

FIGURE 1.2: *Examples of complete binary trees for finding the minimum.*

After having dealt with the problem of finding the minimum, we turn to the following problem: that of finding the *maximum*. Many might now object that this is in fact the same problem as the previous one, while, obviously, it is not. Younger children, for instance, will generally correctly insist that the problems are different. The argument of the more advanced reader for claiming identity is that the problems are so similar that the necessary modifications to get from the one to the other (changing > to <)

---

[1]A *complete* binary tree is one for which every internal node has exactly two children; a *full* binary tree is one having all its leaves on the same level. All the trees in Figure 1.2 are complete, but only the one in Figure 1.2(b) is also full. Unfortunately, there is no consensus about these definitions, and in some texts, they are inversed.

are evident and often judged to be not even worth mentioning. In fact, this transformation from one problem into another is called a *reduction*, and is one of the basic notions and helpful tools in the study and design of algorithms. The reduction from the minimum to the maximum problem is very simple, and we shall encounter many others in this book, including some quite sophisticated ones, especially in Chapter 10, dealing with intractability.

As a consequence of the reduction, we know how to solve the problem of finding the maximum element by applying the algorithm Find-maximum derived from ALGORITHM 1.1 with the obvious amendments. Furthermore, we know that it will take $n - 1$ comparisons, and that this is optimal.

The following challenge is then to find both a minimal *and* a maximal element for a given array A. One can, of course, simply concatenate the two programs Find-minimum and Find-maximum and return the requested pair after $2n-2$ comparisons. Any programmer will realize that two independent loops are not necessary and that they can be merged, as in ALGORITHM 1.3, though this does not reduce the number $2n - 2$ of comparisons.

Find-min-max-1(A, n)

$\quad min \leftarrow A[1] \qquad max \leftarrow A[1]$
$\quad$ for $i \leftarrow 2$ to n
$\quad\quad$ if $min > A[i]$ then $min \leftarrow A[i]$
$\quad\quad$ if $max < A[i]$ then $max \leftarrow A[i]$

ALGORITHM 1.3: *Finding a minimum and maximum in an array A.*

Another technical improvement can be achieved by adding an else between the two if statements, yielding ALGORITHM 1.4, for if an element is smaller than the smallest seen so far, it cannot possibly be larger than the currently largest. Thus, if the condition in the first if is true, there is no need to check the second if, which must return false.

The number of comparisons in the revised algorithm could therefore be lower as some comparisons may be saved, yet in the worst case, for example, when the input array is sorted in increasing order, there are no savings and the number of comparisons is still $2n - 2$. The question is, if it is possible to devise an algorithm which returns both minimum and maximum in less than $2n - 2$ comparisons, even in the worst case.

Find-min-max-2(A, n)

$min \leftarrow A[1]$ $\qquad$ $max \leftarrow A[1]$
for $i \leftarrow 2$ to n
$\qquad$ if $min > A[i]$ then $min \leftarrow A[i]$
$\qquad$ else if $max < A[i]$ then $max \leftarrow A[i]$

ALGORITHM 1.4: *Better min–max algorithm.*

To show that the answer to this question is positive, we shall introduce a family of algorithms that has become known under the intriguing collective name of *Divide and Conquer*, the origins of which may be traced back to the Latin *divide et impera* (divide and rule), a motto attributed to several rulers, ranging from Philip of Macedonia and Julius Caesar to the Kings of the Habsburg dynasty. The idea was that the empire they had to reign over was very vast, as that shown in Figure 1.3. To ensure governance, the territory was split into sub-areas, the borders of which are indicated in Figure 1.3 by the broken lines; each area was assigned its own local administrator, represented by the numbers in circles, who reported directly to the central sovereign. Actually, the problem of these administrators was similar to that of their monarch, just on a smaller scale, so they often subdivided their sub-area into sub-sub-parts, and continued the ruling chain recursively.

The algorithmic equivalent of this paradigm considers a problem which has to be solved for an input of n elements. Since n could be quite large, the *Divide* part calls for splitting the problem into k sub-problems, with $k \geq 2$, often, but not necessarily, all with the same input size $\frac{n}{k}$. The essential feature of this procedure is that one strives to obtain sub-problems which are identical or at least very similar to the original problem, just with smaller input. This enables a recursive solution of the sub-problems. Finally, here, to *Conquer* means collect the solutions of the sub-problems and derive from them the solution of the original problem.

Returning to the earlier mentioned min–max problem, one may start, as will often be done, by splitting the input array A into two halves A_1 and A_2. Here and below, the notation $A[i : j]$ will be used for the sub-array of A starting at the element indexed i up to, and including, the element indexed j. For simplicity we continue to assume that the size n of the input

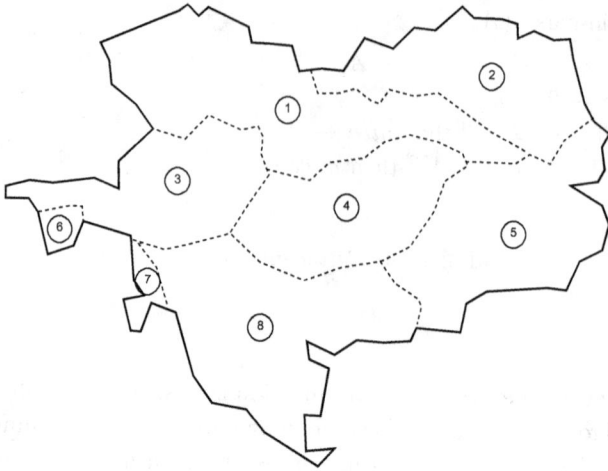

FIGURE 1.3: *Divide and Conquer.*

Find-min-max-3(A, n)

1　if $n = 2$ then
2　　if $A[1] < A[2]$ then return $(A[1], A[2])$
3　　else　　　　　　　　return $(A[2], A[1])$
4　else
5　　$A_1 \leftarrow A[1 : \frac{n}{2}]$　　　　$A_2 \leftarrow A[\frac{n}{2} + 1 : n]$
6　　$(m_1, M_1) \leftarrow$ Find-min-max-3$(A_1, \frac{n}{2})$
7　　$(m_2, M_2) \leftarrow$ Find-min-max-3$(A_2, \frac{n}{2})$
8　　return $\big(\min(m_1, m_2), \max(M_1, M_2)\big)$

ALGORITHM 1.5: *Divide and Conquer min–max algorithm.*

array A is a power of 2, but all the ideas are easily adapted to general n. Applying the procedure recursively on each of the halves yields the two min–max pairs (m_1, M_1) and (m_2, M_2). The sought minimum is then the smaller of m_1 and m_2, and the maximum is the larger of M_1 and M_2. The formal procedure is given in ALGORITHM 1.5 and illustrated in Figure 1.4.

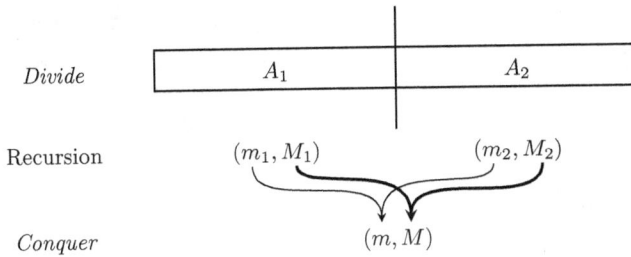

FIGURE 1.4: *Schematic view of the Divide and Conquer min–max algorithm.*

Note that as for all recursive programs, one has to take care of the boundary conditions. In our case, they are reached when the input array contains only two elements, in which case a single comparison suffices to extract both minimum and maximum.

Looking at Find-min-max-3, it is not at all obvious that some improvement has been achieved over the previous attempts to solve the problem. While one may appreciate the elegant partition into similar sub-problems, we know that recursive programs incur some price to pay, the procedure seems much more complicated, and the savings, if there are any, have to be evaluated, which is not trivial. The difficult part is to assess the contributions of the recursive calls, since they are still unknown when they need to be used. The following formal method helps to overcome this hurdle.

Define $T(n)$ as the number of comparisons used by Find-min-max-3 on an input array of size n. For the boundary condition, we have

$$T(n) = 1 \qquad \text{for} \quad n = 2. \tag{1.1}$$

For the general case $n > 2$, there are no comparisons in line 5, which just splits the array into halves. Line 6 is a recursive call, so we do not know how many comparisons are involved, since this is precisely what we are trying to evaluate; but we know how to *represent* the sought value symbolically. We have a recursive call to Find-min-max-3 with parameter A_1, which is an array of size $\frac{n}{2}$, the according number of comparisons is thus, by definition, $T(\frac{n}{2})$. The same is true for line 7, and the min and max functions in line 8 each contribute a single comparison. Summing, we get that

$$T(n) = 2\,T\!\left(\tfrac{n}{2}\right) + 2 \qquad \text{for} \quad n > 2. \tag{1.2}$$

The problem with an expression of the type in Eq. (1.2) is that $T(n)$ is not given explicitly in what is often called a *closed form*, but rather implicitly, since the function T appears also on the right-hand side of the equation.

One way to derive a closed form is by applying the equality of Eq. (1.2) repetitively. The identity is true for any $n > 2$, so also, in particular, for $\frac{n}{2}$, as long as this is still larger than 2. Substituting, we get

$$T(n) = 2\,T\left(\tfrac{n}{2}\right) + 2 = 2\left[2\,T\left(\tfrac{n}{4}\right) + 2\right] + 2 = 4\,T\left(\tfrac{n}{4}\right) + 4 + 2. \qquad (1.3)$$

At first sight, the right-hand side of Eq. (1.3) does not seem more appealing than that of Eq. (1.2). However, the parameter of the function T, $\frac{n}{4}$, is smaller than before, which suggests continuing with our efforts. Indeed, if there is a way to generate a sequence of correct equations, but with successively smaller parameters to the function T on their right-hand sides, we may at some stage reach the boundary condition and thereby escape from the recursive definition by getting rid of the appearance of T.

Applying therefore again the identity (1.2), this time for $\frac{n}{4}$ in Eq. (1.3), we get

$$T(n) = 4\left[2\,T\left(\tfrac{n}{8}\right) + 2\right] + 4 + 2 = 8\,T\left(\tfrac{n}{8}\right) + 8 + 4 + 2,$$

which suggests guessing that a general form of this equation might be

$$T(n) = 2^i\,T\left(\tfrac{n}{2^i}\right) + 2^i + \cdots + 4 + 2 \qquad (1.4)$$
$$= 2^i\,T\left(\tfrac{n}{2^i}\right) + 2^{i+1} - 2. \qquad (1.5)$$

Of course, a guess is always risky and we might err. The correct way to deal with it is by proving the guessed equation by induction. We shall show it only for this first time, but keep in mind that there could be more than one option to extend the first few elements of an unknown sequence.

To prove that Eq. (1.5) holds for all i, note that Eq. (1.2) is the special case $i = 1$. Assuming it holds for i, we show that it also holds for $i + 1$ by applying Eq. (1.2) once again:

$$T(n) = 2^i\left[2\,T\left(\tfrac{n/2^i}{2}\right) + 2\right] + 2^{i+1} - 2 = 2^{i+1}\,T\left(\tfrac{n}{2^{i+1}}\right) + 2^{i+2} - 2,$$

so we get a similar expression as in Eq. (1.5), where each occurrence of i has been replaced by $i + 1$. We may thus conclude that our guess was successful and that Eq. (1.5) holds for all appropriate values of i.

We are interested in getting a closed form for $T(n)$; hence may choose i large enough to let the parameter $\frac{n}{2^i}$ of T reach the border value 2. This yields $n = 2^{i+1}$. Substituting in Eq. (1.5), we get

$$T(n) = \tfrac{n}{2}T(2) + n - 2 = \tfrac{3}{2}n - 2.$$

The surprising result is that the Divide and Conquer strategy helped to reduce the number of required comparisons to not more than $1.5n$, even in the worst case. It should nevertheless be noted that a reduction from $2n$ to $1.5n$ is generally not really considered as an improvement, because our focus will be most of the time only on the *order of magnitude*, as we shall see. The example of Find-min-max-3 was only brought as a first simple illustration of the Divide and Conquer mechanism. In real applications, a simpler program without recursive calls may often run faster, even if there are more comparisons. Actually, it is not necessary to use Divide and Conquer to solve the problem in $1.5n$ comparisons (see Exercise 1.1).

The careful reader could ask why only comparisons have been counted in the above analyses. Other commands clearly also affect the execution time, so why are they ignored? The reason is essentially technical. Not only would it be very hard, if at all possible, to compile a complete inventory of all the commands to be executed within a program, but different commands require different times. For instance, multiplication is much slower than addition, yet still faster than division. Moreover, the execution times, and even their ratios, differ from one hardware platform to another and are constantly changing (generally, getting faster). Statements one could often read in reports such as *"this program takes x milliseconds to execute..."* become almost meaningless after a few years.

A fair comparison between different algorithms solving the same problem is therefore not obtained by means of a stopwatch, but rather by trying to compare the order of magnitude of the number of operations involved, as a function of the size of the input of the algorithm. Such a function is called the *time complexity*, or often simply *complexity*, of the given algorithm.

For many algorithms, and in particular for those presented earlier, the number of commands to be performed between two consecutive comparisons is either constant or bounded by a constant. The number of comparisons is therefore a good estimate for the complexity we are interested in. We shall also see examples of algorithms, for example in Section 1.3, where the basic operations we count are not comparisons.

Background concept: Asymptotic notation

The following notation is widely used in Computer Science, and will replace the rather vague formulations used above referring to the complexity of an algorithm being *of the order of* some function. The notation is generally used to evaluate the execution time required by some algorithm as a function of the size n of its input. For a given integer function g, we define the set $O(g)$, which is read as *big-O of g*, as the set of all integer functions f for which there exist constants n_0 and $C > 0$ such that

$$\forall n \geq n_0 \qquad |f(n)| \leq C\,|g(n)|.$$

This means that, roughly speaking, the function f is bounded above by g, but only approximately, since the bound may be relaxed as follows:

(1) what matters in our estimate is the asymptotic behavior of f, when its argument n tends to infinity, so the values of $f(n)$ for arguments smaller than some predefined constant n_0 are not relevant, and in particular we do not care if they are smaller or larger than $g(n)$;

(2) the requirement is not on the function f itself — it suffices that some constant $C > 0$ multiplied by f be smaller than g.

For example, $f(n) = n^3 + 99n \in O(n^3)$, as can be checked using the constants $n_0 = 10$ and $C = 2$.

Though $O(g)$ is defined as a set and one should write $f \in O(g)$ as in the example above, a prevalent abuse of notation refers to $O(g)$ as if it were a number, and one will often see it appear in equations like $f(n) = O(g(n))$. This is not really an equality; in particular, it is not transitive, as both $n^2 = O(n^3)$ and $n^2 = O(n^4)$ hold, but this does clearly not imply that $O(n^3) = O(n^4)$.

A similar notation exists for lower bounds: $\Omega(g)$, which is read as *big-omega of g*, is the set of all integer functions f for which there exist constants n_0 and $C > 0$ such that

$$\forall n \geq n_0 \qquad |f(n)| \geq C\,|g(n)|.$$

For example, $n^3 - n^2 + \frac{1}{n} \in \Omega(n^3)$.

For a more precise evaluation, a combination of both bounds may be used: $\theta(g)$, which is read as *theta of g*, is the set of all integer functions f for which both $f \in O(g)$ and $f \in \Omega(g)$, obviously with different constants, that is, there are constants n_0, $C_1 > 0$ and $C_2 > 0$ such that

$$\forall n \geq n_0 \qquad C_1\,|g(n)| \leq |f(n)| \leq C_2\,|g(n)|.$$

For example, $\sqrt{n} + \cos(n) \in \theta(\sqrt{n})$.

The asymptotic notation can also be extended to the reals with the argument tending to some constant, often 0, rather than to ∞.

Using the asymptotic notation, the solutions of the min–max problem by algorithms requiring $2n$ or $1.5n$ comparisons belong both to $O(n)$. The examples in the following sections show how the Divide and Conquer approach may also help to reduce the asymptotic complexity.

A final comment concerns the passage from Eq. (1.4) to Eq. (1.5), where the sum $2^i + \cdots + 4 + 2$ has been substituted by $2^{i+1} - 2$. Since similar identities are helpful in many instances, they are treated in the following insert.

Background concept: Simplifying summations

When coming to evaluate the complexity of algorithms, one often encounters implicit summation in which an ellipsis $(\cdots)$ is used to suggest that some general term is supposed to be inferred. Applying the following technique might help to get a closed form of such summations. As example, we shall work here on a finite *geometric progression*,

$$A_n = 1 + \alpha + \alpha^2 + \cdots + \alpha^n. \tag{1.6}$$

To get rid of the ellipsis, we aim at creating another identity that should be similar to, but different from, (1.6). The rationale of such an attempt is that, if we have similar equations, it might be possible to subtract them side by side and because of the similarity, many of the terms may cancel out. We might thereby derive a new equation which gives some useful information. The question is of course how to generate the other equation.

Ellipses are generally used in cases in which the reader may derive some regularity in the sequence of terms from the given context. This is similar to the number sequences that used to appear in IQ tests, in which one has to guess the next element. So the next term in $4, 5, 6, \cdots$ should apparently be 7, and $4 + 7 + 10 + \ldots$ is probably followed by 13. In our case, the regularity seems to be that each term is obtained by multiplying the preceding one by α. One may thus try to multiply the whole identity by α and get

$$\alpha A_n = \alpha + \alpha^2 + \alpha^3 \cdots + \alpha^{n+1}. \tag{1.7}$$

Subtracting both sides of (1.7) from their counterparts in (1.6), one gets

$$A_n - \alpha A_n = 1 - \alpha^{n+1},$$

an equation without ellipses. The left-hand side being $(1 - \alpha)A_n$, two cases have to be considered. If $\alpha = 1$, then the sum (1.6) reduces to

$A_n = 1 + 1 + \cdots + 1 = n + 1$. If $\alpha \neq 1$, we can divide by $(1 - \alpha)$ to get

$$A_n = \frac{1 - \alpha^{n+1}}{1 - \alpha}. \tag{1.8}$$

In particular, if $\alpha < 1$ then $\alpha^{n+1} \to 0$ when n tends to infinity and we get

$$\lim_{n \to \infty} A_n = \frac{1}{1 - \alpha}.$$

For example, for $\alpha = \frac{1}{3}$, one gets

$$1 + \frac{1}{3} + \frac{1}{9} + \cdots + \frac{1}{3^n} = \frac{1 - \frac{1}{3^{n+1}}}{1 - \frac{1}{3}} = \frac{3}{2} - \frac{1}{2 \cdot 3^n},$$

and for $\alpha = 2$, one gets

$$1 + 2 + 4 + \cdots + 2^n = \frac{1 - 2^{n+1}}{1 - 2} = 2^{n+1} - 1,$$

which has been used for the passage from (1.4) to (1.5).

1.2   Mergesort

The following example of the Divide and Conquer family deals with a sorting procedure known as *Mergesort*. The need to keep the elements of some set in sorted order is probably almost as old as mankind, but very simple methods, like *Insertion sort* or *Bubble sort* seemed to satisfy most needs. These methods require $\Omega(n^2)$ comparisons to sort n elements.

To sort an array $A[1:n]$ by Mergesort, we start similarly to what has been done for the min–max problem by splitting the input array into two halves. Mergesort is then applied recursively on the left and right halves of the array, that is on $A[1:\frac{n}{2}]$ and $A[\frac{n}{2}+1:n]$ (this is the *divide* phase); finally, a function merge(i, k, j) is invoked, which supposes that the sub-arrays $A[i:k]$ and $A[k+1:j]$ are already sorted, to produce a merged array $A[i:j]$ in the *conquer* phase.

The merging may be implemented in what we shall call a naïve approach with the help of an auxiliary array B, which temporarily stores the sorted output, before it is moved back to the array A. The formal definition in ALGORITHM 1.6 uses three pointers, two to the parts of the array that are to be merged, and the third to the output array $B[i:j]$. For convenience, the element following the last is defined by $A[j+1] \leftarrow \infty$, which facilitates the processing of the boundary condition in which the second of the sub-arrays is exhausted:

merge(i, k, j)
$\quad$ $p_1 \leftarrow i;$ $\qquad$ $p_2 \leftarrow k + 1;$ $\qquad$ $p_3 \leftarrow i$
$\quad$ while $p_3 \leq j$ do
$\qquad$ if $p_1 \leq k$ and $A[p_1] < A[p_2]$ then
$\qquad\qquad$ $B[p_3] \leftarrow A[p_1];$ $\qquad$ $p_1\text{++};$ $\qquad$ $p_3\text{++}$
$\qquad$ else $B[p_3] \leftarrow A[p_2];$ $\qquad$ $p_2\text{++};$ $\qquad$ $p_3\text{++}$
$\quad$ $A[i : j] \leftarrow B[i : j]$

ALGORITHM 1.6: *Naïve merge algorithm.*

Since after each comparison, the pointer p_3 is incremented by 1, the number of comparisons is clearly the size of the output array, that is $j-i+1$. Is this the best possible way to merge two sorted lists? We shall return to this and similar intriguing questions in a dedicated section at the end of this chapter. The formal definition of mergesort(i, j), which sorts the sub-array $A[i : j]$, is then given in ALGORITHM 1.7.

mergesort(i, j)
$\quad$ if $i < j$ then
$\qquad$ $k \leftarrow \lfloor (i + j)/2 \rfloor$
$\qquad$ mergesort(i, k)
$\qquad$ mergesort$(k + 1, j)$
$\qquad$ merge(i, k, j)

ALGORITHM 1.7: *Mergesort.*

The analysis is similar to the one we did above for min–max. Let $T(n)$ denote the number of comparisons needed to sort an array of size n by mergesort; then $T(1) = 0$, and for $n > 1$,

$$T(n) = 2\,T(\tfrac{n}{2}) + n. \tag{1.9}$$

This recursive definition is quite similar to the one we saw in Eq. (1.2) for the min–max problem, yet the small differences lead to different solutions. If one follows the derivation pattern used to get from Eqs. (1.2) to (1.5), the next equation is

$$T(n) = 4\,T(\tfrac{n}{4}) + 2n,$$

which could lead to the (false) guess that the general form would be

$$T(n) = 2^i\, T(n/2^i) + 2^{i-1}\, n, \qquad \text{for} \quad i = 1, 2, \ldots.$$

This example shows that a premature guess might be misleading, but is not really harmful since a wrong formula can, of course, not be proved by induction. The correct generalization in this case is

$$T(n) = 2^i\, T(n/2^i) + i\, n, \qquad \text{for} \quad i = 1, 2, \ldots. \tag{1.10}$$

We choose i large enough to get to the boundary condition $n/2^i = 1$, from which one can derive $i = \log_2 n$. Substituting in (1.10) then yields

$$T(n) = n\, T(1) + n \log_2 n = n \log_2 n.$$

While the min–max example gave only a reduction of the number of comparisons from $2n$ to $1.5n$, which is not really considered an improvement in terms of order of magnitude, lowering the complexity of sorting from $\Omega(n^2)$, required by most simple methods, to $O(n \log n)$ was a significant step forward. In the next section, we shall see that there is no guarantee for Divide and Conquer methods to achieve any improvement at all!

1.3   Multiplying large numbers

The reader should be intrigued by the title of this section. Indeed, given two input numbers X and Y, the calculation of their product has not been considered a noteworthy issue in our discussion so far, and the "algorithm"

return X * Y

seems to achieve the requested goal. What turns this apparently trivial task into a problem deserving a section on its own is the fact that the size of the involved numbers will not be limited.

The numbers appearing in many, possibly in most, algorithmic questions represent some measurable physical entities such as salaries, distances, weights, etc., and they therefore vary within some reasonable ranges. The 64 bits allotted by most modern computers to represent such numbers are generally more than enough for most of our tasks, being able to represent numbers as large as 16 billion billions, or equivalently, using 20 significant digits. There are, however, problems, in which a number X does not measure anything, but it is X itself that is the subject to be investigated. One such problem is that of *primality*, to be treated later in Chapter 5: given a number X, is X a prime number or not?

Why should one be interested whether a given number is prime? Is that not purely intellectual curiosity? It surely used to be in the many centuries the question was treated by mathematicians around the world. Only recently, in the 1970s, has it turned out that many methods of modern *cryptography* rely on the primality of the numbers involved, as we shall see in Chapter 9. Without the ability to identify large prime numbers, the secure use of the internet and any sort of e-commerce are hardly conceivable.

We are thus given two large numbers X and Y, each of length n bits. Assuming the same length for both does not limit generality, as one can always pad the shorter with leading zeros. The length n of the input strings is an unbounded parameter, and we shall, in particular, not assume that one may store the numbers in any fixed sized location, such as a register. Since any information, be it numbers, characters, instructions and the like, are ultimately converted into binary form to be stored in our computers, we shall consider all these representations as equivalent and use them interchangeably in this section, and several other places in this book. For example, the term Test could be represented, via its ASCII encoding, in any of the following forms:

character string	T e s t
binary string	01010100 01100101 01110011 01110100
decimal number	1,415,934,836

The algorithm for multiplying two numbers is possibly one of the earliest we have been taught in elementary school, and even if our notion of what it meant to be a *large* number was different at that time, the algorithm we are looking for is essentially the same. Figure 1.5 depicts the multiplication of 5708 by 1948 in the standard layout, using both the decimal and binary representations of these numbers: we multiply the upper number by each digit or bit of the lower number, write the results in consecutive rows, shifting each by one position further to the left than the preceding row, and finally add all these rows.

How should one evaluate the complexity of a multiplication algorithm? In the previous sections, we advocated the use of the number of comparisons as a good estimate, but here, there are clearly no comparisons involved. An adequate estimate in the present application could be the number of basic operations, that is, the addition or multiplication of two single bits or digits. If each of the numbers to be multiplied is of length n, be it decimal digits or binary bits, the required number of operations is clearly about n^2 single

$$
\begin{array}{r}
5708 \\
1948 \\
\hline
45664 \\
22832 \\
51372 \\
5708 \\
\hline
11119184
\end{array}
$$

$$
\begin{array}{r}
1011001001100 \\
11110011100 \\
\hline
0000000000000 \\
0000000000000 \\
1011001001100 \\
1011001001100 \\
1011001001100 \\
0000000000000 \\
0000000000000 \\
1011001001100 \\
1011001001100 \\
1011001001100 \\
1011001001100 \\
\hline
1010100110101010 01010 10000
\end{array}
$$

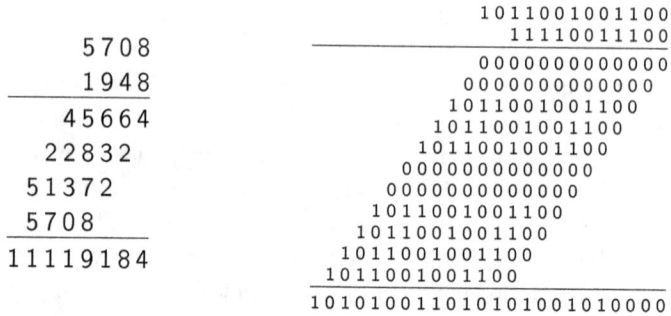

FIGURE 1.5: *Standard multiplication of* 5708 *by* 1948 *in decimal and binary.*

digit or bit multiplications, followed by about n^2 additions, together $\theta(n^2)$. For a simpler discussion, we shall henceforth assume that we are processing bits.

Let us try to apply the Divide and Conquer paradigm to this problem, with the aim of reducing its complexity. Given are $X = x_1 x_2 \cdots x_n$ and $Y = y_1 y_2 \cdots y_n$, where we assume that all $x_i, y_j \in \{0, 1\}$ are bits. Split each of the two strings into two halves and call them X_1, X_2, Y_1 and Y_2, respectively, as shown in Figure 1.6.

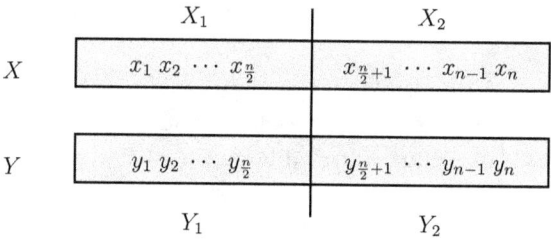

FIGURE 1.6: *Divide and Conquer multiplication algorithm.*

Using the equivalence between the strings X and Y and the numbers they represent, one can write

$$
X = X_1\, 2^{n/2} + X_2
$$
$$
Y = Y_1\, 2^{n/2} + Y_2,
$$

from which one can derive the sought product as

$$
XY = X_1 Y_1\, 2^n + (X_1 Y_2 + X_2 Y_1)\, 2^{n/2} + X_2 Y_2. \tag{1.11}
$$

Even if Eq. (1.11) looks like a mathematical identity, it in fact encapsulates a recursive procedure to calculate the product: to multiply two n-bit strings X and Y, it suffices to (*Divide:*) evaluate X_1Y_1, X_1Y_2, X_2Y_1 and X_2Y_2, which can be done recursively, since each of these pairs is the multiplication of two strings of length $\frac{n}{2}$ each, and then (*Conquer:*) manipulate the results obtained by the recursive calls to obtain the correct answer, in our case, using some shifts and additions. The formal algorithm is shown in ALGORITHM 1.8.

---

mply-1(X, Y, n)

    if $n = 1$ then   return $X * Y$

    else

        $X_1 \leftarrow \lfloor X/2^{n/2} \rfloor$ $X_2 \leftarrow X \bmod 2^{n/2}$

        $Y_1 \leftarrow \lfloor Y/2^{n/2} \rfloor$ $Y_2 \leftarrow Y \bmod 2^{n/2}$

        $A \leftarrow$ mply-1$(X_1, Y_1, \frac{n}{2})$ $B \leftarrow$ mply-1$(X_2, Y_2, \frac{n}{2})$

        $C \leftarrow$ mply-1$(X_1, Y_2, \frac{n}{2})$ $D \leftarrow$ mply-1$(X_2, Y_1, \frac{n}{2})$

        return $A\,2^n + (C + D)\,2^{n/2} + B$

ALGORITHM 1.8: *Multiplying large numbers: a simple approach.*

For the analysis, let $T(n)$ stand for the number of basic binary operations needed to multiply two strings of length n, and note that there are four recursive calls in the procedure. Hence, $T(1) = 1$, and for larger n, which we again assume to be a power of 2,

$$T(n) = 4\,T(\tfrac{n}{2}) + c\,n. \tag{1.12}$$

The constant c in the last term, cn, accounts for all the linear operations like the three additions and the two shifts in the return statement, all of which are applied on strings of length $\theta(n)$ and thus take time $\theta(n)$. Though written as divisions and mod operations, the definitions of X_1, X_2, Y_1 and Y_2 in the first lines of the else part do not really take any time and are just assignments of substrings.

As before, we apply (1.12) repeatedly to get

$$T(n) = 4^2\,T(\tfrac{n}{2^2}) + c\,n\,[2 + 1] = \cdots$$
$$= 4^i\,T(\tfrac{n}{2^i}) + c\,n\,[2^{i-1} + \cdots + 2 + 1] = 4^i\,T(\tfrac{n}{2^i}) + c\,n\,[2^i - 1].$$

Choosing i such that $2^i = n$ and substituting then yields

$$T(n) = n^2 T(1) + c\,n\,[n - 1] = \theta(n^2).$$

We have to conclude that in spite of the sophisticated tools we applied, the final complexity $\theta(n^2)$ is the same as that of the straightforward simple algorithm, and not only is there no gain, but the involved constants seem to be even higher in the recursive version of the multiplication. The lesson to be learned is then that the very fact that we have found a surprising and correct way to solve a problem is not sufficient for the new solution to be really useful. Nevertheless, in the present case, the Divide and Conquer approach may still be improved.

The reason why there might be some hope stems from a closer inspection of our analysis. One difference between the recurrence relations in (1.9) and (1.12) is that the number of recursive calls has doubled from 2 to 4. If one could reduce this number, the complexity could improve. And why should it be possible to achieve such a reduction? Because four recursive calls, each with two numbers as parameters, may process eight different numbers, yet there are only four: X_1, X_2, Y_1 and Y_2, each appearing twice on the right-hand side of Eq. (1.11). There is thus some regularity which has not been exploited so far.

Here is a way to do it. Instead of wasting two recursive calls on the evaluation of C and D in mply-1, we calculate E as the product of $(X_1 + X_2)$ and $(Y_1 + Y_2)$. This should look very strange at first sight. In the decimal equivalent on our running example **5708** and **1948**, it would mean to calculate

$$(57 + 08) \times (19 + 48) = 65 \times 67 = 4355, \qquad (1.13)$$

and none of these numbers seems to have anything to do with the result we are looking for. Note, however, that the sum of two $\frac{n}{2}$-bit numbers is still an $\frac{n}{2}$-bit number (ignoring for the moment the carry, which can turn the result into a $(\frac{n}{2} + 1)$-bit number), hence the evaluation of E can be done in a single recursive call. Moreover, to show the usefulness of E, let us expand its value:

$$E = (X_1 + X_2)(Y_1 + Y_2) = X_1Y_1 + X_1Y_2 + X_2Y_1 + X_2Y_2$$
$$= A + (X_1Y_2 + X_2Y_1) + B,$$

and we recognize in the parentheses the number that should be multiplied by $2^{n/2}$ in Eq. (1.11). We can therefore update the program to get ALGO-RITHM 1.9.

We are thus left with only three recursive calls, and the updated recurrence for the complexity is then

$$T(n) = 3\,T(\tfrac{n}{2}) + c'n,$$

mply-2(X, Y, n)
> if $n = 1$ then $\quad$ return $X * Y$
> else
>> $X_1 \leftarrow \lfloor X/2^{n/2} \rfloor \qquad\qquad X_2 \leftarrow X \bmod 2^{n/2}$
>> $Y_1 \leftarrow \lfloor Y/2^{n/2} \rfloor \qquad\qquad Y_2 \leftarrow Y \bmod 2^{n/2}$
>> $A \leftarrow$ mply-2$(X_1, Y_1, \frac{n}{2}) \qquad B \leftarrow$ mply-2$(X_2, Y_2, \frac{n}{2})$
>> $E \leftarrow$ mply-2$(X_1 + X_2, Y_1 + Y_2, \frac{n}{2})$
> return $\quad A\,2^n + (E - A - B)\,2^{n/2} + B$

ALGORITHM 1.9: *Multiplying large numbers: an improved version.*

where the increased constant $c' > c$ refers also to the additional operations $X_1 + X_2$, $Y_1 + Y_2$ and $E - A - B$, all of which take time linear in n. The idea was therefore to trade one of the (expensive) recursive calls for several (much cheaper) additions and subtractions, expecting an overall reduction of the complexity. Indeed, deriving a closed form, we get

$$T(n) = 3^2\, T(\tfrac{n}{2^2}) + c'n\,\left[\tfrac{3}{2} + 1\right] = \cdots$$
$$= 3^i\, T(\tfrac{n}{2^i}) + c'n\,\left[\left(\tfrac{3}{2}\right)^{i-1} + \cdots + \tfrac{3}{2} + 1\right]$$
$$= 3^i\, T(\tfrac{n}{2^i}) + 2\,c'n\,\left[\left(\tfrac{3}{2}\right)^i - 1\right].$$

If we choose $i = \log_2 n$ as before, we have

$$T(n) = 3^{\log_2 n}\,[1 + 2c'] - 2c'n,$$

where the term in brackets is a constant, and the rightmost term is linear in n. The complexity may thus be written as $\theta(3^{\log_2 n})$, which is

$$3^{\log_2 n} = 3^{[\log_3 n \cdot \log_2 3]} = \left[3^{\log_3 n}\right]^{\log_2 3} = n^{\log_2 3} = n^{1.584}.$$

This might be the first time we see a complexity function with an exponent that is not a decimal number. The result is slightly larger than $n\sqrt{n}$, but significantly smaller than n^2, the complexity of the simple multiplication.

For our running example, we have $A = 57 \times 19 = 1083$, $B = 08 \times 48 = 384$, and E is given in (1.13). The product is therefore

$$1083 \cdot 10^4 + (4355 - 1083 - 384) \cdot 10^2 + 384 = 11119184.$$

Note that to multiply two 4-digit numbers, we have only used three products of two 2-digit numbers, and several additions and subtractions.

Actually, there is no need to derive the closed form of certain recursive formulas from scratch, as many, though not all, are covered by the so-called *Master theorem*, see Exercise 1.4.

1.4 Lower bounds

This section is not an additional example of the Divide and Conquer approach and it deals with lower bounds on the complexity of a problem, which is an important notion to be seen in an introductory chapter on algorithms. The question has been mentioned in Section 1.2 in the discussion of mergesort, where a linear $O(n+m)$ time algorithm for merging two sorted arrays of sizes n and m, respectively, has been suggested.

An obvious question arises, and not only for merging algorithms: how can we know that an algorithm we found for a given problem is reasonable in terms of its time complexity? Sorting, for instance, in time $O(n^2)$, as can be achieved by Insertion sort, might have looked like a good solution before a better one was discovered, like Mergesort in Section 1.2, with time complexity $O(n \log n)$. So, maybe one day, somebody will come up with an even faster merging or sorting procedure?

This question is not just a theoretical one. If your boss is not satisfied with the performance of the solution you came up with, you may invest a major effort in the design of an improved algorithm, which is not always justified. The ability to *prove* mathematically that some suggested solution is already optimal could avoid this waste of time. If it is not possible to show such optimality, as will often be the case, we might settle for finding a so-called lower bound for a given problem.

A *lower bound* for a problem P with input size n is a function $L(n)$ such that it can be shown that no algorithm solving P in time less than $L(n)$ can possibly exist. Note that this does not imply the existence of an algorithm that does solve P in time $L(n)$. It might happen that it is possible to show a lower bound of, say, $O(n \log n)$ for P, whereas the best known solutions require $\Omega(n^2)$. A significant part of computer science research aims at reducing, and hopefully closing, this gap between a lower bound and the complexity of a known algorithm. Sorting, for instance, is one of the problems for which this goal has been achieved.

We show here how to derive a lower bound for the time complexity of merging, but the technique is general enough to be applied also to other problems. We restrict the discussion to merging techniques based on comparisons between the elements, like the simple method merge(i, k, j) mentioned in Section 1.2.

Suppose then that we are given two ordered sequences
$$X = \{x_1 < x_2 < \ldots < x_n\} \quad \text{and} \quad Y = \{y_1 < y_2 < \ldots < y_m\}$$
of n and m elements, respectively, and we assume, for ease of discussion, that all the $n + m$ elements are different.

It would be convenient to reformulate this merging task as finding the subset S of the first $n + m$ integers $\{1, 2, \ldots, n + m\}$ that corresponds to the n indices of the elements of X in the merged set. For example, if $n = 5$ and $m = 4$, then S might be $\{2, 4, 7, 8, 9\}$, implying that the merged set is

1	2	3	4	5	6	7	8	9
y_1	x_1	y_2	x_2	y_3	y_4	x_3	x_4	x_5

where the elements of X appear in gray. Obviously, if this subset is known, then the subset of the indices of the elements of Y is the complementing set, and the actual merging is just a matter of re-arranging the items in a linear scan.

To abbreviate the notation, we shall identify a specific possibility to merge X and Y by the ordered string of the indices of the terms in the merged sequence. For the above example, the merging will be represented by the string **1**1**2**2**3**4**345**, where the indices of the elements of X have been boldfaced.

One should be careful to avoid the common pitfall for showing lower bounds by claiming that the given task can "obviously" only be done in some specific way. This is generally wrong, as there might be infinitely many ways to tackle the problem. The difficulty of showing a lower bound is that the proof must apply to all the, possibly infinitely many, algorithms, including those that have not yet been invented! The proof can thus not restrict the discussion to just some specific methods and has to refer generically to all of them and not just to some specific ones.

We shall therefore describe a merging algorithm using the only information we have about it: that it consists of a sequence of comparisons. Indeed, the merge procedure suggested in Section 1.2 scans the two sets linearly and increments in each iteration the pointer to the smaller element. This seems to be a reasonable approach, yet if one of the sets consists of a singleton, say, $m = 1$, then a linear scan requiring $n + 1$ comparisons is wasteful, and we would be better off by using binary search to find y_1 in X in time $O(\log n)$. However, not all merging methods follow such well-defined and regular comparison patterns.

The tool by means of which the lower bound will be derived is called a *decision tree*. This is a binary tree corresponding to a specific merging algorithm, that is, to a specific sequence of comparisons. Figure 1.7 is an example of a small decision tree, corresponding to the algorithm of merging two elements $X = \{x_1, x_2\}$ with two elements $Y = \{y_a, y_b\}$ using the sequence of comparisons $x_1 : y_a$, $x_2 : y_b$ and either $x_2 : y_a$ or $x_1 : y_b$, depending on the outcome of the first two comparisons. Each level of the tree corresponds to one of the comparisons, in order, and the nodes on level i of the tree, for $i \geq 0$, contain the state of our knowledge after the first i comparisons. In particular, the root, on level 0, reflects the fact that there has not been any comparison yet, so we have not gained any information so far.

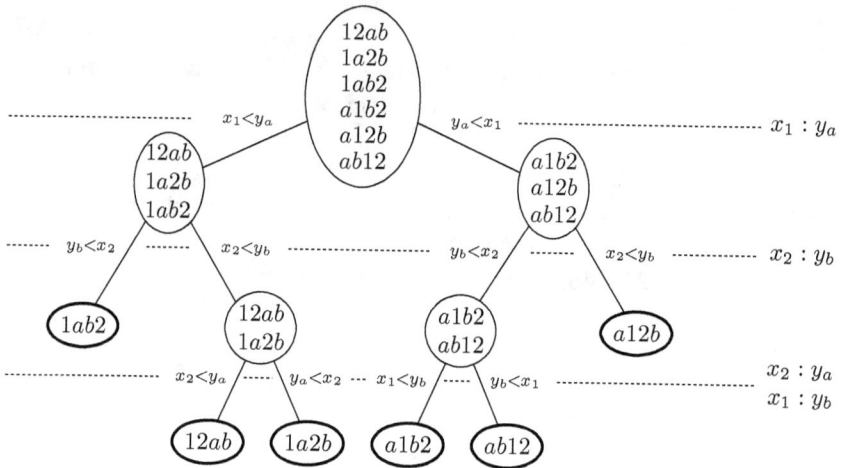

Figure 1.7: *Example of a decision tree for merging two with two elements.*

The knowledge is described as a set of index strings. The root contains all the $\binom{n+m}{m}$ possible choices of selecting a subset of size m from a set of size $n + m$, $\binom{4}{2} = \frac{4!}{2!2!} = 6$ strings of length $2 + 2 = 4$ in our example:

$$\boxed{1\,2\,a\,b}\quad\boxed{1\,a\,2\,b}\quad\boxed{1\,a\,b\,2}\quad\boxed{a\,1\,b\,2}\quad\boxed{a\,1\,2\,b}\quad\boxed{a\,b\,1\,2}$$

If it is not a singleton, the set of strings S_v in a node v on level i is split according to the i-th comparison into two nodes that will be the children of v on level $i+1$: if the comparison is $x_k : y_\ell$, one of these nodes will contain the strings in S_v for which $x_k < y_\ell$, that is, k precedes ℓ in the index string, and the other node will contain the complementing set, those strings of S_v for which $x_k > y_\ell$; the edges leading to these children nodes will be labeled accordingly. For example, let v be the left node on level 1 of the tree in Figure 1.7 that contains the strings $S_v = \{12ab, 1a2b, 1ab2\}$; the second comparison is $x_2 : y_b$; hence S_v is split into $\{1ab2\}$ for which $y_b < x_2$, and $\{12ab, 1a2b\}$, for which $x_2 < y_b$, and these sets are stored in the left and right children of v on level 2, respectively.

The reason for calling this a decision tree is that the state of knowledge at a given node w on level i is a consequence of the *decisions* written as labels on the path from the root to w. For example, let w be the node on level 2 with $S_w = \{a1b2, ab12\}$. The decisions leading to w are $y_a < x_1$ and $y_b < x_2$, which are not enough to completely merge the four elements, because there are two index strings, $a1b2$ and $ab12$, which are compatible with these outcomes of the first two comparisons. If the decisions had been $y_a < x_1$ and $x_2 < y_b$, this would lead us to the node storing just one permutation $a12b$, because we know that $x_1 < x_2$, so in this case, the two first comparisons are sufficient to determine the order.

The aim of a merging algorithm is to perform sufficiently many comparisons such that each leaf of the corresponding decision tree will contain just a single index string. In other words, whatever the outcomes of these comparisons, we must be able to determine the exact order of the $n + m$ elements. Note that the number of necessary comparisons is therefore the depth of the deepest leaf, which is the depth of the tree. As can be seen, for the example in Figure 1.7, three comparisons are sufficient: the tree has six leaves, which are emphasized, each containing a different single string of the six possible index strings. We also see that the first two comparisons alone are not enough: although there are some leaves on level 2, there are also nodes containing more than one index string, so the merging job is not completed.

How can this be generalized to a generic merging algorithm? We obviously do not know which comparisons are used, and in which order. The answer is that we rely on certain properties of binary trees, of which decision trees are a special case. The number of different merging orders of the given $n + m$ elements is $\binom{n+m}{m} = \frac{(n+m)!}{n!m!}$, and each of the corresponding index strings is stored as a singleton in one of the leaves of the decision tree; thus the number of leaves of this tree is $\binom{n+m}{m}$.

As to the depth of the decision tree, there is only one leaf in a tree of depth 0, which is a single node. There are at most two leaves in a binary tree of depth 1, and, by induction, there are at most 2^h leaves in a binary tree of depth h. The number of leaves of a binary tree will thus reach $\binom{n+m}{m}$

only if h is large enough so that

$$2^h \geq \binom{n+m}{m},$$

from which we derive a lower bound on the depth:

$$h \geq \log_2 \binom{n+m}{m}. \tag{1.14}$$

In particular, returning to our example, after 2 comparisons, the tree has at most 4 leaves. Since there are 6 index strings, there must be at least one leaf that contains more than a single string. This shows that not only this specific comparison order fails to merge the input of two sets of two elements in two steps, but that *any* sequence of two comparisons will equally be unsuccessful. To return to our larger example with $n = 5$ and $m = 4$, the number of possible index strings would be $\binom{9}{4} = 126$; after *any* 6 comparisons, the decision tree will only have $2^6 = 64$ leaves, so there is no way to merge 4 with 5 elements in 6 comparisons. Are 7 comparisons enough? It could be, since $2^7 = 128 > 126$, but the technique of using the decision tree gives no clue about whether it is indeed feasible and if so, about how to do it. The procedure merge of Section 1.2 uses $n + m - 1$ comparisons, which is 8 in our case.

---

binary-merge(X, Y, n, m)

$r \leftarrow \lfloor \frac{n}{m} \rfloor; \qquad i \leftarrow r; \qquad j \leftarrow 1$
while $j \leq m$ do
    if $x_i < y_j$ then
        $i \leftarrow i + r$
    else
        binary search of y_j in $[x_{i-r+1} \cdots x_i]$
        $j \leftarrow j + 1$

ALGORITHM 1.10: *Binary merge algorithm.*

For an estimate of the bound in Eq. (1.14), suppose that $n \geq m$ to get

$$\binom{n+m}{m} = \frac{(n+m)(n+m-1)\cdots(n+1)}{m\,(m-1)\,\cdots\,2\,1} \geq \frac{(n)(n)\cdots(n)}{(m)(m)\cdots(m)} = \left(\frac{n}{m}\right)^m,$$

which yields as lower bound $\Omega(m \log(\frac{n}{m}))$. Actually, the lower bound is $\Omega(m \lceil \log(\frac{n}{m}) \rceil)$, because for the case $n < 2m$, one can use another estimate, $\binom{n+m}{m} \geq 2^m$.

There exists a merging algorithm, called *binary merge*, that achieves this lower bound by compromising between the $O(n+m)$ naïve merge and applying binary search m times for all the m elements in Y, yielding $O(m \log n)$. Binary merge scans the larger set X in steps of $r = \lfloor \frac{n}{m} \rfloor$ to find the interval to which y_j belongs; once this interval $[x_{(i-1)r+1} \cdots x_{ir}]$ is located, the exact position of y_j within it is determined by binary search. The formal procedure appears in ALGORITHM 1.10.

In the entire algorithm, i can be incremented at most m times, and there are at most m binary searches over r elements each, so the worst-case complexity is $O(m + m \lceil \log \left(\frac{n}{m} \right) \rceil)$, which is identical to the lower bound since we assume $n \geq m$.

1.5 Exercises

1.1 Find a simple and non-recursive way to solve the min–max problem for n elements using only $1.5n - 2$ comparisons.

1.2 By reducing the number of recursive calls from 4 to 3, the complexity of the multiplication algorithm improved from $\theta(n^2)$ to $\theta(n^{1.584})$. What is wrong with the following "argument" showing how to reduce the number of recursive calls even further by performing

$$F \leftarrow (X_1 \, 2^{n/2} + X_2) \, Y_1 \, 2^{n/2}$$
$$G \leftarrow (X_1 \, 2^{n/2} + X_2) \, Y_2,$$

so that we have calculated $X \times Y = F + G$ using just two recursive calls and some more additions, all of which take linear time? If the argument were correct, what would have been the resulting complexity of multiplying two n-bit numbers?

1.3 Given a non-sorted array A containing n (positive or negative) numbers, use the Divide and Conquer approach to devise an $O(n \log n)$ algorithm which returns a pair of indices (i, j) for $1 \leq i < j \leq n$, such that the sum of the elements $A[i] + A[i+1] + \cdots + A[j]$ is minimized.

1.4 *Master Theorem:* Generalize equations similar to those in (1.2), (1.9), and (1.12) to Divide and Conquer algorithms in which $T(1) = 1$ and for $n > 1$, where we assume n to be a power of s,

$$T(n) = m\,T(\tfrac{n}{s}) + n^d.$$

These are cases in which the input of size n is split into s sections of equal size $\frac{n}{s}$ and the recursion is applied with a multiplicity of m. The exponent d is the degree of the polynomial describing the complexity of reassembling the outcomes of the recursive calls. Show that the closed form of $T(n)$ is given by the following cases:

$$\frac{m}{s^d} < 1 \qquad \rightarrow \qquad T(n) = O(n^d)$$

$$\frac{m}{s^d} = 1 \qquad \rightarrow \qquad T(n) = O(n^d \log n)$$

$$\frac{m}{s^d} > 1 \qquad \rightarrow \qquad T(n) = O(n^{\log_s m})$$

1.5 Consider the following algorithm for finding the *median* of a set of n numbers, given in an unsorted array A.

$median(A, n)$

> if $n \leq 5$ then   find median directly
> else
>> $A_1 \leftarrow A[1 : \lfloor \frac{n}{2} \rfloor]$ $A_2 \leftarrow A[\lfloor \frac{n}{2} \rfloor + 1 : n]$
>> $m_1 \leftarrow median(A_1, \lfloor \frac{n}{2} \rfloor)$ $m_2 \leftarrow median(A_2, \lceil \frac{n}{2} \rceil)$
>> if $m_1 = m_2$ then return m_1
>> else   // assume w.l.g. that $m_1 < m_2$
>>> define $C \leftarrow \{x \in A_1 \mid x \geq m_1\} \cup \{x \in A_2 \mid x < m_2\}$
>>> $m \leftarrow median(C, |C|)$
>>> return m

(1) Show that the algorithm indeed finds the median.
(2) What is the size of the set C?
(3) Evaluate the complexity of the algorithm to reach the conclusion that the algorithm is useless.

1.6 The question deals with sorting algorithms based on comparisons between elements.

    (1) Show that a lower bound for the number of comparisons needed to sort n elements in a comparison-based sorting procedure is asymptotically at least $n \log n$.

    (2) Show how to sort 5 elements with 7 comparisons.

    (3) How are your answers to the preceding questions compatible, since $5 \log_2 5 = 11.6$?

1.7 Twelve apparently identical balls are given and you are told that the weight of one of the balls is different. The problem is to find this ball and to decide whether it is lighter or heavier by means of a two-sided weighing scale.

    (1) Give a lower bound ℓ on the number of necessary weighings to find the different ball.

    (2) Show how to solve the problem in ℓ weighings. *Hint:* The cover page of this book shows a part of the decision tree of one possible solution of this riddle.

    (3) Using the same method as for (1), find a lower bound ℓ' for the same problem, but with 13 instead of 12 balls.

    (4) While you can find a solution matching the lower bound ℓ in (2), show that no solution using ℓ' weighings can exist for the problem with 13 balls.

Chapter 2

Dynamic Programming

2.1 Binomial coefficients

As we saw in the previous chapter, the fact that a problem may be solved recursively using a Divide and Conquer approach is not always sufficient for the solution to be useful. The present chapter deals with a family of algorithms for which a straightforward recursive definition may have a disastrous impact on the execution time, and we shall see how this can often be circumvented.

Consider the problem of evaluating a binomial coefficient $\binom{n}{k}$, for some $n \geq k \geq 0$. As you might have learned, and this is what most people reply when asked what this coefficient stands for,

$$\binom{n}{k} = \frac{n!}{k!\,(n-k)!}, \qquad (2.1)$$

but this is its *value*, which can be reached based on combinatorial arguments, rather than its *definition*. The latter is

$$\binom{n}{k} = \begin{array}{l} \text{number of ways to choose a subset of} \\ k \text{ elements from a set of } n \text{ elements.} \end{array}$$

This definition implies immediately the boundary conditions, namely, $\binom{n}{0} = 1$, since there is only one way to choose an empty subset from a set of n elements, and this holds even if the set itself is empty, that is, $n = 0$, and so $\binom{0}{0} = 1$.

If one tries to plug the extreme values $k = 0$ or $k = n$ into Eq. (2.1), one has to evaluate $0!$, zero factorial, which, as we have been told, is equal to 1. But why should this be so? Is not

$$n! = n\,(n-1)\,(n-2)\cdots 2\,1,$$

the product of n times $n-1$, times $n-2$, etc., down to 1, a definition that does not make any sense for $n = 0$? Defining then $0!$ as equal to 1 seems rather arbitrary. The truth is that the equality in (2.1) does not hold *because* $0! = 1$, but on the contrary, $0!$ has been *defined* as 1 for the equation to hold also in the extreme cases $k = 0$ and $k = n$.[1]

A first step in evaluating a binomial coefficient is to try to establish a recurrence relation. To find the number of possibilities to choose a subset S of size k within a set A of size n, let us split the possible subsets into two classes, according to whether or not they contain a specific element $a \in A$. If a is not in S, then we have to choose the k elements among the $n - 1$ elements of A that are different from a, and this can be done, by definition, in $\binom{n-1}{k}$ ways. If, on the other hand, $a \in S$, then only $k - 1$ elements remain to be chosen from $A \backslash \{a\}$, for which there are $\binom{n-1}{k-1}$ options. Since the two classes form a partition, that is, they are disjoint and their union accounts for all the possibilities, we conclude that

$$\binom{n}{k} = \binom{n-1}{k} + \binom{n-1}{k-1}, \qquad (2.2)$$

which holds for all integers $n \geq k > 0$. Together with the boundary conditions, this is equivalent to the recursive program in ALGORITHM 2.1.

Binom(n, k)

    if $n = k$ or $k = 0$ then return 1
    else return Binom$(n - 1, k)$ + Binom$(n - 1, k - 1)$

ALGORITHM 2.1: *Recursive evaluation of a binomial coefficient.*

This solution might seem elegant, but it is far from being efficient. Let $T(n)$ denote the amount of work needed to evaluate a coefficient for a set of size n, where the exact definition of "work" is not important here, be it comparisons or other operations, and where we only consider the size n of the set and not that of the requested subset. The last line of the program then implies that

$$T(n) \geq 2\,T(n-1),$$

---

[1] The factorial operator can similarly also be extended to non-integer arguments. Actually, a useful definition is connected to the gamma function Γ, which has the property $\Gamma(n) = (n-1)!$ for integer n. For instance, one could define $\frac{1}{2}!$ as $\Gamma(1.5) = \frac{\sqrt{\pi}}{2} = 0.886$.

which, together with the fact that $T(0)$ is some non-zero constant, yields

$$T(n) \geq 2^2\, T(n-2) \geq \cdots \geq 2^n\, T(0) = \Omega(2^n).$$

For a set of size n, ALGORITHM 2.1 thus uses an exponential number of operations, while the number of *different* binomial coefficients is not more than the number of different pairs (i,j), with $n \geq i \geq j \geq 0$, which is $\frac{n(n-1)}{2} = O(n^2)$. The reason for this wasteful performance can be seen if we apply (2.2) once more:

$$\binom{n}{k} = \binom{n-1}{k} + \binom{n-1}{k-1}$$
$$= \binom{n-2}{k} + \binom{n-2}{k-1} + \binom{n-2}{k-1} + \binom{n-2}{k-2}$$

The two terms in the middle of the last line are in fact the same, yet the program does not remember that $\mathrm{Binom}(n-2, k-1)$ has already been evaluated when it encounters its second occurrence.

To cure its amnesiac behavior, the program needs some additional memory in which already calculated coefficients will be recorded for later reference. We shall use a matrix $B[n,k]$, storing the coefficient $\binom{n}{k}$ in the cell at the intersection of the n-th row with the k-th column. Table 2.1 shows the matrix for $n = 5$. In this case, the table is triangular, as only the elements below and including the main diagonal are defined.

TABLE 2.1: *Matrix of binomial coefficients: the* Pascal *triangle.*

n \ k	0	1	2	3	4	5
0	1					
1	1	1				
2	1	2	1			
3	1	3	3	1		
4	1	4	6	4	1	
5	1	5	10	10	5	1

Instead of evaluating the elements in the table on demand in the order implied by the recursive ALGORITHM 2.1, it may be easier to fill the matrix systematically so that when evaluating the element of a given location, those needed for its computation are already in the table. In our case, the first column and the main diagonal are the boundary conditions, and each of the other elements is the sum of the element x just above it and the adjacent element to the left of x. It follows that one may fill the matrix row by row top-down, and within each row in any order, for example, left to right, as shown in ALGORITHM 2.2. The complexity is clearly $O(n^2)$.

Binom-1(n)

     for $i \leftarrow 0$ to n do

         $B[i, 0] \leftarrow 1$ $B[i, i] \leftarrow 1$

         for $j \leftarrow 1$ to $i - 1$ do

             $B[i, j] \leftarrow B[i - 1, j] + B[i - 1, j - 1]$

ALGORITHM 2.2: *Iterative evaluation of binomial coefficients.*

This technique of replacing a wasteful recursive algorithm by an equivalent procedure, filling a table which stores already evaluated results to eliminate duplicate recursive calls, is known as *dynamic programming*, an admittedly not very descriptive name that could be applied to many, if not most, of the families of algorithms. Nonetheless, this is the technical term that has been chosen and it is kept, probably, for historical reasons.

Let us summarize the main ingredients of dynamic programming algorithms:

- The solution of the given problem, often an optimization problem, can be defined recursively as depending on the solutions of sub-problems, usually of the same problem, but applied on proper subsets of the original input. Finding the recursive formulation is often quite tricky and may require some creativity. For example, the solution may involve the introduction of non-obvious auxiliary variables, which complicate the representation but facilitate the definition.

- The solutions can be stored in a table, an array or parts of a higher-order matrix, but care has to be taken to evaluate the elements in the table in a correct order, not relying on elements that have not been defined so far. The dependencies for the binomial coefficient example are shown in Figure 2.1, the current element at position

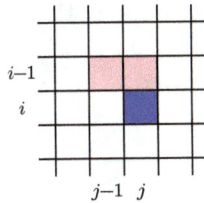

FIGURE 2.1: *Dependencies for the evaluation of binomial coefficients.*

(i, j) being drawn in blue, and the two elements on which it depends in lighter pink.

Accordingly, there are several ways of choosing the order in which the elements of the matrix should be evaluated, some of which are schematically depicted in **Figure 2.2**, along with the corresponding control statements for the nested loops. The leftmost example processes mainly the columns left to right, and within each column top-down. The bullet represents the current element x, and one can see that the elements on which x relies are already defined. The other examples process by row first, top-down, and within the rows, either left to right or right to left.

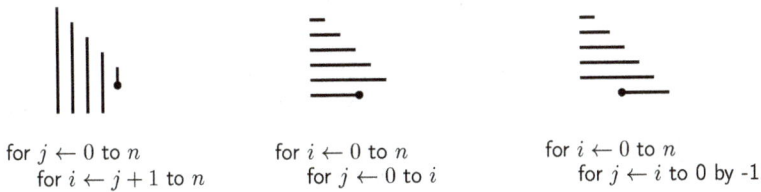

for $j \leftarrow 0$ to n for $i \leftarrow 0$ to n for $i \leftarrow 0$ to n
  for $i \leftarrow j + 1$ to n for $j \leftarrow 0$ to i for $j \leftarrow i$ to 0 by -1

FIGURE 2.2: *Schematic representation of some possible scanning orders.*

- For optimization problems, one may differentiate between a solution s and its associated *cost* or *value* $v(s)$. We seek a solution with minimal cost or maximal value. It will often be convenient to evaluate in a first stage only the value $v(s)$ of an optimal solution, and then adapt the program to give also (one of) the solution(s) yielding this value.

The introductory example of evaluating a binomial coefficient is not an optimization problem, rather a very simple example of replacing recursive calls by constructing a table. In fact, this table is so well known that it deserves a name of its own: it is the *Pascal triangle*. The following examples are more useful and more involved.

2.2 Edit distance

Computers were originally invented as large calculating machines, and many of our algorithms process mainly numbers. Only much later did researchers realize that efficiently manipulated textual data may also have many useful applications, some of which will be handled in the chapters in Part IV of this book.

While the comparison of numbers is straightforward, the *distance* $d(S,T)$ between two character strings $S = S_1 S_2 \cdots S_n$ and $T = T_1 T_2 \cdots T_m$ could be defined in several plausible ways. If the strings are of the same length $n = m$, one possibility would be the *Hamming distance* $\mathsf{HD}(S,T)$, defined as the number of positions i, with $1 \le i \le m$, in which the corresponding characters are different, that is, $S_i \ne T_i$. For example, the Hamming distance between the English and French versions of the name of Austria's capital, VIENNA and VIENNE, is just 1. But how can this be extended to strings of differing lengths? The necessity of such a generalization is obvious if one extends the example also to the German name WIEN.

Another important application of the comparison of strings is *information retrieval*. Given is a large textual database, ultimately the entire World Wide Web, and we are looking for documents including some keywords defining our information needs. Not always are we able to formulate our requests accurately, and both text and keywords are prone to typos and other spelling mistakes. We therefore should not insist on finding only exact copies but also allow *approximate* matches. The mathematical formulation of this vague notion of proximity consists of defining a distance operator d and a threshold value k: the strings S and T are said to match approximately if

$$d(S,T) \le k.$$

The Hamming distance is, however, not always a good candidate for expressing such a distance, because it compares only characters in corresponding positions. ESPAÑA is clearly related to SPAIN, but if both strings are considered left-justified, we would compare E to S, S to P, etc., completely missing the matching substrings SPA.

This leads naturally to defining the distance as the number of *editing* operations needed to transform one string into another. The most commonly used such operations are the following:

rep — *replace* one character by another,
ins — *insert* a new character, and
del — *delete* one of the existing ones.

The following two examples in Figure 2.3 show how to pass from the French name of the city of Brussels to its German equivalent, and from the German preposition DURCH to its English translation THROUGH.

	BRUXELLES		DURCH
rep U–Ü	BRÜXELLES	rep D-T	TURCH
rep X–S	BRÜSELLES	ins H	THURCH
ins S	BRÜSSELLES	del U	THRCH
del L	BRÜSSELES	ins O	THROCH
del E	BRÜSSELS	ins U	THROUCH
del S	BRÜSSEL	rep C-G	THROUGH

FIGURE 2.3: *Two examples of string editing.*

Note that these are not the only possibilities to move from the first to the last term. In general, the number of rep/ins/del operations in different ways to transform S into T does not need to be the same, even if the strings S and T are fixed in advance. We therefore define the *edit-distance* ED(S,T) as the *minimal* number of operations necessary to transform S into T. One could also assign not necessarily equal costs to the different operations and define the edit-distance more generally as the minimal total cost of the transformation.

As mentioned earlier, the key idea allowing the derivation of a recursive formula for the solution of our problem is the introduction of some auxiliary variables. In the case of the edit distance, we are given strings S and T of lengths n and m, respectively, and it will be convenient to consider also the possible *prefixes* of these strings $S[1..i] = S_1 S_2 \cdots S_i$ and $T[1..j] = T_1 T_2 \cdots T_j$, for $1 \le i \le n$ and $1 \le j \le m$. This step is somewhat surprising, because nothing in the statement of the problem indicates that the prefixes of the strings are connected to the requested values, and therefore choosing them as additional parameters is not at all obvious. This is exactly the point where some intuition is needed, and we shall see below more examples of similar decisions. What they all have in common is that while the formulation of the problem gets obviously more complicated, its solution may then be derived more easily.

The dynamic programming approach to solve the edit-distance problem stores the values in a two-dimensional table E in which each dimension corresponds to one of the strings. The entry $E[i,j]$ in the i-th row and j-th column holds the number of required operations to transform the i-th prefix of S into the j-th prefix of T, that is $\mathsf{ED}(S[1..i], T[1..j])$. The first column and the first row correspond to empty strings, denoted by Λ. They can be initialized by

$$E[0,j] = j \quad \text{and} \quad E[i,0] = i,$$

as one may just insert the first j characters one after the other to get from the empty string Λ to $T[1..j]$, or delete all the characters to get from $S[1..i]$ to Λ. For the general step, consider the last characters S_i and T_j, of the currently dealt prefixes:

- if they are equal, we can just ignore them and the best way to transform $S[1..i]$ into $T[1..j]$ is the best way to transform $S[1..i-1]$ into $T[1..j-1]$;
- if the last characters of the prefixes do not match, the solution is the best of the three options

  - replacing the last characters, S_i by T_j,
  - inserting T_j, or
  - deleting S_i,

  and then continuing with the corresponding sub-problem.

This is summarized by defining the elements of the table E recursively by

$$E[i,j] = \begin{cases} E[i-1,j-1] & \text{if } S_i = T_j \\ 1 + \min\left(E[i-1,j-1], E[i,j-1], E[i-1,j]\right) & \text{if } S_i \neq T_j. \end{cases} \quad (2.3)$$

Figure 2.4 shows the dependencies used in the recursive definition of E, the current element indexed (i,j) appearing, as in Figure 2.1, in blue, and the up to three elements on which it depends being drawn in pink. A straightforward top-down, left to right, scan of the elements is therefore sufficient to assure that all the elements are already defined when they are referred to.

Table 2.2 shows the corresponding table for the example strings BRUXELLES and BRÜSSEL. For instance, the value 4 in position $(3,6)$ is the minimal number of operations needed to get from BRU to BRÜSSE: since the

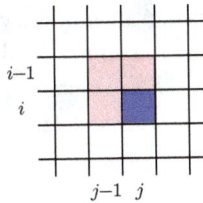

FIGURE 2.4: *Dependencies for the evaluation of the edit distance.*

last characters U and E of these substrings do not match, the value 4 is obtained by adding 1 to the minimum of the three surrounding values, 3, 3 and 4, which have been evaluated in previous iterations.

TABLE 2.2: *Edit distance E.*

E		B	R	Ü	S	S	E	L
	0	1	2	3	4	5	6	7
0	0	1	2	3	4	5	6	7
B 1	1	0	1	2	3	4	5	6
R 2	2	1	0	1	2	3	4	5
U 3	3	2	1	1	2	3	4	5
X 4	4	3	2	2	2	3	4	5
E 5	5	4	3	3	3	3	3	4
L 6	6	5	4	4	4	4	4	3
L 7	7	6	5	5	5	5	5	4
E 8	8	7	6	6	6	6	5	5
S 9	9	8	7	7	6	6	6	6

TABLE 2.3: *Auxiliary pointers P.*

P		B	R	Ü	S	S	E	L
	0	1	2	3	4	5	6	7
0								
B 1		D	L	L	L	L	L	L
R 2		U	D	L	L	L	L	L
U 3		U	U	D	L	L	L	L
X 4		U	U	U	D	L	L	L
E 5		U	U	U	U	D	D	L
L 6		U	U	U	U	U	U	D
L 7		U	U	U	U	U	D	D
E 8		U	U	U	U	U	D	U
S 9		U	U	U	D	D	U	U

The sought minimal value for the number of operations necessary to transform S into T, 6 for our example, can then be found in the right lower corner of the matrix, and it appears boxed in Table 2.2. This, however, is only a first step, since we are also interested in the exact sequence of the operations to be performed, and not just in their number.

It would not be reasonable to try to store the optimal sequence of operations for each of the possible (i, j) pairs. Assuming for simplicity that the strings are of equal length $n = m$, the E matrix has n^2 entries, and for a given pair of indices (i, j), there might be up to n ins/del/rep operations necessary to transform $S[1..i]$ into $T[1..j]$, requiring $\Omega(n^3)$ storage. This can be circumvented by defining an auxiliary matrix $P[i, j]$, which stores at each entry enough local information to enable the construction of an optimal path through the matrix E. We shall define $P[i, j]$ to be either L, U or D, depending on whether the element (i, j) is obtained from the element to its Left, or from the Upper or Diagonal element in Eq. (2.3) defining the elements of E.

The optimal sequence of transformations can then be traced backwards, starting in the right lower corner of the P matrix and proceeding Left, Up or Diagonally, as indicated by the matrix elements, until reaching the left upper corner.

Edit-Distance(S, n, T, m)

    for $j \leftarrow 0$ to m do
       $E[0, j] \leftarrow j$
    for $i \leftarrow 1$ to n do
       $E[i, 0] \leftarrow i$
    for $i \leftarrow 1$ to n do
       for $j \leftarrow 1$ to m do
          if $S_i = T_j$ then
             $E[i, j] \leftarrow E[i - 1, j - 1]$ $P[i, j] \leftarrow \underline{\text{D}}$
          else
             min $\leftarrow E[i, j - 1]$ $P[i, j] \leftarrow$ L
             if $E[i - 1, j] < $ min   then
                min $\leftarrow E[i - 1, j]$ $P[i, j] \leftarrow$ U
             if $E[i - 1, j - 1] < $ min   then
                min $\leftarrow E[i - 1, j - 1]$ $P[i, j] \leftarrow$ D
             $E[i, j] \leftarrow 1 + $ min
    return $E[n, m]$

ALGORITHM 2.3: *Iterative evaluation of the edit distance between two strings.*

The matrix P corresponding to our example is shown in Table 2.3. The elements of the optimal sequence of transformations appear boldfaced in red, and so are the corresponding values of E in Table 2.2. The fact that $P[i, j] = $ D may be caused either by a match of the last characters of the corresponding substrings, $S_i = T_j$, in which case no action is taken, or by a mismatch for which the minimum in Eq. (2.3) is achieved by the Diagonal element, triggering a replacement of S_i by T_j. The former case is indicated by underlining the $\underline{\mathsf{D}}$.

The formal algorithm calculating the edit distance between S and T by filling in the elements of the matrix E is given in ALGORITHM 2.3. There is a constant amount of work for each matrix entry, so both time and space complexities are $O(nm)$. The program also generates the matrix P, which can then be used to derive an optimal sequence of editing steps in time $O(\max(n, m))$, as shown in ALGORITHM 2.4. Applying it to our example in Table 2.3 yields, backwards, the sequence of editing steps in the left column of Figure 2.3.

Editing-Steps(S, n, T, m)
$$
\begin{aligned}
&(i, j) \leftarrow (n, m) \\
&\textbf{while } i > 0 \textbf{ or } j > 0 \textbf{ do} \\
&\quad \textbf{if } P[i, j] = \mathsf{U} \textbf{ then} \\
&\qquad \text{print "delete } S_i\text{"} \quad i \leftarrow i - 1 \\
&\quad \textbf{else if } P[i, j] = \mathsf{L} \textbf{ then} \\
&\qquad \text{print "insert } T_j\text{"} \quad j \leftarrow j - 1 \\
&\quad \textbf{else} \quad /\!/ \ P[i,j] = \mathsf{D} \text{ or } \underline{\mathsf{D}} \\
&\qquad \textbf{if } S_i \neq T_j \textbf{ then} \\
&\qquad\quad \text{print "replace } S_i \text{ by } T_j\text{"} \\
&\qquad (i, j) \leftarrow (i - 1, j - 1)
\end{aligned}
$$

ALGORITHM 2.4: *Evaluating a sequence of editing steps.*

There are several other problems involving strings that can be solved by a similar dynamic programming approach, for example, the *Longest Common Subsequence* problem LCS (see Exercise 2.1).

2.3 Matrix chain multiplication

Our following example deals with matrices. A $p \times q$ matrix, where both p and q are non-zero integers, is a collection of pq numbers, organized in p rows and q columns, the usefulness of which is taught in any first course on linear algebra. We shall assume that the reader is familiar with basic matrix operations, like addition and multiplication of matrices. For instance, a $p \times q$ matrix $A = (a_{ij})$ can be multiplied with an $s \times r$ matrix $B = (b_{k\ell})$ in this order, only if $s = q$, that is, the number of columns of A must equal the number of rows of B. The resulting product is then a matrix $E = (e_{i\ell})$ of dimensions $p \times r$, in which

$$e_{i\ell} = \sum_{t=1}^{q} a_{it} b_{t\ell}. \tag{2.4}$$

Figure 2.5 shows a part of these matrices and their corresponding dimensions. Since by Eq. (2.4), q multiplications and $q - 1$ additions are necessary for the evaluation of a single element of the matrix E, and there are pr such elements, the complexity of multiplying A by B is pqr. In particular, if all dimensions $p = q = r$ are equal, the cost of multiplying two square matrices of side length p is p^3. Using a Divide and Conquer approach similar to the one of multiplying large numbers in Section 1.3, it has been shown by [Strassen (1969)] how to reduce the complexity for square matrices to $O(p^{2.81})$.

$$\underset{p \times q}{\overset{A}{\begin{pmatrix} & & & \\ a_{i1} & a_{i2} & \cdots & a_{iq} \\ & & & \end{pmatrix}}} \underset{q \times r}{\overset{B}{\begin{pmatrix} b_{1\ell} \\ b_{2\ell} \\ \vdots \\ b_{q\ell} \end{pmatrix}}} = \underset{p \times r}{\overset{E}{\begin{pmatrix} & \vdots & \\ \cdots & e_{i\ell} & \cdots \\ & \vdots & \end{pmatrix}}}$$

FIGURE 2.5: *Multiplying two matrices.*

The fact that for general matrices, multiplication is associative, but not necessarily commutative, is well known. Consider three matrices A, B and C, of dimensions 50×3, 3×100 and 100×2, respectively. Their product is a 50×2 matrix, and associativity means that it does not matter whether we first multiply A by B, and then the result by C to get $(A \times B) \times C$, or rather $A \times (B \times C)$, as the product would be the same, which means that the parentheses may be omitted.

Nonetheless, the total number of operations to be performed on the individual elements when multiplying several matrices may vary substantially, depending on the order in which the matrices are processed, even though the final resulting matrix is not affected by this order. Figure 2.6 shows the matrices, their dimensions and the intermediate steps involved when multiplying $(AB)C$ using 25,000 operations, or $A(BC)$ using only 900 operations. In particular, the intermediate matrix AB formed in the former case is of size 50×100, while the corresponding matrix BC in the latter has dimensions 3×2.

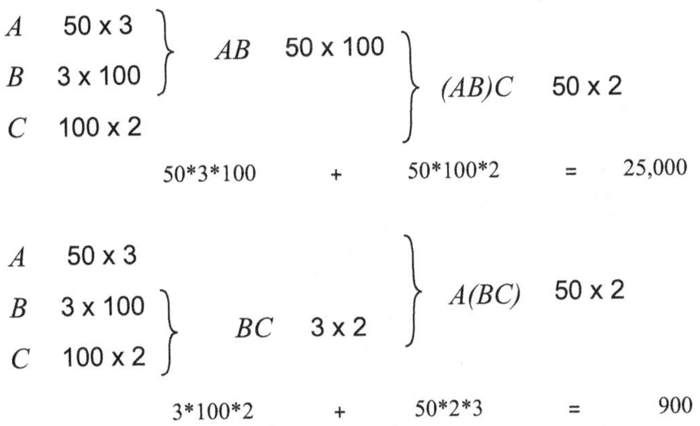

A 50 x 3
B 3 x 100 AB 50 x 100
C 100 x 2 $(AB)C$ 50 x 2

$50*3*100$ $+$ $50*100*2$ $=$ 25,000

A 50 x 3
B 3 x 100 $A(BC)$ 50 x 2
C 100 x 2 BC 3 x 2

$3*100*2$ $+$ $50*2*3$ $=$ 900

FIGURE 2.6: *Number of operations for multiplying matrices A, B and C.*

While one may have guessed for this example that it is preferable to strive for dealing with smaller temporary matrices by eliminating first the largest dimension 100, the generalization of this strategy to a chain of n matrices to be multiplied does not yield an optimal solution.

Consider then the problem of multiplying n matrices in a way that minimizes the total number of individual operations on the matrices' elements. Given is a sequence of $n + 1$ positive integers $p_0, p_1, \ldots, p_n$ and matrices

$$M_1 \, M_2 \, M_3 \, \cdots \, M_n, \tag{2.5}$$

where matrix M_i is of dimensions $p_{i-1} \times p_i$, so that the product $\prod_{i=1}^{n} M_i$ in Eq. (2.5), which yields a $p_0 \times p_n$ matrix, is well defined. The challenge is to find the best way of pairing successively adjacent matrices.

There were only two possibilities to do so for the product of three matrices in the example of Figure 2.6. If four matrices are involved, the five possible combinations are depicted in Figure 2.7, and one may recognize that each pairing sequence corresponds to one of the possible layouts of a

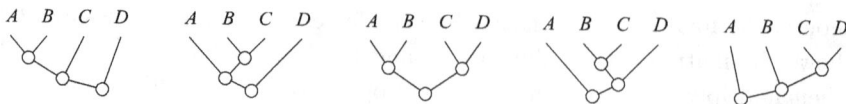

FIGURE 2.7: *Possible evaluation order to multiply four matrices.*

complete binary tree with 4 leaves. More generally, the number of alternatives from which the optimal sequence of matrix pairs has to be chosen is the number of different complete binary trees with n leaves.

Background concept: Number of different tree shapes

To evaluate the number of different possible shapes of a complete binary tree with $m + 1$ leaves, we shall establish some equivalence relations between those trees and related combinatorial objects. Consider the complete binary tree $\mathcal{T}_\mathrm{C}$ in the upper right part of Figure 2.8 as a running example. It has $m + 1 = 16$ leaves which appear as squares and are colored gray. The binary tree $\mathcal{T}_\mathrm{B}$ in the left upper part of Figure 2.8 is not complete, has $m = 15$ *nodes* (not *leaves*), and is obtained from $\mathcal{T}_\mathrm{C}$ by deleting all the leaves of the latter. This transformation is reversible: one can take any binary tree with m nodes and add two children to all its leaves and the missing child to all its nodes having only one child, and one ends up with a complete binary tree with $m + 1$ leaves, so there is a bijection between the two sets and the number of possible trees of the two kinds must be the same.

On the other hand, every binary tree corresponds to a unique ordered forest of general trees, in which every node may have any number of children, not just two as for binary trees. The three small trees forming a forest $\mathcal{F}$ in the lower part of Figure 2.8 correspond to $\mathcal{T}_\mathrm{B}$.

The details of the transformation are as follows: the root of $\mathcal{T}_\mathrm{B}$ will be the root of the leftmost tree in $\mathcal{F}$. The left child of a node in $\mathcal{T}$ will be the first (leftmost) child of that node in $\mathcal{F}$, and the right child of a node v in $\mathcal{T}$ will be the sibling immediately to the right of v in $\mathcal{F}$, if there is such a sibling at all. The roots of the trees in $\mathcal{F}$ are special cases, and are considered as siblings. The corresponding nodes in Figure 2.8 are indicated by matching colors.

As above, the construction process is reversible, implying that the number of different such forests with m nodes is equal to the number of different binary trees with m nodes, and also to the number of complete binary trees with $m + 1$ leaves.

It is convenient to use the following compact representation $\mathcal{R}(T)$ of a general tree T as a sequence of opening and closing parentheses: if the sub-trees of T are $T_1, T_2, \ldots, T_k$, in this order, then
$$\mathcal{R}(T) = (\, \mathcal{R}(T_1)\, \mathcal{R}(T_2) \cdots \mathcal{R}(T_k)\,).$$
The sequence for a forest is just the concatenation of the sequences of the corresponding trees. The representation of forest $\mathcal{F}$ appears in the right lower part of Figure 2.8, using the same color coding.

Since every opening parenthesis has a matching closing one, their number must be equal. Denoting a sequence of a opening parentheses and b closing parentheses by an (a, b)-sequence, the sequence corresponding to a forest with m nodes has thus to be an (m, m)-sequence.

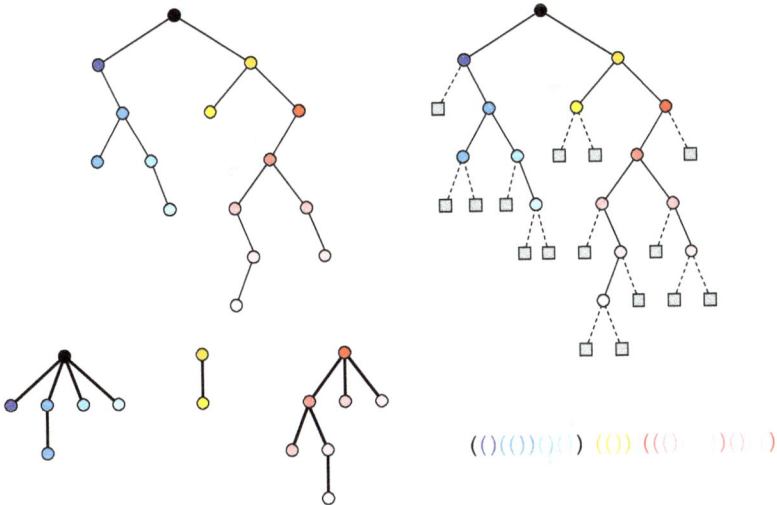

FIGURE 2.8: *Equivalence between sets of complete binary trees, binary trees, ordered forests of general trees and nested parentheses.*

That this is not a sufficient condition can be seen when considering, e.g., the sequence $S = (\,)\,(\,(\,)\,)\,)\,(\,($, which has the correct number of parentheses, but is not properly nested. So the following nesting condition has to be added: when scanning the sequence left to right, at each point, the number of closing parentheses seen so far is not larger than the number of opening parentheses. Let us call *complying* (m, m)-sequences those satisfying the nesting condition. To evaluate their number, we consider the complementing set of *not complying* (m, m)-sequences.

In any noncomplying (m, m)-sequence, the number of closing parentheses exceeds the number of opening ones, so there must be a leftmost position in a left-to-right scan at which this happens for the first time. In the example of the sequence S above, it is at position 7, and the violating closing parenthesis is boldfaced. Consider then the function g transforming such noncomplying (m, m)-sequences by inverting all the parentheses following this first occurrence. For our example, the resulting sequence would be $g(S) = ( ) (( ))) \mathbf{(} \mathbf{)} \mathbf{)}$, in which all inverted parentheses are boldfaced. The resulting sequence will always be an $(m - 1, m + 1)$-sequence, and since the function is reversible by applying it again, that is, $g^{-1} = g$, the function g is a bijection, implying that the number of noncomplying (m, m)-sequences is equal to the total number of $(m-1, m+1)$-sequences, which is $\binom{2m}{m-1}$. We can thus conclude that

number of forests with m nodes	=	number of complying (m, m)-sequences	=	total number of (m, m)-sequences	−	number of $(m - 1, m + 1)$-sequences

$$= \binom{2m}{m} - \binom{2m}{m-1} = \frac{1}{m+1}\binom{2m}{m}.$$

These numbers are known as *Catalan* numbers, the first of which are $1, 2, 5, 14, 42, \ldots$. Asymptotically, the m-th Catalan number is $O(\frac{4^m}{m^{3/2}})$. In particular, the number of complete binary trees with $m + 1 = 4$ leaves in Figure 2.7 is $\frac{1}{4}\binom{6}{3} = 5$.

Returning to our problem of multiplying a sequence of n matrices, we recognize the importance of analyzing the number of possibilities: if it were just growing, say, as $O(n^2)$, we could generate all possible combinations, evaluate the cost of each and choose the optimum, as we did for $n = 3$ in Figure 2.6. This strategy has to be ruled out for the exponentially increasing number of alternatives. For instance, for $n = 30$, this number would be more than 10^{15}, a million billions!

We shall therefore try a recursive approach, as advocated in this chapter. As was done for the Edit Distance in Section 2.2, we consider in a first stage only the *cost* of an optimal solution, defined here as the minimal number of required operations on the individual matrix elements. Once this is found, we shall adapt the algorithm to give not only the cost, but also a specific way of grouping the matrices to achieve this optimal cost.

To enable a recursive formulation, first the definitions have to be generalized to deal with sub-problems: let $\mathsf{mult}(i,j)$, for $i \leq j$, be the cost of the sub-problem defined by the matrices with indices $i, i+1, \ldots, j$, that is, the minimal number of individual operations to multiply the chain of matrices $M_i M_{i+1} \cdots M_j$. The resulting matrix has dimensions $p_{i-1} \times p_j$. In particular, the solution to our initial problem is then given by $\mathsf{mult}(1,n)$.

The key idea for the solution of this problem, and many other similar ones, is that for the order of operations to multiply the n matrices to be optimal, all sub-sequences of these matrices have to be multiplied in an optimal way. We start with a sequence of $j - i + 1$ matrices and gradually decrement this number in each step until we are left with a single matrix.

$\mathsf{mult}(i,j)$

    if $i = j$ then return 0

    else

        $x \leftarrow \infty$

        for $k \leftarrow i$ to $j - 1$ do

            $y \leftarrow \mathsf{mult}(i,k) + \mathsf{mult}(k+1,j) + p_{i-1} p_k p_j$

            if $y < x$ then

                $x \leftarrow y$

        return x

ALGORITHM 2.5: *Recursive evaluation of the order to multiply matrices.*

If we start with a single matrix, that is, $i = j$, there is no cost involved; thus $\mathsf{mult}(i,i) = 0$. Otherwise, consider the situation just before the last multiplication. There are then exactly two matrices left, the first being the product of $M_i \cdots M_k$, the second of $M_{k+1} \cdots M_j$, for some index k such that $i \leq k < j$. If we were somehow to know which index k corresponds to the indices i and j defining the range of the matrices in our problem, then the optimality of the solutions of the sub-problems would imply that

$$\mathsf{mult}(i,j) = \mathsf{mult}(i,k) + \mathsf{mult}(k+1,j) + p_{i-1} p_k p_j,$$

where the last term $p_{i-1} p_k p_j$ is the cost of multiplying the two last matrices, which are of sizes $p_{i-1} \times p_k$ and $p_k \times p_j$. Lacking any knowledge about k, mult(i, j) can nevertheless be evaluated by checking all possible alternatives, since there are only $j - i$:

$$\text{mult}(i, j) = \min_{i \le k < j} \Big(\text{mult}(i, k) + \text{mult}(k + 1, j) + p_{i-1} p_k p_j \Big).$$

This yields the recursive ALGORITHM 2.5.

To evaluate the complexity of ALGORITHM 2.5, set $T(m)$ as the number of operations to evaluate mult(i, j), where $m = j - i + 1$ is the number of matrices dealt with in the sub-range at hand. We can then state that

$$T(m) = \sum_{r=1}^{m-1} \Big(T(r) + T(m - r) + c \Big),$$

where we have taken into account the two recursive calls in the assignment to y, and where the constant c is a bound on the amount of work to add the terms and to multiply $p_{i-1} p_k p_j$ for any k. This is equivalent to

$$T(m) = (m - 1)c + 2 \sum_{r=1}^{m-1} T(r), \tag{2.6}$$

which, in particular, also holds for $m - 1$, yielding

$$T(m - 1) = (m - 2)c + 2 \sum_{r=1}^{m-2} T(r). \tag{2.7}$$

Subtracting side by side (2.7) from (2.6), one gets

$$T(m) - T(m - 1) = 2\, T(m - 1) + c,$$

from which one can derive that

$$T(m) \ge 3\, T(m - 1) \ge \cdots \ge 3^{m-2}\, T(2) = \Omega(3^m). \tag{2.8}$$

So, meanwhile, it seems that we have reduced the complexity from about 4^n to 3^n, which is still exponential and does not really improve the feasibility of the solution. The reason for the large number of required steps in Eq. (2.8) is, however, again the fact that each sub-problem is resolved many times, as for the binomial coefficients of Section 2.1. Indeed, there are only $O(n^2)$ different pairs (i, j) with $1 \le i \le j \le n$, not justifying a complexity of $\Omega(3^n)$.

We shall therefore replace again the recursive calls by table lookups, and define a table Mult$[i, j]$ in which each entry holds the corresponding value mult(i, j). The question again is in which order to fill the table, so that no element that has not yet been defined is addressed. Figure 2.9 shows the

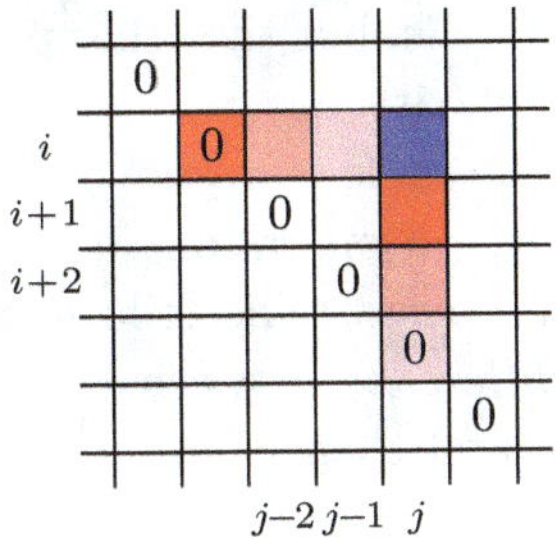

FIGURE 2.9: *Dependencies for the elements in the matrix multiplication table.*

dependencies of an element on previously defined ones. The blue element at position (i, j) is a function of elements (i, k) and $(k+1, j)$ for $i \leq k < j$, where elements belonging to the same value of k appear in matching tones of pink.

As can be seen, the value of an element of the table depends on elements to its left as well as on elements below it. This, however, disqualifies a straightforward top-down left-to-right scan, such as the schematic representation that has been crossed out at the left of Figure 2.10: the elements to the left of the bullet, representing the current element, are already known, but not those below it. To correct this, one could use a left-to-right, bottom-up scan, or inverting the order, bottom-up and internally left-to-right, as shown in the second and third examples from the left of Figure 2.10, respectively. An interesting alternative is the rightmost drawing in Figure 2.10, suggesting to process the elements mainly by *diagonals*, and within each diagonal, top-down.

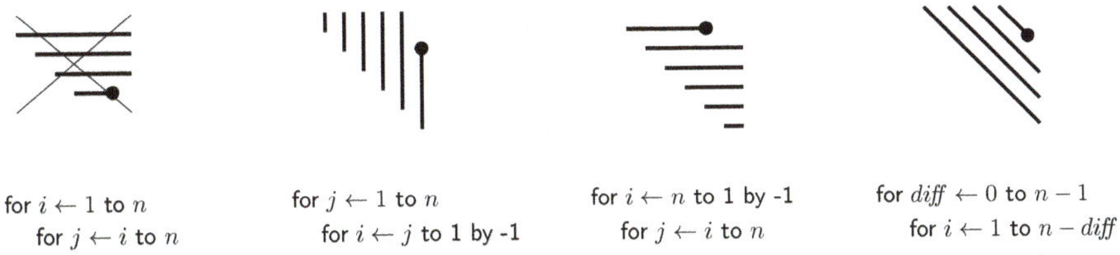

| for $i \leftarrow 1$ to n | for $j \leftarrow 1$ to n | for $i \leftarrow n$ to 1 by -1 | for $diff \leftarrow 0$ to $n-1$ |
| for $j \leftarrow i$ to n | for $i \leftarrow j$ to 1 by -1 | for $j \leftarrow i$ to n | for $i \leftarrow 1$ to $n - diff$ |

FIGURE 2.10: *Scanning orders for matrix chain multiplication.*

For each of the correct orders, the element containing the sought optimal value $\mathsf{Mult}(1, n)$ is the last to be evaluated. We shall give here the details of the approach using diagonals, but the other possibilities are just as good and achieve the same complexity.

The main diagonal of the table corresponds to $i = j$, that is, sub-ranges containing a single matrix; no operations are then needed, so the main diagonal is initialized with zeros. The following diagonals correspond to increasing values of the *difference* between j and i, meaning that we first process pairs of matrices, and then triplets, quadruples, etc. The formal description is given in ALGORITHM 2.6.

---

mult-order$(p_0, p_1, \ldots, p_n)$

    for $i \leftarrow 1$ to n do   Mult$[i, i] \leftarrow 0$
    for $diff \leftarrow 1$ to $n - 1$ do
        for $i \leftarrow 1$ to $n - diff$ do
          $j \leftarrow i + diff$
          $x \leftarrow \infty$
          for $k \leftarrow i$ to $j - 1$ do
              $y \leftarrow$ Mult$[i, k] +$ Mult$[k + 1, j] + p_{i-1} p_k p_j$
              if $y < x$ then
                  $x \leftarrow y$ $P[i, j] \leftarrow k$
          Mult$[i, j] \leftarrow x$

ALGORITHM 2.6: *Iterative evaluation of the order to multiply matrices.*

The first line initializes the main diagonal and is followed by the nested loops processing the other diagonals. The commands are similar to those of the recursive ALGORITHM 2.5, in which recursive calls to the function mult have been replaced by corresponding accesses to the table Mult. At the end, the sought optimal cost is stored in Mult$[1, n]$. One of the optimal partitions of the sequence $M_1 M_2 \cdots M_n$ yielding this optimal cost can then be reconstructed by means of the values stored in the auxiliary table P, similarly to what has been done with Table 2.3 for the Edit distance.

Finally, given a sequence of n matrices $M_1 M_2 \cdots M_n$, an optimal way to process them can be expressed by inserting pairs of nested parentheses into the string of matrix names, by applying recursive-print$(1, n)$, where the recursive function is defined in ALGORITHM 2.7.

recursive-print(i, j)

---

  if $j = i$ then print $'M_i'$
  else
      $k \leftarrow P[i, j]$
      print '('   recursive-print(i, k) recursive-print$(k + 1, j)$ print ')'

ALGORITHM 2.7: *Recursively printing the optimal multiplication order.*

The complexity of finding the optimal partition in ALGORITHM 2.6 is $O(n^3)$, much better than that of our first attempt of a recursive approach, and the complexity of printing the optimal solution in ALGORITHM 2.7 is $O(n)$. In particular, for the example given in Figure 2.6, the solution would be printed as $(A\ (B\ C))$.

These were only a few examples of dynamic programming solutions, and we shall see more examples later in this book, in the next chapter on graphs and in the chapter on intractability.

2.4   Exercises

2.1 Given are two character strings $T = T_1 T_2 \cdots T_n$ and $S = S_1 S_2 \cdots S_m$, where all characters T_i and S_j belong to some alphabet Σ. The problem is to find the *longest common subsequence* LCS of T and S, the characters of which are not necessarily adjacent in T or in S. For example, the longest common subsequence of FRIENDS and ROMANS is RNS and has length 3. Formally, we seek the longest possible sequences of indices

$$1 \leq i_1 < i_2 < \cdots < i_t \leq n \qquad \text{and} \qquad 1 \leq j_1 < j_2 < \cdots < j_t \leq m,$$

such that $T_{i_r} = S_{j_r}$ for all $1 \leq r \leq t$. In case S and T have no common characters, we might set $t = 0$. This problem has, among others, applications in bioinformatics and linguistics. Give a dynamic programming solution that is similar to the edit problem solution.

2.2 An alpinist plans a trek leading from a starting point $S = A_0$, passing through a chain of mountain chalets $A_1, A_2, \ldots, A_n$ that can be used as resting stations, to a target point $T = A_{n+1}$. The chalets are connected by trails of similar length so that it takes one unit of time to get from A_i to A_{i+1} for $0 \le i \le n$, but if the alpinist does not halt and tries to skip some rest stations, the time to traverse the second and subsequent trails is multiplied by $\alpha > 1$. Not all the chalets have the same comfort, and it takes M_i time units to fully recuperate in station A_i, for $0 < i \le n$.

For example, $(M_1, \ldots, M_6) = (2.1, 0.6, 1.8, 0.9, 1.7, 1.3)$ and $\alpha = \frac{4}{3}$. If the alpinist decides to rest only in A_2 and A_4, the total time spent on the trek will be

$$(1 + \alpha) + 0.6 + (1 + \alpha) + 0.9 + (1 + \alpha + \alpha^2) = 10.87,$$

but there are 2^n possible ways (since each chalet can be chosen as rest station or not), which cannot be checked exhaustively. Give a $O(n^2)$ dynamic programming solution for finding the fastest trek.

2.3 Given are positive integers $p_0, p_1, \ldots, p_n$ and a sequence of matrices $M_1 M_2 \cdots M_n$ such that matrix M_i is of dimension $p_{i-1} \times p_i$ for $0 < i \le n$. If we assume that all dimensions are of the same order of magnitude, that is, $O(p) = O(p_0) = \cdots = O(p_n)$, then what is, as a function of n, the largest order of magnitude of p for which the time required to multiply the n matrices will be at most equal to the time to find the optimal order for this multiplication?

2.4 Given are positive integers $p_0, p_1, \ldots, p_n$ and a sequence of matrices $M_1 M_2 \cdots M_n$ such that matrix M_i is of dimensions $p_{i-1} \times p_i$ for $0 < i \le n$. The following greedy algorithm is suggested for finding the optimal processing order: try, in each iteration, to get rid of the largest dimension. Formally, if, at some stage, the current dimensions are $q_0, q_1, \ldots, q_k$, we find an index r such that $q_r = \max\{q_j \mid 0 < j < k\}$, and the matrices to be multiplied in the following iteration are of dimensions $q_{r-1} \times q_r$ and $q_r \times q_{r+1}$. Prove that this approach yields an optimal solution, or disprove this claim by building a counter-example.

2.5 Let $\mathcal{A} = \{A_1, \ldots, A_n\}$ be a set of cities connected by a network of roads, such that there is a direct road leading from A_i to A_j for all $1 \leq i, j \leq n$. Because tunnels and bridges may limit the height of a vehicle on a given road, we are also given a matrix $h(i, j)$ of height constraints, indicating the maximal height permitted on the direct road from A_i to A_j. For a path $A_{i_1}, A_{i_2}, \ldots, A_{i_k}$, the height restriction is the minimum of the corresponding h values, that is, $\min_{1 \leq r < k} h(i_r, i_{r+1})$.

We wish to generate a matrix of global height limitations, giving, for each pair of cities in $\mathcal{A}$, the maximal possible height of a vehicle, taking into account that there may be several options to get from one city to the other. We shall use dynamic programming and introduce the number of roads used on the path as an auxiliary parameter. Define $H(u, v, k)$ as the maximal permitted height on a path of length at most k roads from u to v, with $u, v \in \mathcal{A}$. We shall ultimately be interested in $H(u, v, n - 1)$.

(1) Why are values of $H(u, v, k)$ for $k \geq n$ not interesting?
(2) What is $H(u, v, 1)$?
(3) Suppose the values of $H(u, z, 1)$ and $H(z, v, s-1)$ are already known. Explain why

$$H(u, v, s) \geq \min(H(u, z, 1), H(z, v, s - 1)).$$

(4) Build the recursive relation for $H(u, v, k)$ for all $u, v \in \mathcal{A}$ and $1 \leq k < n$. Define the boundary conditions $H(u, v, 1)$ and show that

$$H(u, v, k) = \max \Big(H(u, v, k - 1) ,$$
$$\max_{z \neq u, v} \big[\min \big(H(u, z, 1), H(z, v, k - 1) \big) \big] \Big).$$

(5) What is the complexity of the procedure?

PART 2
Graph Algorithms

Chapter 3

Minimum Spanning Trees

3.1 Graphs

This second part of the book deals with algorithms that are connected with *graphs*. I shall assume that the reader has already some basic notions on graphs, since they are usually presented in introductory courses on discrete mathematics or data structures. Nevertheless, some of the essential definitions are repeated here.

Let me start with a warning: the drawing in Figure 3.1 is often referred to as a graph, but, strictly speaking, it is just one of its many possible *representations*. Mathematically, a graph G is defined as a pair of sets $G = (V, E)$, with no restriction on the set V, called its *vertices*, whereas the set of *edges* E satisfies $E \subseteq V \times V$ and describes whether some binary relation exists between certain pairs of vertices.

The number of possible representations of the same graph G is infinite, since there are almost no restrictions on the layout of the drawing: the elements representing the vertices (circles, squares, ovals, dots, etc.) may be scattered in any visually appealing way, the edges may be straight lines or arcs or have any other form, they may cross each other, and so on. The formalism of not identifying a graph with one of its representations is thus necessary, because one can easily get drawings that are completely different, yet representing the same graph.

Indeed, the problem, known as *Subgraph Isomorphism*, of finding a given query graph G_1 within another graph G_2, is apparently a very difficult one, as we shall see in Chapter 10. This is in contrast to a textual environment, in which the equivalent problem of finding a given query string within a large repository is a relatively easy task, and we shall deal with it in Chapter 7 on pattern matching.

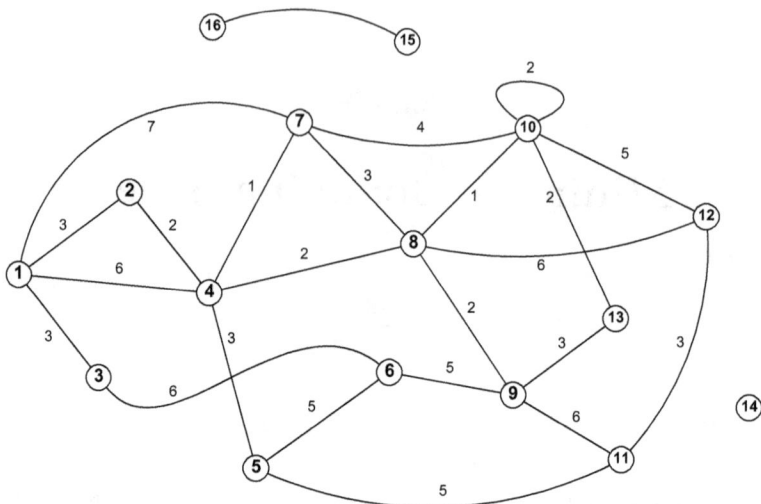

Figure 3.1: *An example graph* $G = (V, E)$ *with* $V = \{1, 2, \ldots, 15, 16\}$ *and*
$E = \{(1, 2), (1, 3), (1, 4), (1, 7), (2, 4), (3, 6), (4, 5), (4, 7), (4, 8), (5, 6), (5, 11), (6, 9), (7, 8),$
$(7, 10), (8, 9), (8, 10), (8, 12), (9, 11), (9, 13), (10, 10), (10, 12), (10, 13), (11, 12), (15, 16)\}.$

The vertices are usually denoted by $V = \{v_1, v_2, \ldots, v_n\}$, or simply
$V = \{1, 2, \ldots, n\}$. As to the edges, the extreme cases are $E = \emptyset$, the
empty graph, and $E = V \times V$, the *full* graph. Note that these notions refer
only to the number of edges, so a graph may be empty and yet have many
vertices. The complexities of algorithms involving graphs are often given
as a function of the sizes of V and E, for example, $O(|V| + |E|)$, but in
practice, one simplifies this notation by omitting the vertical bars, thereby
identifying a set with its size. It is thus acceptable to write expressions like
$O(V + E)$ when no confusion can arise.

In spite of the above warning calling us to be aware of the fact that
a drawing is *not* a graph, it is customary, and often also very helpful
as a visual aid, to represent a graph by a drawing in which the vertices appear as dots or circles, and an edge (a, b) as an arrow or some
line from the circle corresponding to vertex a to that of vertex b. Figure 3.1 is such a drawing for a graph with $V = \{1, 2, \ldots, 15, 16\}$ and
$E = \{(1, 2), (1, 3), (1, 4), (1, 7), (2, 4), \ldots, (11, 12), (15, 16)\}$, and will serve
as a running example in this chapter.

A graph in which we distinguish between edges (a, b) and (b, a) is called *directed*, and its edges are usually drawn as arrows, as we shall do in the following section. In other applications, the only relevant fact is whether there exists a connection between the elements a and b, but the direction of the edge (a, b) or (b, a) is not given any importance; such graphs are called *undirected* and their edges are shown as lines rather than arrows, like in Figure 3.1. In a strictly mathematical sense, we should then denote edges as sets $\{a, b\}$ rather than ordered pairs (a, b), but it has become common practice to use the pair notation (a, b) even for the undirected case.

A *loop* is an edge of the form (a, a), like the one emanating from and pointing to vertex 10. A sequence of edges

$$(v_{i_1}, v_{i_2}),\ (v_{i_2}, v_{i_3}),\ (v_{i_3}, v_{i_4}),\ \ldots, (v_{i_{s-1}}, v_{i_s}),$$

in which the starting point of one edge is the endpoint of the preceding edge, is called a *path* from v_{i_1} to v_{i_s}. For example,

$$(4, 8),\ (8, 9),\ (9, 13),\ (13, 10)$$

is a path from 4 to 10. If $v_{i_1} = v_{i_s}$, that is, if the last edge of a path points to the beginning of the first edge, the path is called a *cycle*. For example,

$$(3, 6),\ (6, 5),\ (5, 4),\ (4, 1),\ (1, 3)$$

is a cycle. If there is a path from every vertex to any other, the graph is *connected*. Our example graph is not connected, because, e.g., vertex 15 cannot be reached from vertex 9.

The *degree* of a vertex v is the number of edges touching it, for example, the degrees of vertices 1, 10 and 16 are 4, 5 and 1, respectively. If there is no edge touching the vertex at all, like for vertex 14, its degree is 0 and it is called *isolated*.

An important special case of a graph is a *tree*, defined as a graph that is connected but has no cycles. Our graph is not a tree, first because it is not connected; even if we remove the loop on vertex 10 and then restrict the intention to the sub-graph G' induced by the vertices $V' = \{1, 2, \ldots, 13\}$, that is, ignoring the vertices 14, 15 and 16 and the edges $(10,10)$ and $(15,16)$, the graph would still not be a tree, because it contains cycles. Our running example will be restricted in the sequel to the graph G'.

Trees have many important and useful properties. For example, the number of edges in a tree built on n vertices is exactly $n - 1$, regardless of the shape of the tree. In particular, any tree for our example graph G' has 12 edges.

It is rather regretful that there is another, quite different, notion in computer science which is also called a *tree* and represents a well-known data structure. There are certain similarities between a tree as a data structure and a tree as a special kind of graph, but there are also many differences, and applying the same nomenclature to essentially different entities may lead to confusion. We shall see an example of an algorithm using trees of both kinds in the next section in Figure 3.4.

For certain applications, a *cost* or *weight* $w(a, b)$ is assigned to each edge (a, b) of a graph, as shown in our example of Figure 3.1. The weight of a path will be defined as the sum of the weights of the edges forming the path. Depending on the context, the weights may represent the price to pay or the time necessary for using the edge, or the distance between its endpoints, and there are many other possible interpretations.

Of the many basic notions and algorithms related to graphs and trees, we have chosen to present only BFS and DFS, two standard graph exploration methods, as **Background concept** inserts.

Background concept: Exploring graphs — BFS and DFS

In many contexts, the need arises to scan the vertices of a graph in some systematic way. The most popular scan orders are known as *Breadth-first search* (BFS) and *Depth-first search* (DFS), which is most easily understood considering a tree: starting at some chosen vertex v, the vertices are explored in BFS by layers, first the immediate neighbors of v, then the neighbors of the neighbors, etc.; the order for DFS is by branches rather than layers, exploring entirely one branch before dealing with another one. Both algorithms can, however, be applied to any connected graph and not only to trees.

ALGORITHM 3.1 shows the formal algorithms for BFS and DFS, both of which get as parameters a connected graph $G = (V, E)$ and a chosen vertex $v \in E$ from which the graph exploration should emanate. The purpose of the function *mark* is to avoid the processing of a vertex more than once.

To emphasize their similarity, the algorithms are displayed side by side. Indeed, the only difference between the two is the data structure for the temporary storage of the newly encountered vertices: for BFS a *queue* Q is used, whose FIFO paradigm assures that the vertices are processed by layers, whereas the elements are stored in a *stack* S for DFS. Before being inserted into Q or S, the vertices are marked as having already been

BFS(G, v)
- mark v
- insert v into the queue Q
- while $Q \neq \emptyset$ do
    - $w \leftarrow$ extract element from Q
    - visit w
    - for all neighbors x of w
        - if x is not marked then
            - mark x
            - insert x into queue Q

DFS(G, v)
- mark v
- insert v into the stack S
- while $S \neq \emptyset$ do
    - $w \leftarrow$ extract element from S
    - visit w
    - for all neighbors x of w
        - if x is not marked then
            - mark x
            - insert x into stack S

ALGORITHM 3.1: *Breadth-first search and depth-first search.*

processed, and immediately after being extracted from Q or S, the vertices are *visited*, that is, given the desired treatment for the current context. The sequences produced by visiting our example restricted graph G' of Figure 3.1, starting at vertex 8 and visiting the neighbors of a vertex in increasing order of their indices, are:

$$\text{BFS}(G', 8) - 8, 4, 7, 9, 10, 12, 1, 2, 5, 6, 11, 13, 3$$
$$\text{DFS}(G', 8) - 8, 12, 11, 5, 6, 3, 1, 2, 10, 13, 9, 7, 4$$

The order of visiting the vertices depends on the order in which the neighbors of a given vertex are inserted into Q or S. The insertion order for our example has been chosen by increasing indices for both BFS and DFS, for example, the neighbors 4, 7, 9, 10, 12 of vertex 8 have been inserted in this order in both the queue Q and the stack S.

The complexity of both algorithms is $O(V + E)$. For certain applications, BFS is preferable, while for others, DFS might be better. Often the exact order does not matter and BFS or DFS can be applied indifferently, the goal being just an exhaustive scan without duplicates.

3.2 The Minimum Spanning Tree problem

Suppose the vertices of the undirected graph $G = (V, E)$ represent computers and that we wish to construct a communication network reaching each vertex $v \in V$. The aim is to allow all the computers to communicate with each other, possibly with the help of one or more intermediate stations. The weight $w(a, b)$ may stand for the cost of building a direct wired line connection from computer a to computer b. For topographical or other reasons, there may be vertex pairs that are not connected by an edge, so the graph is not necessarily full. The problem is to find the cheapest layout of the network of connection lines. Alternatively, one could think of the vertices as locations in an electricity network, of the edges as high-voltage transmission lines and of the weights as the costs to build the lines.

There must be a possibility to allow communication, or to transfer electricity, from every point to any other; therefore the graph G has to be connected. On the other hand, ignoring security threats, there is no reason to build a network with more than one possibility to get from a vertex a to a vertex b. The sought network should thus not include cycles, so what we are looking for is in fact a tree. A tree touching all the vertices in V is called a *spanning tree*. The challenge is to find, among the potentially large number of different possible spanning trees, one for which the cumulative cost of its edges is as small as possible.

The problem can thus be reformulated mathematically as follows: given an undirected connected graph $G = (V, E)$ with weights $w(a, b)$ for each $(a, b) \in E$, find a subset $T \subseteq E$ of the edges such that T is a spanning tree, and such that $\sum_{(a,b) \in T} w(a, b)$ is minimized over all possible choices of such trees T.

The following sections present several algorithms that have been suggested for the construction of such Minimum Spanning Trees (MST). One might object to the choice of presenting so many alternatives to answer a single, specific, graph theoretical question, while many other not less important problems will not even been mentioned in this book. The idea is that by treating the same problem from different aspects, we get insights into the considerations and the decision process leading to the final choice of the method to be applied.

3.3 Kruskal's algorithm

The following simple algorithm is due to Joseph Kruskal [Kruskal (1956)] and bears his name. It is built on a *greedy* approach, which means that one tries at each step to perform an action that seems locally to be the best, but there is no guarantee that these steps indeed lead to a globally optimal solution. In this case, however, it *does* yield an MST.

Starting with an empty set of edges S, Kruskal's algorithm inspects the edges in V by the order of non-decreasing weights, considering them as candidates to be adjoined to S, which should ultimately include all the edges of the MST. To form a spanning tree, exactly $|V| - 1$ edges have to be adjoined to S, so one tries to choose those with the smallest possible weights first to get overall a minimum. The reason for talking about candidates is that not every edge may actually be adjoined, only those that do not close any cycle with the edges already in S.

The edges inserted into S appear in Figure 3.2 in various shades of red, starting with darker ones for the two edges of weight 1. The four edges with weight 2 can be adjoined in any order. The following edges to be processed are six edges of weight 3, but two of them, $(7,8)$ and $(9,13)$, would close cycles; they are therefore skipped by Kruskal's method, which is why they appear as broken lines in Figure 3.2. Similarly, the edge $(7,10)$ with weight 4 is skipped because it would close a cycle, and of the four edges with weight 5, only two can be chosen to complete the choice of 12 edges for the MST. More precisely, the choice is restricted to one of the edges $(5,6)$ or $(6,9)$, and one of the pair $(5,11)$ or $(10,12)$.

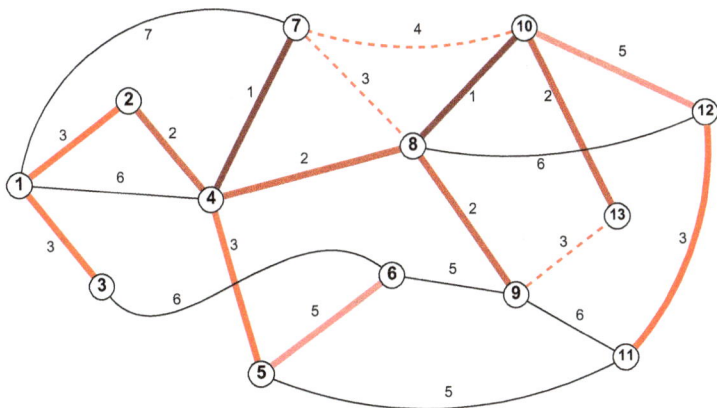

Figure 3.2: *Applying MST/Kruskal on the example graph.*

The formal algorithm appears in ALGORITHM 3.2. As a first step, the edges $e \in E$ are sorted in time $O(E \log E)$ by their weights $w(e)$ to allow access to the next edge to be processed in constant time. Actually, this might be wasteful, because only $|V| - 1$ edges are needed, so many may not even be considered. It might be more economical to build a *heap* of the weights in time $O(E)$, and then extract each element in time $O(\log E)$. In the worst case, this yields also $O(E \log E)$, but if the number of iterations is substantially lower, say of the order of $|V|$, then the global time spent on accessing the relevant edges is only $O(V \log E)$.

MST-Kruskal(V, E)

> sort edges in E by weights into non-decreasing order $e_1, e_2, \ldots, e_{|E|}$
> $S \leftarrow \emptyset \qquad\quad j \leftarrow 1$
> while $|S| < |V| - 1$ do
> > if e_j does not close a cycle with the edges in S
> > > $S \leftarrow S \cup \{e_j\}$
> >
> > $j \leftarrow j + 1$
>
> return S

ALGORITHM 3.2: *Minimum spanning tree: Kruskal.*

The main iteration of the algorithm processes the edges in the given order, and checks for each whether it closes a cycle or not. The key for a fast implementation is to know how to check this fact quickly.

Kruskal's algorithm processes the edges by weight and not by location. The set S being built is therefore not necessarily connected. For example, after having inserted three edges, one of the possibilities would be to get the set $S = \{(2, 4), (4, 7), (8, 10)\}$, consisting of two disjoint connected components. Actually, it will be more convenient to define connectivity on the set of vertices V rather than for edges. Starting with a set of $|V|$ singleton sets, each containing a single vertex $v \in V$, the addition of a new edge $(v, w) \in E$ to S causes one of the following:

(1) If both v and w belong to the same connected component, then the edge (v, w) must close a cycle, so the edge is simply ignored by the algorithm.

(2) If v and w belong to different connected components A and B, then including the edge (v, w) also in S acts as a bridge between A and B, which should henceforth be considered as a single component, so we need to merge A and B.

In terms of our running example, the partition of V after the insertion of the three edges mentioned above would be

$$\{2\text{–}4\text{–}7,\ 8\text{–}10,\ 1,\ 3,\ 5,\ 6,\ 9,\ 11,\ 12,\ 13\}.$$

If $(4,8)$ is the following edge to be adjoined, the two components 2–4–7 and 8–10 are merged, resulting in

$$\{2\text{–}4\text{–}7\text{–}8\text{–}10,\ 1,\ 3,\ 5,\ 6,\ 9,\ 11,\ 12,\ 13\},$$

with a single connected component beside the singleton vertices. If, instead of $(4,8)$, the edge $(10,13)$ is chosen, the component 8–10 has to be merged with the singleton 13, which yields the partition

$$\{2\text{–}4\text{–}7,\ 8\text{–}10\text{–}13,\ 1,\ 3,\ 5,\ 6,\ 9,\ 11,\ 12\}.$$

The question is therefore how to check whether the two endpoints of a given edge (v, w) belong to the same connected component. A first attempt to solve this problem would be to apply one of the graph exploration methods, BFS or DFS: starting from one of the vertices, say v, visit all the vertices belonging to the same connected component A as v; if w is reached at some stage, the edge (v, w) may be skipped; otherwise, w belongs to another component B, and A and B should be merged.

The problem with this approach is its complexity. BFS or DFS may require $O(E)$ steps to be applied for each edge considered by the algorithm, which could yield a complexity of $O(E^2)$, possibly as bad as $O(V^4)$. A much better solution uses a data structure called *rooted trees*, which are a part of a set of procedures known as *Union–Find* algorithms.

Background concept: Union–Find

The following scenario can be found in many applications involving sets. We are given a set $\mathcal{X} = \{x_1, \ldots, x_n\}$ of n elements and its partition into ℓ disjoint subsets $X_1, \ldots, X_\ell$, that is,

$$\mathcal{X} = \bigcup_{i=1}^{\ell} X_i \quad \text{and} \quad X_i \cap X_j = \emptyset \quad \text{for} \quad i \neq j.$$

The following two operations should be supported:

(1) Find(x)—given an element $x \in \mathcal{X}$, find the index of the only subset $X_i \subseteq \mathcal{X}$ that includes x, i.e., $x \in X_i$;
(2) Union(X_i, X_j)—merge the subsets X_i and X_j into a single subset.

We associate each element of the set $\mathcal{X}$ with a node in some tree. Each subset in the partition will be represented by a *rooted tree*, which is a tree in which every node has only one outgoing pointer, to its parent. Figure 3.3 displays an example of three such rooted trees, the roots of which are the colored nodes. Note that the pointers, symbolized by the solid arrows, are pointing upwards, so one can easily reach the root from any node in its tree, but there is no way of passing from a root to any of its descendants.

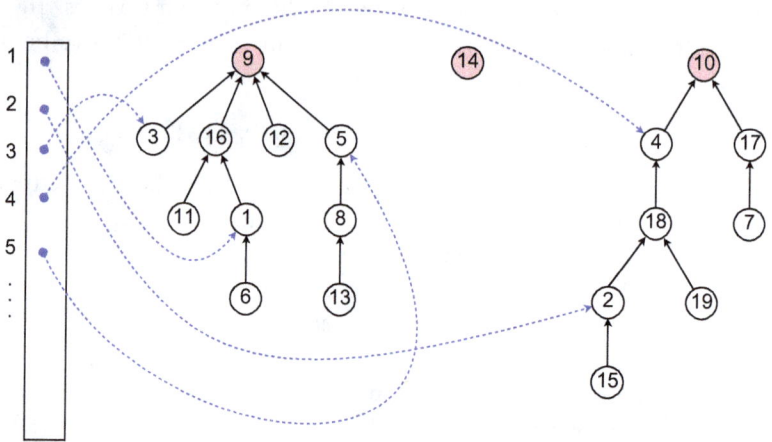

Figure 3.3: *Representing the partition of a set by rooted trees.*

By convention, the name of a set is the label of the root in its associated tree. The three trees in Figure 3.3 therefore represent the sets X_9, X_{14} and X_{10}. The Find operation is implemented by means of an array of pointers giving direct access to each of the nodes, and is shown as blue arrows in the figure. For example, to process Find(2), the array is accessed at entry 2 containing a pointer to the node v_2, where we use the notation v_x to design a node containing the value x. From v_2, one follows several parent pointers, passing through v_{18} and v_4, until reaching v_{10}, which must be the root, since it has no outgoing pointer. We conclude that $2 \in X_{10}$. The number of steps needed for Find(x) is thus related to the depth of x in its tree.

To merge two sets in $O(1)$ time for the Union(X_i, X_j) command, let the root of the smaller tree point to the root of the larger one. Even though our goal is to bound the depth of the trees for efficient Find commands, the criterion for calling a tree smaller or larger in this context will be the number of nodes in the tree, and not its depth. Thus, to perform Union(X_9, X_{10}), we set $parent(v_{10}) \longleftarrow v_9$, since the tree rooted by v_9 has 10 nodes and that rooted by v_{10} has only 8.

It can be shown that if one starts with a partition into n singletons and forms the trees by repeatedly applying Union according to the rules above, the depth of any tree, and thereby the complexity of the Find command, is bounded by $O(\log n)$. There are techniques to reduce this even further, to an average of $O(\log^* n)$, where the function $\log^*$ is called *iterative logarithm* or *log star*, and defined as the number of times the function $\log_2$ can be applied iteratively until the result is ≤ 1. For example, $\log^*(100) = 4$. This function grows extremely slowly: for all reasonable numbers n (up to 2^{65536}), $\log^*(n) \leq 5$.

Returning now to Kruskal's algorithm, if the connected components of the MST being built are managed by Union–Find, it will take at most $O(\log V)$ comparisons to check if a given edge closes a cycle; therefore the overall complexity for all the edges to be checked is bounded by $O(E \log V)$. The dominating part is therefore the sorting of the edges and we conclude that Kruskal's MST construction takes time $O(E \log E)$.

Figure 3.4 shows the forests that have been built by the algorithm after processing the edges $(4, 7)$, $(8, 10)$, $(10, 13)$, $(2, 4)$, $(4, 8)$, $(8, 9)$, $(9, 13)$, $(7, 8)$, $(11, 12)$ and $(1, 3)$, in this order, where the crossed out edges are those that closed cycles and thus did not trigger a Union operation. The upper part of the figure shows the MST forest built so far, whereas the lower part displays the rooted trees generated by Union–Find. Corresponding trees appear in matching colors. One can see that the structures of the trees may be completely different, e.g., the trees in the MST forest have no roots. Nevertheless, the partition of the vertices V into subsets remains the same.

Theorem 3.1: *Kruskal's algorithm constructs a minimum spanning tree.*

Proof: We prove, by induction on the size of the set of edges S, that at each point in ALGORITHM 3.2, there exists a minimum spanning tree T of the given graph that includes S. This is clear at the beginning, since every MST includes the empty set, and if it is true at the end, when $|S| = |V| - 1$, then S itself must be an MST.

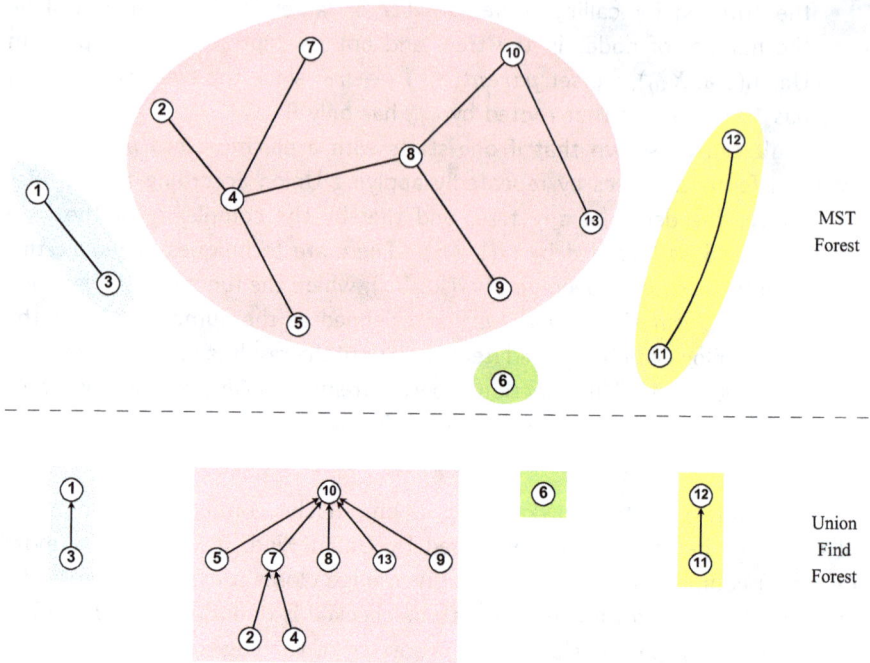

Figure 3.4: *MST versus Union–Find forests.*

For the inductive step, refer to the example in Figure 3.5 in which the edges of S are emphasized as solid or broken bold red lines. Let T_1 be an MST that includes S and also all the other red edges, solid non-bold (f,g) and (e,h), and broken non-bold (a,b).

Consider the next edge e to be adjoined to S in the following iteration. If it is one of the red edges, then it indeed belongs to T_1. Suppose then that e is not in T_1. As for any tree, the addition of an edge to T_1 closes a cycle in $T_1 \cup \{e\}$. Not all of the edges of this cycle may belong to S since e does not close a cycle in S, only in T_1. Let e' be one of the edges of the cycle not belonging to S. If we add e to T_1 but then remove e' we get another spanning tree $T_2 = T_1 \cup \{e\} \setminus \{e'\}$. But the weight of e cannot be larger than that of e', otherwise e' would have been added to S before e. The total weight of T_2 can therefore not be larger than that of T_1, so it must also be an MST. We have thus shown that there exists an MST including $S \cup \{e\}$, which concludes the proof.   □

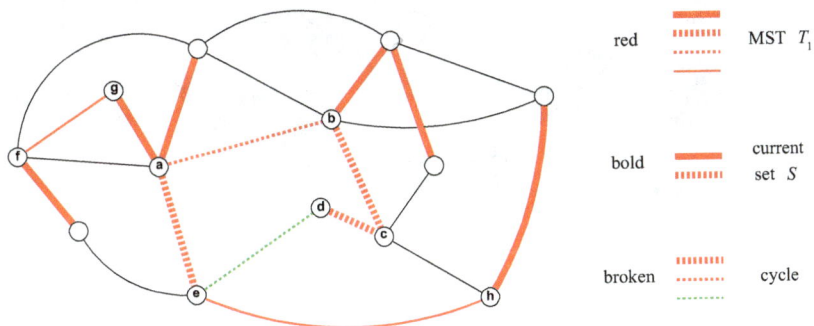

Figure 3.5: *Proof for Kruskal's MST algorithm.*

To illustrate the proof on our example, the edge $e \notin T_1$ could, for example, be the green edge $e = (e,d)$. The corresponding cycle is a, b, c, d, e, a and appears as green or red broken lines. Finally, an edge on the cycle but not belonging to S is $e' = (a,b)$.

3.4   Borůvka's algorithm

Consider a very large input graph in which one of the vertices v has only a single neighbor u, and for which the weight $w(u, v)$ of the edge connecting them is the largest among all the weights. In this case, the edge (u, v) has to be a part of any MST, as there is no other possibility to reach v, yet it will be the last edge to be considered according to Kruskal, even if an *almost* spanning tree with $|V| - 2$ edges has already been formed long before processing (u, v).

The following algorithm, named after its inventor Otakar Borůvka [Borůvka (1926)], alleviates the problem by basing the main loop on processing the vertices rather than dealing with the ordered list of edges as suggested by Kruskal. The minimum spanning tree will be built incrementally in stages, each such stage adding more edges until a tree is formed.

For the first stage, the vertices of V are processed in some order, and each vertex v selects an edge of minimal weight among those edges touching it. An edge can therefore be selected twice, if it happens to have minimal weight for both of its endpoints. All the selected edges are added, as a set, to an initially empty set S, which will ultimately contain the MST. If there are more than one such minimal weight edges, the decision about which one to choose between those may be arbitrary, as long as priority is given

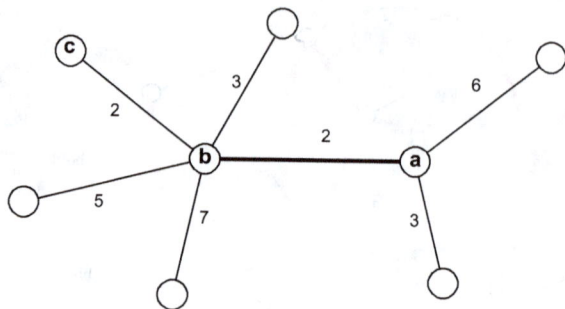

Figure 3.6: *Example of choice by priority for Borůvka's algorithm.*

to an edge that has already been selected by another vertex, if there is such an edge.

For example, if vertex v_a is processed first in Figure 3.6, it must select vertex v_b because $w(a, b) = 2$ is minimal among the edges emanating from v_a. But v_b has then, *a priori*, two plausible choices, v_a or v_c. By the priority rule, v_b has to select v_a. Figure 3.7 displays the graph of our running example after the first stage; the selected edges appear in red.

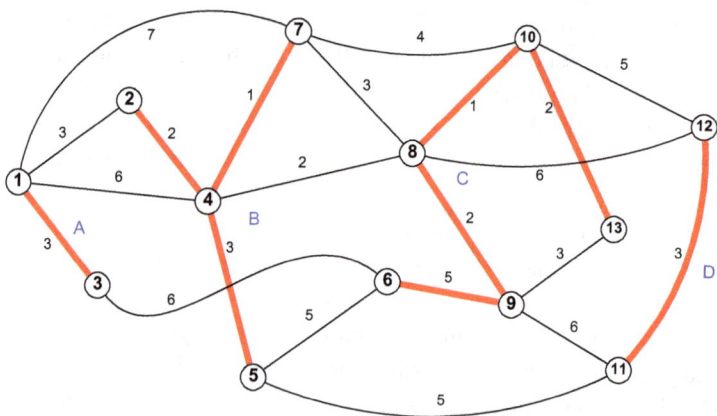

Figure 3.7: *Applying MST-Borůvka — first stage.*

If the graph is implemented by means of an *adjacency list*, that is, for every vertex $v \in V$, a list E_v of edges having v as one of their endpoints is given, then the number of comparisons in the first stage for finding the minima for all the vertices is

$$\sum_{v\in V}(|E_v|-1) = 2|E| - |V| = O(E),$$

since every edge in E is considered exactly twice.

Since every vertex chooses a minimal weight edge, the set S consists of connected components of overall minimal cumulative weight at the end of the first stage. We claim that each of these connected components must be a tree.

To see why this is true, assume that while processing the vertices in the given order, a cycle is formed as, for example, shown in Figure 3.8: v_a selected vertex v_b, which in turn chose v_c, etc., continuing the chain up to v_x, which selected v_b and thereby closed a cycle. According to the priority rule, the weight $w(v_b, v_c)$ must be strictly smaller than $w(v_a, v_b)$, because if it is larger or even equal, v_b would have chosen v_a. Similarly, $w(v_c, v_d) < w(v_b, v_c)$, and continuing the chain of strictly decreasing weights, one finally gets that

$$w(v_x, v_b) < \cdots < w(v_c, v_d) < w(v_b, v_c).$$

But this is a contradiction to the very existence of such a cycle, because if indeed the weight $w(v_x, v_b)$ is strictly smaller than $w(v_b, v_c)$, then v_b would have chosen v_x in the selection process rather than v_c. The conclusion is that the connected component has to be cycle free, so it must be a tree.

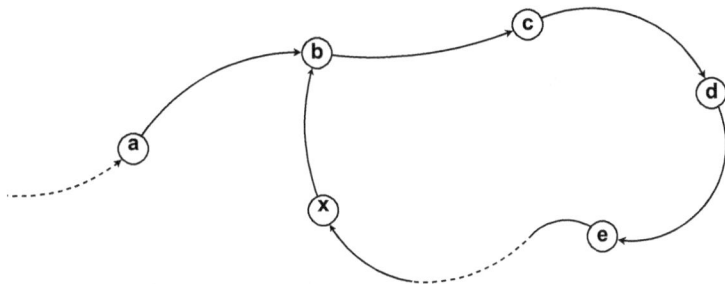

Figure 3.8: *Why the connected components are cycle free.*

If at the end of this first stage, the number of connected components is just one, then this is a tree and not a forest, so it is an MST. In many cases, however, the number of connected components will be larger, as in our running example in Figure 3.7, for which it is four.

A key observation for Borůvka's algorithm is that we are faced, at the end of the first stage, with a problem similar to that in the beginning: the exact internal structure of the trees forming the connected components is not relevant any more, and all that is needed is to find the cheapest ways to connect the components, regardless of which vertices within the trees are actually used for building the connection. We can thus conceptually consider each of the connected component at the end of the first stage as if it were a single vertex in some new graph $G' = (V', E')$, as the one shown in Figure 3.9. The second stage of Borůvka's algorithm is then similar to the first stage, just working on G' instead of on G: if the new vertices are scanned in alphabetic order, then A chooses the edge with weight 3, B chooses (B, C) with weight 2 and so does C, finally D has two choices, the cheapest edge having weight 5, and we have chosen (B, D). Figure 3.9 shows these edges in green.

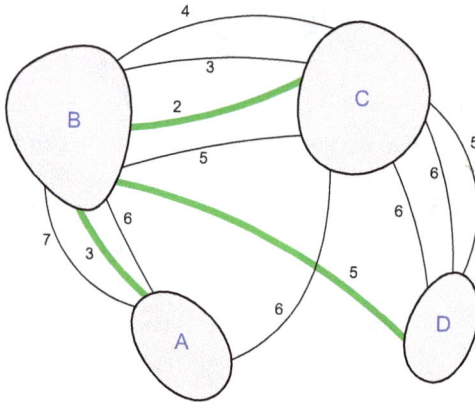

Figure 3.9: *Borůvka's algorithm: redefining the graph.*

Formally, G' is not really a graph. Any pair of vertices defines at most one edge, which is why the number of edges in a graph is at most $|V|^2$, and having such a restriction is often convenient for bounding the complexity of graph algorithms. If more than a single edge is allowed between two given vertices, we talk about a *multi-graph*, similarly to the extension of the notion of a set (to which an element can only belong once) to a *multi-set*. For example, the set $\{1, 2, 2, 1, 1\}$ contains only two elements; but regarding it as a multi-set, it contains five.

In our case, however, there is no such disadvantage in the newly defined multi-graph, because the set of edges E' is a subset of the original set E, so the same bound applies. In practice, the multi-graph can be implemented by considering sets of vertices, with the help of the Union–Find procedures seen earlier. The edge of minimum weight emanating from a connected component X can thus be found by iterating over the vertices forming X. The complexity of the second stage, excluding the cost of the bookkeeping commands to handle the sets, is therefore $O(E') \leq O(E)$, as for the first stage. For the technical details of working with a multi-set instead of a set, without exceeding the bound $O(E)$ per stage, see Exercise 3.1.

How many stages do we need? For our example, two stages did suffice to end up with an MST, as can be seen in the summarizing Figure 3.10, in which the red or green edges form the MST. But this may not be true in general, where three and more stages might be needed. All these stages start by forming new connected components (applying Union) and then running over all of them to select minimum weight outgoing edges. The complexity of each of the stages is thus $O(E)$.

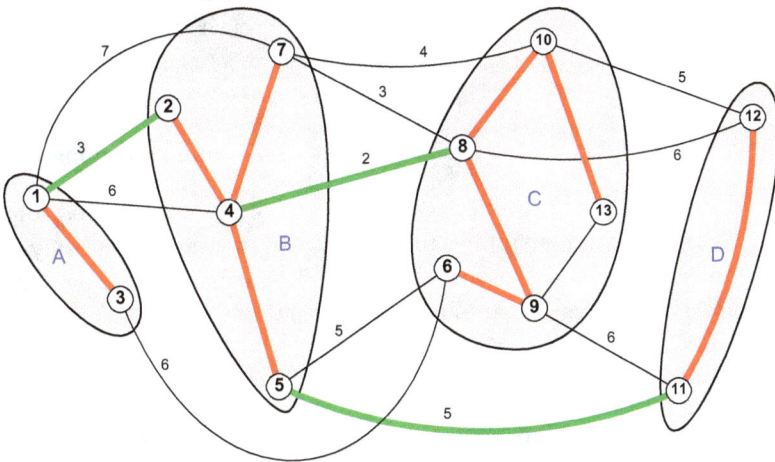

Figure 3.10: *Borůvka's MST algorithm.*

ALGORITHM 3.3 shows the formal algorithm. The set $\mathcal{C}$ contains the connected components, initially $n = |V|$ singletons, one for each of the vertices. The MST is built in the set S and is initially empty. Each iteration of the main while loop corresponds to one of the stages. For each stage, every

MST-Borůvka(V, E)

   $\mathcal{C} \leftarrow \{\{v_1\}, \{v_2\}, \ldots, \{v_n\}\}$
   $S \leftarrow \emptyset$

  while $|\mathcal{C}| > 1$ do
     for each set $C \in \mathcal{C}$ do
        $e \leftarrow$ minimum weight edge with one endpoint in C
          priority to edges already in S
        $S \leftarrow S \cup \{e\}$
     redefine $\mathcal{C}$ as the connected components in graph (V, S)
        merging components from previous iteration
  return S

ALGORITHM 3.3: *Minimum spanning tree: Borůvka.*

component chooses the cheapest outgoing edge, handling ties as explained to avoid the formation of cycles. At the end of each stage, the partition of V into the set $\mathcal{C}$ of connected components is redefined with the help of Union–Find commands.

Theorem 3.2: *Borůvka's algorithm constructs a minimum spanning tree.*

Proof: The fact that at the end of each stage, the current connected components form a minimum spanning forest is straightforward, since only minimum weight outgoing edges have been used to connect the components defined in the previous stage. When the number of components is reduced to 1, the forest has become a tree. □

To get a bound on the number of stages, consider the worst possible scenario, in which the components are the smallest and contain just two vertices. This will happen if repeatedly some x chooses y and then y chooses x, as depicted in the example in Figure 3.11.

In this case, the number of components halves at the end of each stage. Since we start with $n = |V|$ vertices, we have at most $\frac{n}{2}$ components at the end of the first stage, and in general $\frac{n}{2^i}$ components at the end of the i-th stage. The algorithm stops when a single tree is reached, that is, $\frac{n}{2^i} = 1$, so the number of stages cannot exceed $\log_2 n$. The worst-case complexity of Borůvka's algorithm is therefore bounded by $E \log V$.

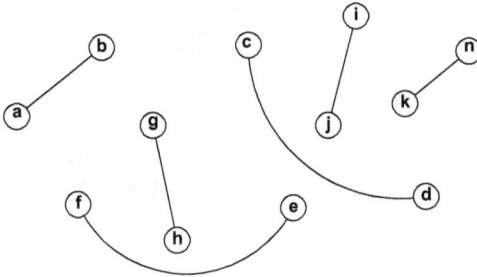

Figure 3.11: *Worst case of the number of components in each stage.*

Compare this with the algorithm of Kruskal in the preceding section.

- Kruskal's complexity is $O(E \log E)$, so there is *a priori* an improvement, as we consider only graphs for which $|E| > |V|$; otherwise, there is no connectivity and no MST could exist.
- On the other hand, $|E| \leq |V|^2$, implying that $O(E \log E) = O(E \log V)$.
- A closer look shows that one of the reasons for getting $O(E \log E)$ for Kruskal is the need to sort the edges by weight, and we know that no faster sort by comparisons exists. But for Borůvka, the $\log V$ factor is the number of stages in the worst case, so in practice, this can be faster than Kruskal's algorithm.
- Another advantage of Borůvka's method is that it is the basis of Yao's algorithm, to be seen next.

3.5 Yao's algorithm

Borůvka's algorithm seeks, for each vertex v, an edge of minimum weight emanating from v, but this operation has to be repeated, on only slightly modified graphs, for every stage. This suggests that it might possibly be worth to *sort* the edges in the adjacency list E_v of v by their weights in a pre-processing phase, which will subsequently enable access to the minima in constant time. However, sorting these lists can itself require, in the worst case, a time of

$$\sum_{v \in V} |E_v| \log |E_v| = \log |V| \sum_{v \in V} |E_v| = O(E \log V), \tag{3.1}$$

so nothing would be gained. An elegant trade-off has been suggested in [Yao (1975)]: considering the weights in the adjacency lists, if not sorting them at all or fully sorting them results in the same complexity, albeit for different reasons, maybe some *partial* sorting could improve these extreme cases?

Many alternatives exist for the definition of what could be called a partial sort. Here is what Yao suggested. We first choose some integer K as a parameter, and keep it fixed throughout the algorithm. We shall decide on an advantageous value of K only during the complexity analysis. Given a set A of n elements, the partial sort partitions A into K disjoint sets $A_1, A_2, \ldots, A_K$, all of the same size $\lceil \frac{n}{K} \rceil$ (except, possibly, the last), so that the smallest elements are in A_1, the next smallest are in A_2, etc., and the largest are in A_K. The sorting is partial because the sets A_i are not necessarily ordered internally. Formally,

$$A = A_1 \cup A_2 \cup \cdots \cup A_K, \qquad A_i \cap A_j = \emptyset \text{ for } i \neq j \qquad (3.2)$$

$$\forall a \in A_i \quad \forall b \in A_j \quad i < j \ \rightarrow \ a \leq b.$$

As an example, consider the infinite expansion of π and define the elements of the sequence A as the 25 first pairs of digits after the decimal point. The first line of Figure 3.12 displays the original sequence, whereas in the second line, the elements are grouped, from left to right, into five parts $A_1, \ldots, A_5$ of five elements each.

We see that the sets A_i are not sorted internally; in our example, after having determined the partition into subsets, the order within each A_i is induced from the order in the original sequence A.

14 15 92 65 35 89 79 32 38 46 26 43 38 32 79 50 28 84 19 71 69 39 93 75 10

14 15 26 19 10 │ 35 32 38 32 28 │ 46 43 38 50 39 │ 65 79 71 69 75 │ 92 89 79 84 93

Figure 3.12: *Example of partial sort for Yao's algorithm.*

One can now apply Borůvka's algorithm, but considering, in the first stage, only the first set in the partition of each adjacency list, since it is there where the minimum weight will be found. At the end of each stage, the elements in the adjacency lists that are not useful anymore are tagged and will not participate in the search for the minimum weight edge. The tagged elements are those edges having both of their endpoints in the same connected component. Figure 3.13 shows an example for $K = 3$, in which the tagged elements appear as black dots.

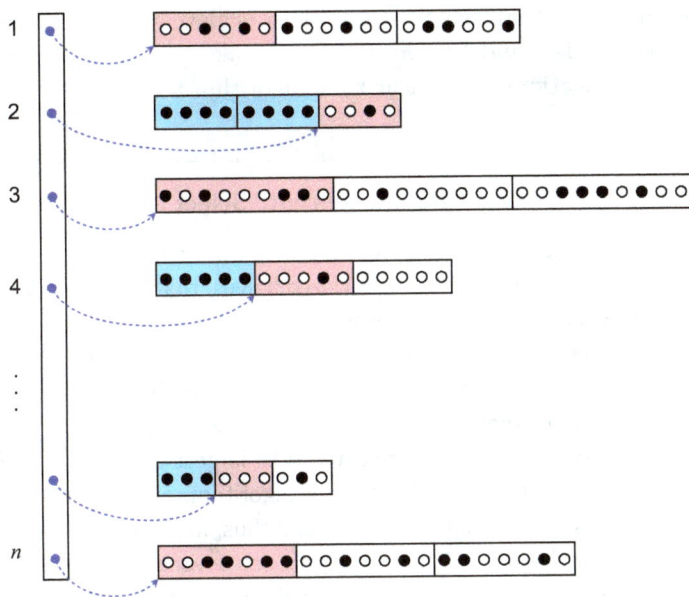

Figure 3.13: *MST–Yao.*

When all the edges in the first set of the adjacency list of a vertex v are tagged, like for vertex 4 in the example, we shift our attention to the second set of the partition. When the second set also contains only tagged elements, we turn to the third set, etc, like for vertex 2 in the example. Yao's algorithm therefore maintains an array of pointers p_v giving, in each stage and for each vertex v, the currently "active" set of size $\frac{|E_v|}{K}$ in the partition of the adjacency list of v. These sets are colored light red in Figure 3.13, the sets including only tagged elements appear in light blue, and the pointers p_v are shown as blue broken line arrows.

The main difference between Yao's and Borůvka's algorithms is then that a minimum is sought for a set of size $\frac{|E_v|}{K}$ rather than $|E_v|$, and this remains true for each vertex v and at each stage. The time spent on each stage is then

$$\sum_{v \in V} \frac{|E_v|}{K} = \frac{1}{K} \sum_{v \in V} |E_v| = \frac{2|E|}{K},$$

and we get $O(\frac{E \log V}{K})$ for the $\log V$ stages together. We ignore in this simplified analysis the processing of the connected components (see Exercise 3.2 for details).

The question that remains is how to achieve the partial ordering of the adjacency lists and how long this will take. The solution is based on repeated applications of a linear time algorithm for finding the *median* of a set.

Background concept: Select and median

The median of a set A of size n is an element $m \in A$ that could partition the set into about equal sized halves, that is, if A would be sorted, then m would be placed in position $\lfloor \frac{n}{2} \rfloor$. A straightforward algorithm to find the median is therefore to sort the set A, which requires $\Omega(n \log n)$ time. But this might be an overkill, since for the same price we would get much more than just the median, for example, one could access any required percentile and, in fact, the k-th largest element of A, for any k.

In fact, there exists a linear time algorithm, often called $Select(A, k)$, for finding such a k-largest element and thus, in particular, also the median by applying $Select(A, \frac{n}{2})$. The details are beyond the scope of this book but can be found in most books on data structures, like [Klein (2016)]. We just mention here that the algorithm is an extension of the *Quicksort* sorting procedure belonging to the Divide and Conquer family discussed in Chapter 1, and that its running time is $O(n)$ in the worst case.

To see how to achieve the requested partial sorting into K subsets for a given set A of size n, consider increasing the values of K, starting with $K = 2$. For this special case, one may find the median m in time $O(n)$, and then build the sets A_1 and A_2 defined in Eq. (3.2) by comparing each element of A with m.

For $K = 4$, all we need is to repeat this process on each of the subsets A_1 and A_2. If the partition of A into two parts took time cn for some constant c, the partitions of each of A_1 and A_2 into two parts will take time $c\frac{n}{2}$, with the same constant c. The partition into four parts thus takes time $c\left(n + \frac{n}{2} + \frac{n}{2}\right) = 2cn$. Similarly, a partition into eight parts will take $c\left(n + 2\frac{n}{2} + 4\frac{n}{4}\right) = 3cn$, and generally, a partition into 2^{i+1} parts will take $c\left(n + 2\frac{n}{2} + \cdots + 2^i\frac{n}{2^i}\right) = (i+1)cn$. It thus takes $O(n \log K)$ to partition a set A into K parts, and summing for all adjacency lists E_v, the pre-processing part of Yao's algorithm takes time $O(E \log K)$.

If we add the pre-processing part of the partitioning to the time taken to find the minimum weight edges, we get as overall complexity of Yao's algorithm $O(f(K))$, where f is a function of the parameter K defined by

$$f(K) = E \log K + \frac{E}{K} \log V. \tag{3.3}$$

The first term is an increasing and the second a decreasing function of K, and the extreme values are

- $K = 1$, for which there is no pre-processing and we get just the complexity of Borůvka's method;
- $K = |V|$, which means that the subsets A_i of the partition in Eq. (3.2) are all singletons, so that in fact, all the adjacency lists are fully, and not just partially, sorted. The pre-processing time alone is then already $O(E \log V)$, as seen earlier in Eq. (3.1).

To find a good trade-off, consider f as if it were a continuous function and find a minimum by taking its derivative:

$$f'(K) = \frac{1}{\ln 2} \left(\frac{E}{K} - \frac{E \ln V}{K^2} \right).$$

One gets $f'(K) = 0$ for $K = \ln V$, and checking the second derivative, we see that $f''(\ln V) = \frac{1}{\ln 2} \frac{E}{\ln^2 V} > 0$, so that $K = \ln V$ is a minimum. Plugging this value of K into Eq. (3.3), we conclude that the complexity of Yao's MST algorithm is

$$f(\log V) = O(E \log \log V + E) = O(E \log \log V). \tag{3.4}$$

This is almost a linear function of the number of edges, because the $\log \log V$ factor grows very slowly: it will be at least 6 only for graphs with more than 16 billion billions of nodes!

Since Yao's algorithm only adds a pre-processing phase to improve the time complexity, without changing the approach of Borůvka's method of choosing minimum weight outgoing edges for each connected component, the correctness of Yao's algorithm follows from that of Borůvka.

3.6   Prim's algorithm

The last method we shall see for the construction of an MST is known as Prim's algorithm [Prim (1957)]. It starts with an arbitrary vertex and grows a single tree by adjoining a minimum weight edge leaving the currently built single connected component.

Figure 3.14 shows the result of applying Prim's algorithm to our running example, starting with vertex v_4. The edges of the MST appear in various

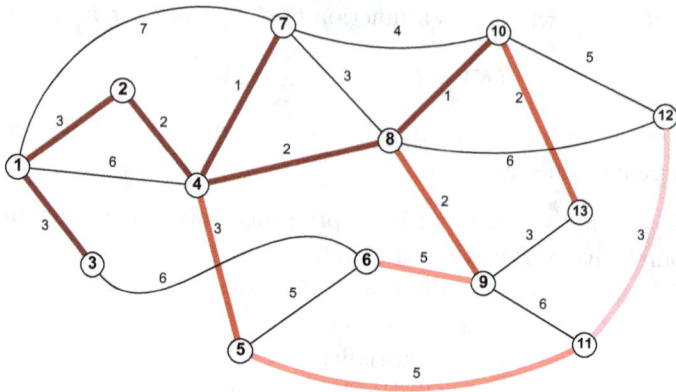

Figure 3.14: *Prim's MST algorithm.*

shades of red, from darker ones for the first edges to be added, to the lighter ones for those adjoined later to the tree.

The details, including the formal description, proof of correctness and complexity analysis, are deferred to the next chapter dealing with the shortest paths in a graph. The reason is that we shall see the algorithm of Dijkstra in Chapter 4, and the similarity between Dijkstra's and Prim's methods is so strong that the algorithms will appear side by side to highlight their differences. We shall thus treat both algorithms together, in spite of the fact that they solve different problems.

3.7   Exercises

3.1 Given a graph $G = (V, E)$, Borůvka's algorithm spends $O(E)$ on each stage for finding the minimum weight outgoing edges from each vertex, for a total of $O(E \log V)$. Only in the first stage each vertex is considered on its own, while in subsequent stages, the "vertices" are in fact connected components. Show how to implement the necessary Union and Find commands to manage the components, so that the total complexity of the MST algorithm remains bounded by $O(E \log V)$.

3.2 Extend Exercise 3.1 to Yao's algorithm, discussing how to implement the updates of the connected component and the resulting complexities.

3.3 Yao's algorithm is based on using a linear time method to find the median of an unordered set. What would be the complexity of Yao's method if a $\theta(n \log n)$ time procedure is used for finding the median?

3.4 A *bitmap* is a simple data structure often used to represent a set, for example in information retrieval systems. If the maps are sparse, that is, the number of 0-bits is much larger (or much smaller) than that of 1-bits, the maps can be compressed efficiently. Suppose that we are given a set of n bitmaps $\mathcal{B} = \{B_1, \ldots, B_n\}$, all of the same length ℓ and suppose that certain bitmaps A and B are correlated and include similar bit-patterns. Instead of compressing A and B individually, we take advantage of their similarity to produce a new map C by recording only the differences between A and B. This can be done by defining $C = A$ XOR B, which may be sparser than both A or B, and thereby more compressible. We keep A and C instead of A and B, since B can be recovered by simply XORing again: $B = A$ XOR C.

The exercise consists of generalizing this idea to n bitmaps. The problem is to find a partition of the set of bitmaps $\mathcal{B}$ into two disjoint subsets: $\mathcal{I}$, the bitmaps that will be compressed individually, and $\mathcal{X}$, those that will be XORed before being compressed. We also look for a function g from $\mathcal{X}$ to $\mathcal{B}$, assigning to each bitmap $B \in \mathcal{X}$ a bitmap $B' \in \mathcal{B}$ with which it should be XORed, subject to the constraints that:

(1) for all bitmaps $B \in \mathcal{X}$, the sequence $B, g(B), g(g(B)), \ldots$, as long as it is defined, should not contain any cycle;
(2) the total number of 1-bits in the set of compressed bitmaps,

$$\sum_{B \in \mathcal{I}} popc(B) + \sum_{B \in \mathcal{X}} popc(B \text{ XOR } g(B)),$$

should be minimized over all possible partitions and choices of g, where $popc(B)$ is a function known as *popcount*, returning the number of 1-bits in the bitmap B.

Hint: There is a good reason why this exercise appears in this chapter.

3.5 Suppose an electricity network is a graph of high-voltage transmission lines connecting n vertices. The probability for the direct line from a to b to work properly, and not being subject to some technical failure, is $p(a, b)$. We wish to build a network with a minimal number of edges, that is, a spanning tree of the graph. In this case, a single failure at one of the edges of the tree will disconnect the graph. How can the tree be chosen, so that the probability of keeping a connected graph is maximized? You may use any MST algorithm as a black box, without adapting it.

Chapter 4

Shortest Paths

This chapter deals with one of the fundamental graph problems for which the translation of a real-life situation to an abstract description in mathematical terms seems most natural. Given is a map (a graph $G = (V, E)$) with locations (vertices V) and paths or roads (edges E) connecting them, and we wish to find a path from some source location, a special vertex we call s, to a target vertex t, in the fastest, or cheapest, or shortest, of the possible ways. Though the formulation of these optimization problems might look different at first sight, depending on the units in which the *cost* or *weight* $w(a, b)$ of an edge $(a, b) \in E$ is expressed, be it minutes, dollars or kilometers, the reader will realize that all these problems are essentially equivalent. We shall adopt the language of shortest or cheapest paths, but these should be understood in a broader sense encompassing also similar challenges.

There is no need to stress the importance of finding efficient solutions to such problems, which we encounter on a daily basis, for example, in our GPS-guided navigation software. We shall see that the usefulness of shortest path algorithms goes beyond these straightforward applications, and may extend to areas that seem to have little in common with graphs.

4.1 Classification of path problems

The generic formulation of problems for finding shortest paths can be classified into three conceptual layers. All the problems work on a given directed input graph $G = (V, E)$, with weights $w(a, b)$ assigned to its edges.

- The lowest layer is the **individual** one, in which, in addition to the graph G, also a *source* vertex s and a *target* vertex t are given, and the question is to find a shortest path from s to t. Formally, we seek a sequence of vertices $X = (x_0 = s, x_1, \ldots, x_{r-1}, x_r = t)$, with $x_i \in V$, such that

$$\sum_{i=0}^{r-1} w(x_i, x_{i+1})$$

is minimized over all possible choices of the sequence X. This is the most basic level, corresponding to somebody who wants to travel from s to t.

- The intermediate layer, called the **local** one, has only a source vertex s as parameter and looks for shortest paths from s to all other vertices in V. This layer corresponds to a travel agency in our local neighborhood: most of its clients are likely to start their journey from s, but with various possible destinations, so the travel agency might prepare a list of optimal paths in advance.

- In the highest layer, the **global** one, no source or target is determined, and the aim is to derive shortest paths from any point to any other in the graph. This corresponds to a global travel agency with no specific location, such as those working on the Internet, with clients (and thus travel origins) from anywhere in the world.

Obviously, a solution to the problem of a higher layer solves also instances of lower layers, and if we knew how to deal with the individual version, one could derive solutions for the local and global ones by applying a single or two nested loops. It would thus seem reasonable to start our study with the individual layer, but surprisingly, no good algorithm solving it has been found so far. The best approach for the lowest layer is to consider the second, local, layer, and to stop the execution of the procedure as soon as the sought for target vertex t has been reached. In the worst case, this will happen at the end, so the time complexity for the first layer will be the same as for the second.

4.2   Single source, multiple destinations

The following algorithm solving the shortest path problem is due to Edsger W. Dijkstra (1959) and uses the additional constraint that all the involved weights are non-negative, that is, $w(a, b) \geq 0$ for all $(a, b) \in E$. This does not seem to be a severe limitation as far as negative distances or negative time would be concerned, but negative costs, actually earnings instead of expenses, are certainly conceivable. Therefore, if one needs also negative weights, an alternative algorithm should be used, for instance, the one to be seen in the following section for the third layer.

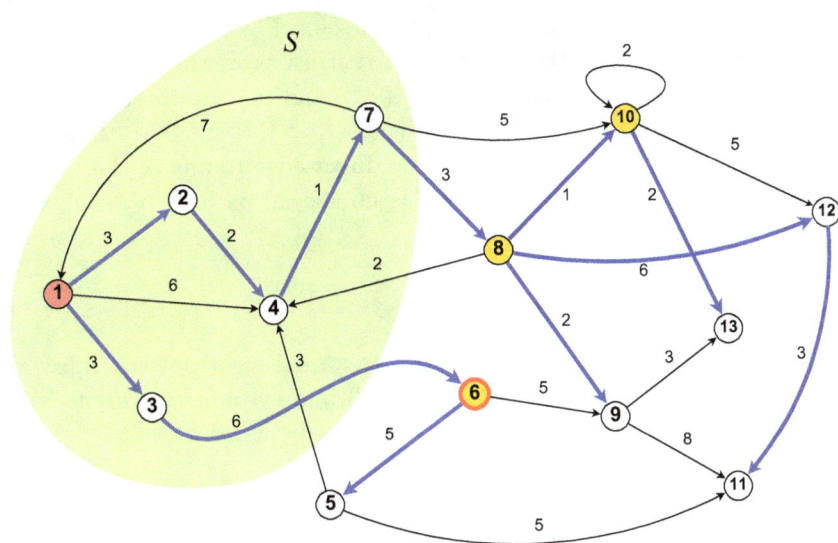

Figure 4.1: *Dijkstra's shortest path algorithm with source v_1.*

Refer to our slightly adapted running example in Figure 4.1, in which the pink vertex v_1 has been chosen as source. Dijkstra's algorithm shares certain features with the dynamic programming methods we have seen in Chapter 2, in particular we shall, in a first stage, derive only the *costs* of the shortest paths, and then see in a second stage how to adapt the algorithm to get also the associated *paths* themselves.

The expected output is thus not just a single value, but the cost of the shortest path from the source v_1 to v, for all $v \in V \setminus \{v_1\}$. These costs will be stored, at the end of the execution of the algorithm, in an array $D[v]$, called in tribute to its inventor or because its elements store distances from v_1 to v.

In addition, the algorithm builds incrementally a set S of vertices, starting with $S = \{v_1\}$ and adjoining a single vertex in each iteration, until $S = V$ at the end. The set S contains at each step those vertices v for which the cost of the shortest path from the source to v is already known and stored in $D[v]$; for the other vertices $y \in V \setminus S$, the set of vertices not yet included in S, $D[y]$ will be the cost of the shortest *special path* from the source to y.

A special path from the source to $y \in V \setminus S$ is defined as a path all of whose vertices belong to S, except y itself. Figure 4.1 displays the set S in light green, at the stage when it consists of the vertices $S = \{v_1, v_2, v_3, v_4, v_7\}$. The yellow vertices v_6, v_8 and v_{10} are those to which there are special paths. No special paths exist, at this stage, to the white vertices outside of S, because there is no direct edge to any of them from some vertex in S (in particular, there is no special path to v_5, since the edge (v_5, v_4) is directed from v_5 to v_4).

4.2.1   Algorithm

The left part of ALGORITHM 4.1 shows the formal algorithm of Dijkstra, whose input is a directed graph $G = (V, E)$ and a source vertex $s \in V$. Prim's algorithm on the right-hand side will be discussed later.

	Dijkstra(V, E, s)	Prim(V, E, s)
1	for $v \in V \setminus \{s\}$	for $v \in V \setminus \{s\}$
2	if $(s, v) \in E$ then $D[v] \leftarrow w(s, v)$	if $(s, v) \in E$ then $W[v] \leftarrow w(s, v)$
3	$\quad\quad\quad\quad\quad\quad\quad P[v] \leftarrow s$	$\quad\quad\quad\quad\quad\quad\quad P[v] \leftarrow s$
4	else $\quad\quad\quad\quad D[v] \leftarrow \infty$	else $\quad\quad\quad\quad W[v] \leftarrow \infty$
5	$S \leftarrow \{s\}$	$S \leftarrow \{s\}$ $\quad\quad T \leftarrow \emptyset$
6	while $S \neq V$ do	while $S \neq V$ do
7	$\quad x \leftarrow z \in V \setminus S$ with minimal $D[z]$	$\quad x \leftarrow z \in V \setminus S$ with minimal $W[z]$
8	$\quad S \leftarrow S \cup \{x\}$	$\quad S \leftarrow S \cup \{x\}$
9		$\quad T \leftarrow T \cup \{(x, P[x])\}$
10	$\quad$ for all $y \in V \setminus S$ do	$\quad$ for all $y \in V \setminus S$ do
11	$\quad\quad$ if $D[x] + w(x, y) < D[y]$ then	$\quad\quad$ if $w(x, y) < W[y]$ then
12	$\quad\quad\quad D[y] \leftarrow D[x] + w(x, y)$	$\quad\quad\quad W[y] \leftarrow w(x, y)$
13	$\quad\quad\quad P[y] \leftarrow x$	$\quad\quad\quad P[y] \leftarrow x$

ALGORITHM 4.1: *Dijkstra's shortest path and Prim's minimum spanning tree.*

Since at initialization, the set S contains only the source vertex s, there are special paths only to the direct neighbors of s, which are the vertices v for which $(s, v) \in E$. The cost of the path to v is then simply the weight of the only edge in the path, so we can set $D[v] \leftarrow w(s, v)$ in line 2. The values in array D for all the other vertices are initialized as ∞ in line 4.

In the main loop 6–13, the new vertex x of $V \setminus S$ to be adjoined to S is chosen in line 7 as one whose current value in D is the lowest. Then the array D has to be updated in lines 10–13 for all vertices y remaining in $V \setminus S$. Two possibilities arise as to whether or not the addition of vertex x to S has an effect on $D[y]$:

(1) if yes, this means that there is a new special path from the source to y passing through x, and that the cost of this path is lower than the current value of $D[y]$; in this case, $D[y]$ is updated in line 12;

(2) if not, the value of $D[y]$ does not change, and no action needs to be taken.

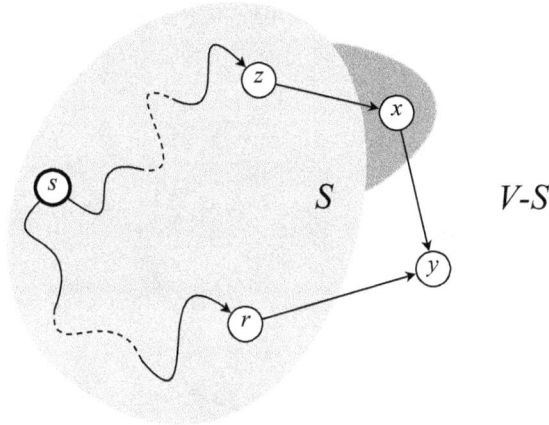

Figure 4.2: *Adjoining a new vertex x to S in Dijkstra's algorithm.*

This is schematically depicted in Figure 4.2, showing two paths from the source s to a vertex y. The set S is shown in light gray and is extended to include also the darker gray area by adjoining the vertex x. Before adding x to S, there is only one special path to y, the one passing through vertex r, and its cost is stored in $D[y]$. The inclusion of x in S creates a new special path to y, passing through x; its cost is $D[x] + w(x, y)$. By comparing these values in line 10, the algorithm decides on the new value of $D[y]$. It will be convenient to define the weight of non-existing edges as infinity, that is,

$$(x, y) \notin E \qquad \longrightarrow \qquad w(x, y) \leftarrow \infty.$$

As mentioned earlier, we have so far only evaluated the *cost* of a shortest path from the source vertex s to any target vertex v. If one is interested in the actual shortest paths from s to any other vertex in the graph, it seems that the size of the output may be up to $\theta(V^2)$, since a path in the graph can consist of up to $|V|$ vertices. There must, however, be quite some overlap between all these shortest paths, as implied by the following:

Lemma 4.1: *For a weighted directed graph $G = (V, E)$ and any pair of vertices $a, b \in V$, let $P(a, b)$ denote the shortest path from a to b. Let x and y be two vertices belonging, in this order, to this path. Then restricting $P(a, b)$ to the segment between x and y yields a shortest path $P(x, y)$ from x to y.*

Proof: Consider Figure 4.3 in which $P(a, b)$ consists of the solid lines A, B and C. Suppose that there is a shorter path D from x to y, shown as the broken line. Then A,D,C would be a path from a to b that is shorter than $P(a, b)$—a contradiction. □

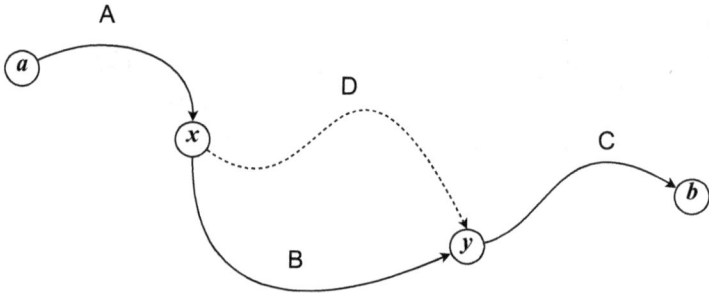

Figure 4.3: *Any part of a shortest path is itself a shortest path.*

The fact that these shortest paths are not disjoint can be exploited as follows: the algorithm will not return explicitly the paths, and rather provide enough information so as to enable us to build a shortest path from s to v on demand. This is implemented by means of an additional array P, giving, for each vertex $v \in V \setminus \{s\}$, the vertex immediately preceding v in a shortest path from s to v, similarly to what has been done in the dynamic programming examples of Chapter 2. The shortest path is therefore constructed backwards as v, $P[v]$, $P[P[v]]$, etc., until getting to the source s, as shown in ALGORITHM 4.2. A stack can be used if one prefers getting the path in its natural forward direction.

Shortest-Path(v)

$x \leftarrow v$
while $x \neq s$ do
    print x
    $x \leftarrow P[x]$
print s

ALGORITHM 4.2: *Shortest path from source v_1 to vertex v (backwards).*

Table 4.1 summarizes a run of Dijkstra's algorithm on our example graph in Figure 4.1. The leftmost column shows the vertices successively inserted into S, and the other columns give the values of the array D for all the vertices in $V \setminus \{v_1\}$. The i-th row shows the values of D at the end of the i-th iterations, where $i = 1$ corresponds to the initialization of lines 1–4. In each row, the smallest value is boxed, and the corresponding vertex is the one to be inserted into S in the following iteration. The set S displayed in light green in Figure 4.1 corresponds to the fifth row; a minimal value is 9 and obtained for vertex v_6, emphasized by the red circle in the figure, so in the sixth line, v_6 is added as sixth element to S.

Values left unchanged from the previous iteration appear in gray, and updated values appear in black. Every time a value in D is changed in line 11, the corresponding vertex is recorded in the array P in line 12. The last row of Table 4.1 shows the array P at the end of the execution of the algorithm. To find the predecessor of a vertex v in the shortest path from v_1 to v, one locates the last row in which the column of v contains a black value. For example, for v_{12}, the final value $D[v_{12}] = 15$ has been updated in the row for which $x = v_8$; thus $P[v_{12}] = v_8$.

All the shortest paths are shown as blue edges in Figure 4.1. They form a tree, which should not be confused with the minimum spanning tree seen in the previous chapter. For example, the shortest path from the source vertex v_1 to v_{12} can be constructed backwards by ALGORITHM 4.2 as

$$v_1 \quad v_2 \quad v_4 \quad v_7 \quad v_8 \quad v_{12},$$

as can also be seen in Figure 4.1. Since ties are broken arbitrarily, the shortest path is not necessarily unique: an alternative for the last example could be

$$v_1 \quad v_2 \quad v_4 \quad v_7 \quad v_8 \quad v_{10} \quad v_{12}.$$

Table 4.1: *Execution of Dijkstra's algorithm on the example graph.*

x	v_2	v_3	v_4	v_5	v_6	v_7	v_8	v_9	v_{10}	v_{11}	v_{12}	v_{13}
v_1	3	3	6	∞	∞	∞	∞	∞	∞	∞	∞	∞
v_2		3	5	∞	∞	∞	∞	∞	∞	∞	∞	∞
v_3			5	∞	9	∞	∞	∞	∞	∞	∞	∞
v_4				∞	9	6	∞	∞	∞	∞	∞	∞
v_7				∞	9		9	∞	11	∞	∞	∞
v_6				14			9	14	11	∞	∞	∞
v_8				14				11	10	∞	15	∞
v_{10}				14				11		∞	15	12
v_9				14						19	15	12
v_{13}				14						19	15	
v_5										19	15	
v_{12}										18		
v_{11}												
P	v_1	v_1	v_2	v_6	v_3	v_4	v_7	v_8	v_8	v_{12}	v_8	v_{10}

4.2.2   Complexity

As to the complexity of Dijkstra's algorithm, the initialization in lines 1–5 is clearly bounded by $O(V)$. In the main while loop, a minimum value in the array D is repeatedly retrieved in line 7. One needs $O(V)$ for a single minimum, and so a total of $O(V^2)$. A single iteration of the internal for loop starting at line 9 takes just $O(1)$, but at first sight it seems that up to $|V|$ vertices y may be accessed for each given x, which yields again a total of $O(V^2)$. A closer look, however, reveals that $D[y]$ can only change if there is an edge from x to y, so no cases are skipped if we restrict the scope of the for loop in line 9 only to the direct neighbors of x. We can thus replace line 10 by

$$\text{for all } \ y \in (V \setminus S) \cap E_x \ \text{do}$$

where $E_x = \{y \mid (x, y) \in E\}$ is the set of vertices to which there are edges starting from x.

    This amendment lets us reevaluate the complexity of the for loop: every edge in E is treated at most once, so the bound is $O(E)$, which can be better than the $O(V^2)$ we had so far. But this does not change the cost of the main while loop starting in line 6, because of the repeated extraction of minima in line 7, which still requires $\Omega(V^2)$. Nonetheless, having resolved one of the bottlenecks helps to focus our efforts on the other one. We were confronted with a similar situation in Section 3.5, with Yao's algorithm trying to lower the number of evaluations of minimal values in Borůvka's method. Yao's solution was some partial sorting. Here we may just use a more sophisticated data structure.

Background concept: Heaps

A *heap* is a full binary tree storing values in its nodes, and is often used to implement priority queues. The main property of a heap is that for each node x, the value stored in x is at least as large as the values in each of its left and right child nodes, as long as they exist. This immediately implies that the maximum value is stored in the root of the heap, which is why it may be referred to as a *max-heap*. Similarly, one can also define a *min-heap*.

If a non-ordered array of n elements is given, it can be transformed into a heap in $O(n)$ time. Access to the maximal element is then possible in constant $O(1)$ time, and insertion of a new element, or extraction of a given element, in particular, the extraction of the maximal element, can all be performed in time $O(\log n)$.

Actually, the definition of a heap as a tree is just a conceptual tool, and heaps can be implemented as simple arrays without any pointers. The key to such an implementation is storing the root at the element indexed 1, and storing at indices $2i$ and $2i + 1$ the left and right children, if they exist, of the element at index i.

The suggested alternative is to handle the elements of the array D as a min-heap. The initialization phase thus takes time $O(V \log V)$. Finding the minimum in line 7 can be done in $O(1)$, yielding a total of $O(V)$ for the entire while loop. The updates of D in line 11 are not in constant time any more and require $O(\log V)$ for each, and since there are at most $|E|$ such updates, the complexity of the while loop is bounded by $O(E \log V)$, which is the dominant part for the complexity of the entire algorithm.

We are therefore in the interesting situation in which the complexity of the algorithm, $O(E \log V)$ or $O(V^2)$, depends on the chosen data structure, with or without heaps, respectively. One can enforce the lower of the two by deciding on the data structure according to the relation, in the given input graph, between E and V: only if $E < \frac{V^2}{\log V}$ is it worth implementing the array D as a min-heap, otherwise it is preferable to process D just as an non-sorted array. We conclude that the complexity of Dijkstra's algorithm is $\min(O(E \log V), O(V^2))$.

4.2.3   *Application to Prim's MST algorithm*

As mentioned earlier, Prim's algorithm belongs in fact to the previous chapter on Minimum Spanning Trees, but it has been placed here because of the similarity to Dijkstra's method. Refer again to ALGORITHM 4.1, this time to both sides.

For Prim, the parameter s is in fact not needed: we can start from any vertex as starting point. The set S contains those vertices for which the MST is already known. Instead of an array D maintaining the costs of the currently derived shortest paths, an array W is defined, such that, for $y \notin S$, $W[y]$ is the lowest cost for connecting y by a direct edge to the MST being built. Both algorithms adjoin the vertex x with the lowest such cost to S, and then update the values of $D[y]$ or $W[y]$ that have changed, for all vertices y outside of S.

A major difference between the algorithms is that Dijkstra calculates only the costs in D and provides the array of predecessors P to generate the paths on demand. For Prim's algorithm, the cost of the MST is not required, and the tree itself is constructed with the help of the array P in the commands in lines 5 and 9, colored in gray.

The complexity of Prim is thus identical to that of Dijkstra as seen in Section 4.2.2, including also the option whether or not heaps should be used. For the correctness of Prim, one can show by induction on $|S|$ that at the end of each iteration, T is a minimum spanning tree of the currently defined set S. Note, in particular, that there are no restrictions on the weights $w(x, y)$ as for Dijkstra, and that Prim's algorithm also works in the presence of negative weights. The correctness of Dijkstra is more involved as shown in the next subsection.

4.2.4   *Correctness of Dijkstra's algorithm*

The fact, that the values in the array D at the end of the execution of Dijkstra's algorithm are indeed the minimal possible costs for the required paths, is by no means self-evident.

Theorem 4.1: *The following claims hold at any stage of the execution of Dijkstra's algorithm:*

(1) for any $x \in S$, $D[x]$ is the cost of a shortest path from the source vertex s to x, and all the vertices on this path belong to S;

(2) for any $y \in V \setminus S$, $D[y]$ is the cost of a shortest special path from the source vertex s to y.

Since the claims hold at any stage, this is true in particular at the end when $S = V$, which proves the optimality of Dijkstra's method. It is then rather intriguing to find such a complicated formulation rather than the expected statement *Dijkstra's algorithm is optimal*. Moreover, since the proof will be by induction on the stages, that is, on the number of elements in the set S, why are the claims not separated into two independent theorems, each of which can be proven on its own? And why, as a matter of fact, is claim (2) needed at all, as in the last stage, for $S = V$, which is the case we are interested in, the claim is empty?

There is a single answer to all these questions, namely that the claims cannot be separated because the truth of each of the claims (1) and (2) at stage i relies on the truth of *both* claims at stage $i - 1$.

Proof: We shall refer to the values in D at the end of the i-th iteration, for $i = 1, 2, \ldots, |V|$, where $i = 1$ is the initialization in lines 1-5. One could define $D[s] = 0$, so that claim (1) holds also for the only element in S for $i = 1$. As to claim (2), $D[x]$ is initialized as the cost of the direct edge from s to x, if it exists, so this is the cost of the shortest path since it is the only path in the initialization stage.

Suppose the claims hold at the end of iteration $i - 1$, that is, when $|S| = i - 1$. Let x_0 be the element that is adjoined to S in line 7 in the i-th iteration. The truth of claim (1) for the elements in S, besides x_0, follows from the inductive assumption (1), so it remains to be shown that $D[x_0]$ is the cost of a shortest path from s to x_0. Assume then, on the contrary, that this is not true. We know from the inductive assumption (2) that at the end of the previous iteration, $D[x_0]$, whose value has not been changed, was the cost of a *special* shortest path P from s to x_0. Hence if this does not correspond now to a shortest path, there must be another, non-special, path P' from s to x_0, for which $c(P') < c(P) = D[x_0]$, where $c(X)$ denotes the cost of a path X, i.e., the sum of the weights of its edges. The path P' starts at s, passes through vertices in S and then leaves S, directly to a vertex $y \notin S$, from which it continues, possibly reentering S once or more times, before reaching finally x_0, as shown in Figure 4.4.

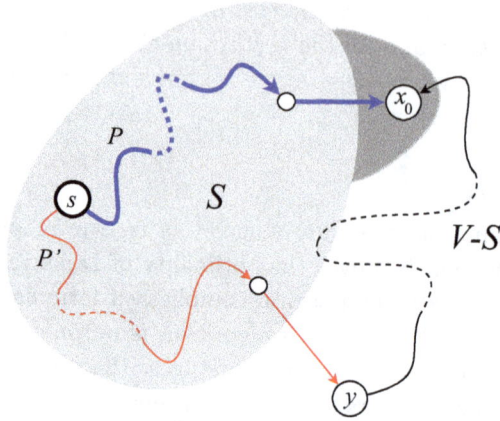

Figure 4.4: *Correctness proof of Dijkstra's algorithm.*

The figure shows the set S at the end of the iteration $i-1$ in light gray, the path P from s to x_0 emphasized as blue lines, and the alternative path P' as thinner red or black lines. If y is the first vertex on the path P' that does not belong to S, then $y \neq x_0$, otherwise P' would have been special. Denote the part of P' from s to y as P'', shown as the red lines in Figure 4.4. We then know that $D[y] \leq c(P'')$, because P'' is a special path to y, not necessarily the shortest one. On the other hand, the cost of the partial path P'' cannot be larger than the cost of the entire path P', because all the (black) edges on the remaining path from y to x_0 have non-negative weights. We can thus conclude that

$$D[y] \leq c(P'') \leq c(P') < c(P) = D[x_0],$$

but this is in contradiction: if indeed $D[y] < D[x_0]$, then y would have been chosen instead of x_0 in line 7 of the i-th iteration.

Since at the end of iteration $i-1$, the path P to x_0 was a special path, and x_0 has been adjoined to S, the entire path belongs to S at the end of iteration i. This concludes the proof of the inductive step for claim (1).

For claim (2), refer again to Figure 4.2 and some element $y \in V \setminus S$. There are two possibilities for the vertex preceding y on a shortest special path from s to y: either it is x, the element added to S in this i-th iteration (on the dark gray background), or it is some other vertex r that belonged to S even in iteration $i-1$. In the former case, the cost of the path to y is $D[x] + w(x,y)$, and in the latter, we claim that $D[y]$ did not change. Indeed, by the inductive assumption (2), $D[y]$ was the cost of the shortest special path from s to y at the end of iteration $i - 1$, and this path did not change.

We only have to assure that the addition of x to S in iteration i did not decrease the cost of getting to r, but this follows from the fact that $D[r]$ has been fixed as the cost of a shortest path to r *before* x has been added to S in iteration i, and that, by the inductive assumption (1), a shortest path to r did not include the vertex x. As the algorithm chooses the smaller of $D[x] + w(x, y)$ and $D[y]$ in lines 11–12, claim (2) follows. $\square$

Note in particular that the proof depends critically on the additional assumption that all the weights are non-negative. In fact, not only would the proof be wrong when edges with negative weights are allowed, but also it is easy to build counter-example graphs for which the theorem itself does not hold. This may lead to the suggestion to try to change the graph when there are negative weights, but this is not straightforward, see Exercise 4.1.

Dijkstra's algorithm is not the only one dealing with single-source, multiple-destination, shortest paths. In particular, we mention an alternative approach known as the Bellman–Ford algorithm [Bellman (1958); Ford (1956)], which has the advantage of working also in the presence of negative weights. Its complexity is, however, $O(VE)$, which is worse than Dijkstra's.

4.3 Multiple sources, multiple destinations

We now turn to the third, global, layer mentioned in Section 4.1, dealing with the problem of finding a shortest path from anywhere to everywhere. One can, of course, repeat the algorithms of Dijkstra or Bellman–Ford in a loop for all possible starting vertices $s \in V$, which would yield complexities of $\min(O(EV \log V), O(V^3))$ and $O(V^2 E)$, respectively. An elegant alternative, which is advantageous when the graph is dense, and more precisely when the number of edges in E is larger than $\frac{V^2}{\log V}$, has been suggested by Robert Floyd [Floyd (1962)]. Moreover, Floyd's algorithm works also in the presence of negative weights and forbids only the presence of negative cycles. A *negative cycle* is a sequence of vertices $x_1, x_2, \ldots, x_r \in V$, such that the sum of weights on the edges of the cycle is negative, i.e.,

$$w(x_1, x_2) + w(x_2, x_3) + \cdots + w(x_{r-1}, x_r) + w(x_r, x_1) < 0.$$

There is a fundamental difference between Dijkstra's condition prohibiting any negative weight and Floyd's restriction concerning cycles. The first is a weakness of the algorithm itself: not that negative weight edges do not appear in real-life situations, they do, but the algorithm does not work properly in their presence. On the other hand, if we would allow negative cycles, not only Floyd's algorithm would not be able to handle them, but the mere problem of finding shortest paths would then be ill-defined. This is so, because if one can reach a negative cycle C from a vertex v_i, and also get from C to some vertex v_j, then one could choose a path from v_i to v_j passing several times through C, and each such passage would reduce the global cost of the path. Thus no minimal cost path from v_i to v_j exists.

Floyd's algorithm is one of the classical examples of dynamic programming we have seen in Chapter 2. Like for the other examples, we shall split the problem into first finding, for each pair of vertices $v_i, v_j \in V$, the *cost* of a shortest path, and only then adapt the program to enable also the generation of an actual path, and not just the evaluation of its price.

The expected output for the first part is a two-dimensional cost matrix $D[i,j]$. Similarly to what we did for Dijkstra, the program will not return a list of shortest paths: the length of a single path can be $\theta(V)$, which could require, for all possible pairs together, an output of size $\theta(V^3)$. The algorithm will therefore provide, just as has been done for the problem of matrix chain multiplication in Section 2.3, enough information to enable the generation of the shortest paths on demand, in time proportional to the length of the path.

As for other problems solved by a dynamic programming approach, the solution of the multiple-source, multiple-target shortest path problem is based on the introduction of an additional parameter, namely the set S of vertices that are permitted as intermediate nodes in all the currently managed shortest paths. We have mentioned already that this is the subtle point in the design of a dynamic programming solution, where intuition plays a crucial role. Since the solution has to be built incrementally, one could have expected to allow, at initialization, no intermediate nodes at all; then only a single one, then two, etc., until ultimately having no restrictions on the number of vertices in the shortest paths. The surprising idea of Floyd's algorithm is to not only consider the *size* of the set S, but to confine S to a specific sequence of vertices. Actually, the order by which the vertices of V are adjoined to S does not matter, and we shall assume, for convenience of notation, that S is initialized as $S = \emptyset$ and will consist, after k iterations, of $S = \{v_1, v_2, \ldots, v_k\}$. At the end of the algorithm, $S = V$, meaning that there are no restrictions on the optimal paths.

Let $D_k[]$ be the square matrix of size $|V| \times |V|$ handled in the k-th iteration, so that $D_{k-1}[i,j]$ is the cost of a shortest path from v_i to v_j at

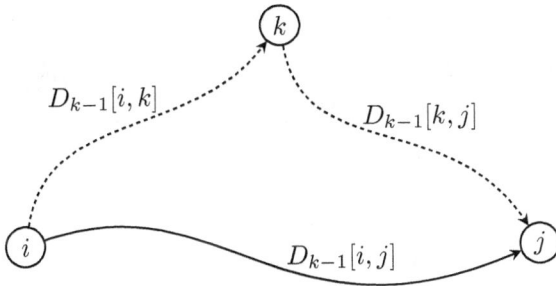

Figure 4.5: *General step in Floyd's shortest-path algorithm.*

the end of the $(k-1)$-th iteration under the constraint that only elements of $S = \{v_1, v_2, \ldots, v_{k-1}\}$ are allowed as intermediate nodes in those shortest paths, for all i and j. At initialization, all shortest paths consist of single edges, so $D_0[i,j] = w(v_i, v_j)$ if $(v_i, v_j) \in E$ and $D_0[i,j] = \infty$ otherwise.

For the general step, refer to the diagram in Figure 4.5. The only difference between iteration k and the preceding one is that a single vertex, v_k, has been added to the set of potential intermediate nodes. All that needs to be done is then to check, independently for each pair (i,j), whether or not this vertex v_k is useful for finding a shorter path from v_i to v_j than currently known, that is, with cost smaller than $D_{k-1}[i,j]$.

(1) If v_k helps to reduce the cost of a shortest path, then it belongs to some path P from v_i to v_j. We may assume that it appears only once in P; otherwise, P would include a cycle. We know that the total cost of the edges of the cycle is not negative, and if it is positive or zero, we would gain or at least not lose by removing the cycle from the path P. According to Lemma 4.1, the path P consists of a shortest path P' from v_i to v_k, followed by a shortest path P'' from v_k to v_j, and neither P' nor P'' contain the vertex v_k as intermediate nodes. The conclusion is that we can use the cost of P' and P'' evaluated in the previous iteration; hence

$$D_k[i,j] = D_{k-1}[i,k] + D_{k-1}[k,j].$$

(2) If v_k does not help to reduce the cost of a shortest path, then there is no change relative to the previous iteration, thus

$$D_k[i,j] = D_{k-1}[i,j].$$

Floyd's algorithm can thus be summarized in a single statement as

$$D_k[i,j] \leftarrow \min \left(D_{k-1}[i,j], \ D_{k-1}[i,k] + D_{k-1}[k,j] \right), \qquad (4.1)$$

which has to be embedded in three nested loops: the two innermost of these loops go over i and j to deal with each element of the matrix D_k, the outer loop on k iterates over the new vertices adjoined to the set S of potential intermediate nodes in the shortest paths. The formal algorithm appears in ALGORITHM 4.3, which includes also the assignments to a matrix P necessary to enable the construction of the shortest paths themselves, not just their costs.

$\underline{\text{Floyd}(V, E)}$

```
1    for all vᵢ ∈ V              D[i,i] ← 0
2    for all (vᵢ, vⱼ) ∈ V × V with i ≠ j
3        if (vᵢ, vⱼ) ∈ E then D[i,j] ← w(vᵢ, vⱼ)
4                              P[i,j] ← −1
5        else                  D[i,j] ← ∞
6                              P[i,j] ← −2
7    S ← ∅
8    for k ← 1 to |V|
9        for i ← 1 to |V|
10           for j ← 1 to |V|
11               if D[i,k] + D[k,j] < D[i,j] then
12                   D[i,j] ← D[i,k] + D[k,j]
13                   P[i,j] ← k
```

ALGORITHM 4.3: *Floyd's shortest path algorithm.*

Recall that for Dijkstra's algorithm, an array P was used, recording in $P[w]$ the vertex just preceding w in a shortest path from the source s to w. Here a two-dimensional matrix $P[i,j]$ is needed, and it suffices to store the index of any element v_k on a shortest path from v_i to v_j, not necessarily the vertex immediately preceding the target v_j. The optimal path can then be reconstructed recursively as shown in ALGORITHM 4.4. The procedure Print-Path(i,j) prints a shortest path from v_i to v_i, including the origin v_i but excluding the target v_j.

The attentive reader will have noticed that lines 11–12 in ALGORITHM 4.3 do not exactly match their explanation in Eq. (4.1): the indices k and $k-1$ of the matrices D are missing, suggesting that a single matrix is used

Print-Path(i, j)
_____
if $P[i, j] = -2$ then print *'there is no path from v_i to v_j'*
if $P[i, j] = -1$ then
    print *'$v_i -$ '*
else
    $k \leftarrow P[i, j]$
    Print-Path(i, k)
    Print-Path(k, j)

ALGORITHM 4.4: *Printing a shortest path from v_i to (and not including) v_j.*

throughout, rather than a different one for each iteration. The following lemma explains why this simplification is justified.

Lemma 4.3: *The elements in the k-th row and in the k-th column do not change in the k-th iteration, for all $k = 1, 2, \ldots, |V|$.*

Proof: The elements of the k-th row are those for which $i = k$ in Eq. (4.1). One gets

$$D_k[k, j] = \min \left(D_{k-1}[k, j], \ D_{k-1}[k, k] + D_{k-1}[k, j] \right).$$

But $D_{k-1}[k, k]$ is an element on the main diagonal, and is therefore zero, so that $D_k[k, j] = D_{k-1}[k, j]$. Similarly, the elements of the k-th column are those for which $j = k$, implying that

$$D_k[i, k] = \min \left(D_{k-1}[i, k], \ D_{k-1}[i, k] + D_{k-1}[k, k] \right) = D_{k-1}[i, k]. \quad \square$$

We have repeatedly addressed, in Chapter 2, the issue of the correct order of processing the elements in a dynamic programming table. The main challenge is to choose an order implying that the elements of the table on which the current calculation relies have to be correctly defined. In our case, this is most easily assured by basing our calculations, for all the elements of the matrix D_k, on those of the matrix D_{k-1}, which has been evaluated in the previous iteration. This is the solution suggested by Eq. (4.1). Note, however, that in iteration k, the element at index (i, j) either does not change, or, when it does, it is the sum of elements belonging to the k-th row or the k-th column, and these, according to Lemma 4.3, are not altered in this iteration. Nothing will therefore be lost if only a single matrix D is used, as shown in ALGORITHM 4.3.

Figure 4.6 is an example graph G with four vertices and Figure 4.7 shows the contents of the matrices D and P for G. D_0 is the matrix D at initialization and D_k is D at the end of the k-th iteration, for $k = 1, 2, 3, 4$.

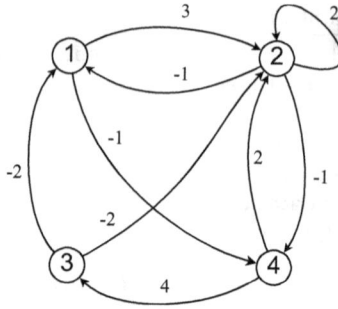

Figure 4.6: *Example graph for Floyd's algorithm.*

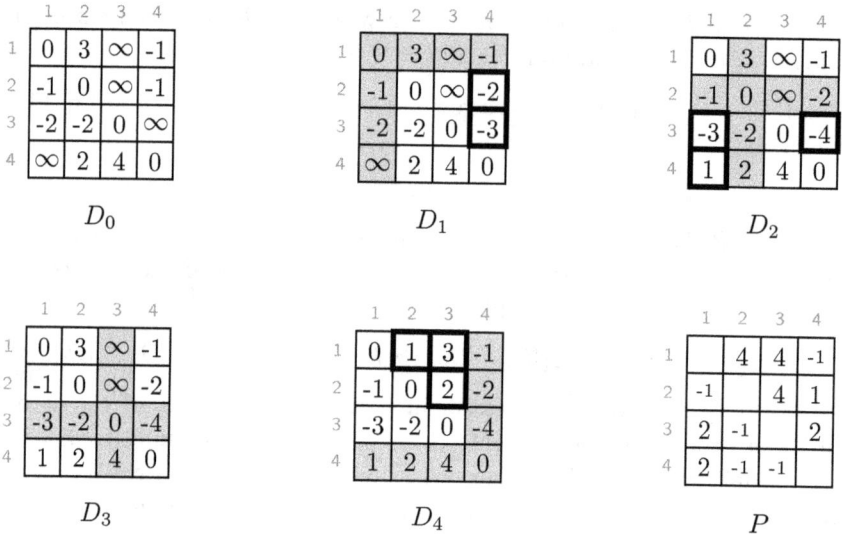

Figure 4.7: *The matrices D and P in Floyd's shortest path algorithm.*

The elements on the diagonal of D are zero, even for $D[2, 2]$, in spite of the self-loop (v_2, v_2), since it is cheaper to stay at v_2 than to use an edge from v_2 to itself (recall that there are no negative weight cycles, and thus no negative weight self-loops). The cells in the k-th row and the k-th column of D_k appear in gray: the evaluation of the other cells of the matrix depends only on them, and their values are equal to the corresponding values in D_{k-1}. Values that have changed are shown in emphasized boxes. For example, the cost of the shortest path from v_3 to v_4 is ∞ at initialization, since there is no direct edge (v_3, v_4) (though there is one in the opposite direction); it is reduced to -3 in D_1 and again reduced to -4 in D_2.

The matrix P in Figure 4.7 shows the values at the end of the execution of the algorithm. The value -1 at (i, j) is used to indicate that the shortest path from v_i to v_j is the direct edge (v_i, v_j). If no such edge exists, it is indicated at initialization by the value -2; if at the end of the algorithm, the value has not been changed, this means that there is no path at all from v_i to v_j. On our example graph, there are paths from every vertex to every vertex.

Applying ALGORITHM 4.3 to find, for instance, a shortest path from v_3 to v_4, yields

$$\text{Print-Path}(3, 4) = \text{Print-Path}(3, 2) \quad \text{Print-Path}(2, 4)$$
$$= \text{Print-Path}(3, 2) \quad \text{Print-Path}(2, 1) \quad \text{Print-Path}(1, 4)$$
$$= v_3 - v_2 - v_1 -$$

The complexity of Floyd's algorithm is clearly $\theta(V^3)$, just as if we had used Dijkstra's method using the implementation without heaps, and repeated it within a loop on all possible sources s. Floyd's method will still be preferable in this case since it works also in the presence of negative weight edges.

The correctness follows from the following theorem, the truth of which has been shown above:

Theorem 4.2: *For each $k \in [0, |V|]$, and for each $i, j \in [1, |V|]$, at the end of the k-th iteration, $D[i, j]$ is the cost of a shortest path from v_i to v_j allowing intermediate nodes to belong only to $\{v_1, v_2, \ldots, v_k\}$, where $k = 0$ corresponds to the initialization phase.*

4.4   Exercises

4.1 Given a graph $G = (V, E)$ with weights $w(a, b)$, some of which are negative. Let $m = \min\{w(a, b) \mid (a, b) \in E\}$ be the smallest weight in the graph, so that, given the assumption of the existence of negative weights, m is negative. Define a graph $G' = (V, E)$ with the same vertices and edges as G, but with different weights; for all $(a, b) \in E$, define the weight w' by $w'(a, b) = w(a, b) - m$, that is, we add $-m$ to all the weights, turning them into non-negative numbers. We can therefore apply Dijkstra's algorithm to the new graph G', find the shortest paths for G' and deduce from it the shortest paths also for G.

Prove by building a counter-example that this reduction does not always give a correct shortest path. More precisely, give an example of a graph G and a shortest path in G which does not correspond to a shortest path in G'. In fact, there is no need to know the details of Dijkstra's algorithm to answer this question.

4.2 A travel agency knows about all the flights from and to all possible airports, including the exact schedules of all the flights. Assume, for simplicity, that the schedule repeats itself periodically every day. The following questions are variants of the shortest path problem. You should, however, not try to adapt Dijkstra's algorithm, but rather use it as a *black box* and modify the input accordingly. What can be done to find the following:

(a) a cheapest path from a source s to a target t, under the constraint that airports A, B and C should be avoided;

(b) a cheapest path from s to t, but preferring, whenever possible, the use of either airline X or Y, because you could get frequent flyer points there;

(c) a path from s to t with the smallest number of takeoffs and landings;

(d) a path from s to t with the smallest number of airline changes (consecutive flights with the same airline does not count as a change);

(e) a path from s to t minimizing the overall travel time from departure from s to landing at t.

4.3 A communication network manages n computers, and a message may be sent from a source computer s to a target computer t via several intermediate stations. There is a small probability $1-p(a,b)$ for a message to get corrupted on its way from a to b. The problem is thus to choose a path in the network from s to t, so that the probability of receiving the correct message is maximized? You may use Dijkstra's algorithm, but you may not change it, like in the previous exercise.

4.4 The following problem will be treated later in Section 6.2. You are given a text $T = x_1 x_2 \cdots x_n$ of length n characters, which all belong to some alphabet Σ. To encode the text, we assume the existence of a *dictionary* D, consisting of a collection of strings of characters of Σ. The goal is to efficiently encode the text T by means of the dictionary D, replacing substrings of T by (shorter) pointers to elements of D.

The difficulty stems from the fact that the elements of D might be overlapping, and that the number of different possible parsings can be too large to allow an exhaustive search over all possibilities. In addition to T and D, there is also a given encoding function $\lambda(w)$ defined for all $w \in D$, producing binary strings of variable length. The length (in bits) of the string encoding w is denoted by $|\lambda(w)|$. The problem is to find a decomposition of the text T into a sequence of elements w_i of the dictionary, that is, $T = w_1 w_2 \cdots w_k$, with $w_i \in D$, minimizing the total length $\sum_{i=1}^{k} |\lambda(w_i)|$ of the encoded text.

Show how to reduce the problem to one using graphs and find an optimal solution.

4.5 The following questions refer to Floyd's algorithm ALGORITHM 4.3:

(a) What happens if we change the control sentence of the main loop in line 8 to:

```
for k ← |V| to 1 by -1   ?
```

(b) What happens if we move the control sentence of the main loop in line 8 after line 10, so that the nested loops become:

```
for i ← 1 to |V|
    for j ← 1 to |V|
        for k ← 1 to |V|
```

PART 3
Probabilistic Algorithms

Chapter 5

Primality

5.1 Probabilistic algorithms

The methods described in this part of the book may seem surprising. The algorithms seen so far are all *deterministic*, in the sense that they consist of a well-defined sequence of commands depending solely on the given input parameters, if any. Such a sequence of commands can be quite complex, including a blend of assignments, function calls and tests that can be used in the design of loops. Nevertheless, the exact effect of each of the commands, as well as the order in which they are executed, can be assessed precisely and is often simulated, for debugging purposes, in what programmers call a *dry run*. Moreover, this property of being at each point well defined is often considered as the definition of what an algorithm should be.

In this part of the book, we consider algorithms which include an element of randomness, so that the above statement about our ability to trace the execution path of the resulting program is not necessarily true any more. The standard way to introduce randomness is by means of a *random number generator*, which exists in most programming languages and is often called rand( ) or the like.

For certain applications, as when playing a card game against a computer, randomly chosen elements are a natural choice: a game dealing time and again the same set of cards to the players would not be very exciting. Another important class of programs would be simulations. For example, in order to schedule the traffic lights at the intersection of two multi-lane roads, one may observe the arrival patterns of the vehicles from the different directions leading to the crossing point; then computer simulations using the gathered information may be applied to study the possible buildup of traffic congestion for various settings of the traffic lights.

We shall, however, concentrate rather on different applications, for which the utilization of elements, which have been generated in a process governed by chance, seems intriguing at first sight. What these algorithms have in common is that they allow the virtual tossing of a coin as one of their possible commands. The outcome of such a tossing, implemented as a comparison of a randomly generated number with some constant, is not foreseeable, so dry runs are not possible any more.

This raises the question: what advantage could possibly be gained by this restriction of our ability to be in control of what exactly happens within our program? Intuitively, it would rather seem that the more painstakingly we supervise the small details, the better we expect the outcome to serve our intended goals. Nevertheless, we shall see not only examples in which the introduction of randomness will *improve* the solution of certain problems, but also instances for which a solution without using randomness is not necessarily *possible* at all.

5.2 Primality testing

The problem we shall concentrate on is that of checking whether or not a given input integer n is a prime number. A number is called *prime* if it can be divided without remainder only by itself and by 1. By convention, 1 is not considered prime. The other integers, those larger than 1 and not prime, are called *composite*. Prime numbers have been studied for thousands of years, and many of their properties are known, for example the fact that there are infinitely many of them.

Note that the problem of primality is different from most other problems we have seen so far. For instance, the input parameter n is generally used to designate the *size* of the problem, like the n numbers from which we wanted to extract the minimum in ALGORITHM 1.1, or the array of size n given as input to mergesort in Section 1.2. Here, the number n is the input itself, not its size! We should also remark that the problem is interesting only for quite large values of n, because the simple loop in ALGORITHM 5.1 solves it.

Clearly, if some divisor $1 < i < n$ of n has been found, then n is not prime by definition, and if it is composite, say, $n = n_1 n_2$, then one of the factors must be smaller or equal to $\sqrt{n}$, otherwise we would get that $n > \sqrt{n}\sqrt{n} = n$, that is, n is greater than itself! Hence, it suffices to check potential divisors up to $\sqrt{n}$. Therefore, even if we consider numbers of the order of billions $\approx 2^{32}$, not more than $2^{16} < 100{,}000$ divisions are needed,

Primality-1(n)

> for $i \leftarrow 2$ to $\lfloor \sqrt{n} \rfloor$
> > if i divides n then
> > > print 'composite'   stop
> > print 'prime'

ALGORITHM 5.1: *Checking primality by exhaustive search.*

which can be checked in a fraction of a second on today's devices.

Actually, the standard input on which one would like to apply primality test algorithms is rather for numbers the representation of which requires hundreds if not thousands of bits, like $n \simeq 2^{400}$ or so. It is not possible to apply ALGORITHM 5.1 for such large numbers: to perform $\sqrt{n} \simeq 2^{200}$ divisions, even if one can do a trillion (2^{40}) operations per second and could split the work among a trillion such powerful processors, it would take

$$\frac{2^{200}}{2^{40} \times 2^{40} \times 3600 \times 24 \times 365} \simeq 2^{95},$$

that is, more than 32,000 trillion trillions of years to accomplish the task.

The question is therefore: why should we care? There is of course nothing wrong to conduct investigations out of pure intellectual curiosity, and only in the 20th century has the usefulness of very large prime numbers been discovered, in particular for cryptography, as we shall see later in Chapter 9. The secure usage of the internet, the possibility to keep a part of our information confidential, even though it may have been transmitted wirelessly, would all be unthinkable without proper encipherment. Many of the coding methods yielding these ciphers depend critically on very large primes and thus on our ability to identify a given number as such.

The question about the compositeness or primality of a given input number n is a *decision problem*. These are problems to be solved by algorithms whose output is either Yes or No. Due to this simplicity, decision problems are often easier to handle, which is why they have become an important tool in the analysis of algorithms, as we shall see in Chapter 10. There is, nevertheless, nothing lost by this simplification, as it can be shown that many other problems, and in particular optimizations, can be reformulated into equivalent decision problems by a mechanism called a *reduction*.

The algorithm presented below has been invented by Michael Rabin [Rabin (1980)]. It is a probabilistic algorithm, in the sense that it uses the outcome of virtual coin tosses as a part of its commands. Similar to many other algorithms of this kind, the scenarios of the possible answers, Yes = prime or No = composite, are not symmetric: if the algorithm returns No, then the input n is definitely not a prime number, but when the algorithm returns Yes, there is no certainty that this is indeed true. There is a chance that our algorithm mistakes, albeit the probability p of such erroneous behavior may be controlled and p can be forced, by adjusting certain parameters of the algorithm, to be as small as we wish it to be, though still positive.

5.2.1 *Rabin's algorithm*

A conspicuous fact about any prime number p can be shown as follows: choose an arbitrary integer a that is not a multiple of p and consider its multiples modulo p.

$$a, \quad 2a, \quad 3a, \quad \ldots, \quad (p-2)a, \quad (p-1)a \quad (\bmod\ p). \tag{5.1}$$

If p were not prime, there could be repetitions in this sequence, but for prime p, if, for $i, j < p$ we have that $ia \equiv ja \ (\bmod\ p)$, then $(i-j)a \bmod p = 0$, implying $i = j$, so that all the $p-1$ elements in the sequence (5.1) must be distinct. In other words, they are just some permutation of the integers $1, 2, \ldots, p-1$. By taking the product of all the elements in the sequence, we get therefore that

$$a^{p-1}\,(p-1)! = (p-1)! \quad (\bmod\ p).$$

Dividing both sides by $(p-1)!$ yields what has been known since the 17th century as

Fermat's Little Theorem:
For any prime number p and any integer a which is not a multiple of p,

$$a^{p-1} \equiv 1 \quad (\bmod\ p). \tag{5.2}$$

For example, $6^{12} = 2{,}176{,}782{,}336 = 13 \times 167{,}444{,}795 + 1$. This surprising fact can be exploited to devise a simple mechanism for checking the compositeness of an input integer n:

(1) pick an arbitrary number b that is not a multiple of n;
(2) raise b to the power $n-1$ and let r be the remainder modulo n, that is, $r = b^{n-1} \ (\bmod\ n)$;
(3) if $r \neq 1$ then n cannot be prime, as it would contradict Fermat's Little Theorem.

Suppose, for example, that we want to check whether $n = 16$ is a prime number or not. Let us choose an arbitrary number, say 5, to apply the heuristic above. One gets $5^{15} \bmod 16 = 13 \neq 1$, so we know that 16 must be composite, even though the process does not give the slightest clue about any of the proper divisors of the number n, besides the fact that they must exist.

The difficulty of mobilizing Fermat's Little Theorem for primality testing lies in the fact that it is one sided: if p is prime, then the theorem holds, but this does not mean that if p is composite, we shall always get that $a^{p-1} \bmod p \neq 1$. Indeed, for $a = 1$, all its powers are also 1, regardless of the modulus p; for composite p, Eq. (5.2) may hold for certain values of a and not hold for other values. Moreover, there are some rare integers c, called *Carmichael* numbers, that behave like primes in the sense of Fermat, i.e., $c^{p-1} \bmod p = 1$ for all primes p. The smallest Carmichael numbers are 561 and 1105, which are not prime since the first is a multiple of 3 and the second of 5.

There are certain parallels between the problem of primality testing and the indictment of somebody accused of a crime in court. In our case, it is the integer n playing the role of the accused, and the crime it allegedly committed is being composite. An irrefutable confirmation for the guiltiness of n would be the appearance in court of a factor q of n: it can easily be checked whether indeed q divides n. We are, however, in the situation in which no such obvious witnesses are known, and the prosecution desperately tries to convince an arbitrary passerby b to come and testify against n.

The declaration of b would be: "*Please raise me to the power $n - 1$ and reduce the outcome modulo n.*" If the obtained result is not 1, then n can be convicted and b turned out to be a witness for n's compositeness. On the other hand, if $b^{n-1} \bmod n = 1$, then b did in fact not add any information about the innocence or guiltiness of n, so it is just not appropriate to call b a witness.

A single witness b is sufficient to declare n to be composite even though b's testimony is circumstantial. However, even the existence of several non-witnesses does not weaken the case against n, like if there are no eyewitnesses to a crime under investigation, it does not mean that it did not happen. Unless, of course, we have some additional information about the number of spectators at the scene: if it was perpetrated on the main street of a small town in the middle of the day, it would not seem reasonable that not even a single witness is detected in repeated random choices.

Rabin's algorithm therefore repeats the test k times, each with a new randomly chosen integer b. If one of those turns out to be a witness, the loop can be aborted and n is declared, with certainty, to be composite. If, however, not a single witness has been found, n is declared to be prime, even though there will always remain a slight chance that this decision is the wrong one. ALGORITHM 5.2 gives the formal description.

Primality-Rabin(n)

   repeat k times
      $b \leftarrow$ randomly chosen integer $< n$
      check whether b is a *witness*
      if yes then print 'composite'   stop
   print 'prime'

ALGORITHM 5.2: *Rabin's probabilistic primality test.*

The problem is that in order to know how many iterations k to perform, we need an estimate of the proportion of witnesses within the population from which they are drawn. Regretfully, no good estimate of this proportion is known for the definition of a witness given so far, which is based on Fermat's Little Theorem alone. In particular, Carmichael numbers are a matter of concern, as they are composite and yet no witnesses exist for them.

What is needed is a broader definition of the notion of a witness. To continue with the analogy with a legal case, the integer b is given a second chance to convince the court that it is eligible to be called a witness.

Given two integers $a \geq b \geq 0$, their *Greatest Common Divisor*, denoted by $\mathsf{GCD}(a, b)$, is the largest integer that divides both a and b without a remainder. Clearly, if an integer x can be found for which

$$1 < g = \mathsf{GCD}(x, n) < n, \qquad (5.3)$$

then n cannot possibly be prime. In this case, not only is there a proof of the *existence* of a proper divisor as in the case when $b^{n-1} \bmod n \neq 1$, but one such divisor, namely g, has actually been found. One could therefore try to calculate the GCD of n and x for a large set of randomly chosen values of x, and if even a single x is found for which Eq. (5.3) holds, this would prove that n is composite. The problem remains as before: also if n is not prime, it might have so few factors that a random selection can miss all of them, even when repeated millions of times.

A major achievement of Rabin's algorithm is to have succeeded in drastically reducing the size of the set of potential candidate integers x that have to be checked with Eq. (5.3), while still being able to bound the probability of failure. This is materialized in the following more relaxed definition of a witness.

Definition 5.1: An integer b is called a *witness* for the compositeness of n if one of the following conditions holds:

(1) $b^{n-1} \bmod n \neq 1$, or

(2) $\exists\, i, m \in \mathbb{N}$ $\quad \frac{n-1}{2^i} = m \;\; \wedge \;\; 1 < \mathsf{GCD}(b^m - 1, n) < n$.

The first condition is Fermat's Little Theorem. If it does not give conclusive evidence, GCD tests are performed, but only for a very specific set of integers. One first checks if there exist integers i and m such that $\frac{n-1}{2^i} = m$, that is, whether $n-1$ can be divided by some powers of 2. There is at least one such value of i, as we may assume that the given input integer n is odd, so $n - 1$ is divisible by 2. The involved integers are easy to detect if one considers the standard binary representation of $n - 1$, denoted by $\mathcal{B}(n-1)$. The number $n - 1$ is divisible by the consecutive powers of 2: $2, 4, 8, \ldots, 2^d$ if and only if the d rightmost bits of $\mathcal{B}(n - 1)$ are zero. For example, if $n = 50,881$, then

$$\mathcal{B}(n - 1) = 1100\ 0110\ 11\mathbf{00\ 0000},$$

in which the $d = 6$ rightmost 1-bits have been emphasized. The exact number d of integers i for which condition (2) has to be checked is therefore equal to the number of consecutive zeros at the right end of $\mathcal{B}(n-1)$, which cannot exceed $\log n$. The corresponding d values of m are $\frac{n-1}{2}, \frac{n-1}{4}, \ldots, \frac{n-1}{2^d}$, and for each of them, an integer x is defined as $x = b^m - 1$ for which the GCD test of Eq. (5.3) is performed. Only if b fails the test of condition (1), and none of the up to $\log n$ values of m yields a successful result in condition (2), will b be declared as a non-witness and the algorithm will select another random number b.

For this extended definition of a witness, Rabin was able to validate the following claim, which is given here without its proof.

Theorem (Rabin): *If n is composite, then there are at least $\frac{3}{4}n$ witnesses smaller than n for its compositeness.*

If n is prime, then obviously it is not possible to find any witness, and the probabilistic test in ALGORITHM 5.2 will print the correct answer 'prime' after k unsuccessful trials to find a witness by random selection.

On the other hand, if n is composite, the probability for a randomly chosen integer $b < n$ of not being a witness is at most $\frac{1}{4}$. By repeating the test for k independent trials, the probability of not finding at least one witness dwindles to not more than $p = 2^{-2k}$. This is then the probability of Rabin's algorithm to return erroneously 'prime' while the true answer would have been 'composite', but by choosing k large enough, this error probability can be compelled to be arbitrarily small. For example, choosing k as low as 10 is sufficient to reduce this error probability to below 1 in a million; with $k = 30$, it is smaller than a billionth of a billionth.

5.2.2 *Space complexity*

Though the idea of applying probabilistic features to solve difficult problems might seem appealing at first sight, a closer look at the tests required by the conditions of the definition of a witness reveals that a straightforward implementation will not be possible for the potentially very large input numbers we wish to handle. Consider, for example, the number checked by Fermat's Little Theorem $b^{n-1} \bmod n$. If, as we suggested, n might be of the order of 2^{400}, then even if b is chosen as small as possible, say, $b = 2$, the number $b^{n-1} \approx 2^{2^{400}}$ would require 2^{400} bits for its representation, which is more than there are electrons in the universe! Such a number would be impossible to store even on all the computers of the world together, not to speak of the time this would take.

What comes to our rescue is the convenient properties of modular arithmetic, as it turns out that one can evaluate $b^{n-1} \bmod n$ without passing first by the evaluation of b^{n-1}.

Background concept: Modular arithmetic

Let m be a fixed integer larger than 1 that we shall call the *modulus*. We define $u \bmod m$, for any natural number u, as the remainder of the division of u by m, that is, $u \bmod m = r$ for the unique integer r satisfying that $0 \leq r < m$ and u can be written as $u = km + r$, for some integer k. In a context when the modulus is assumed to be fixed, we may abbreviate the notation by setting $\bar{u} = u \bmod m$, and call it the *modulo* function, without explicit reference to m. The following properties follow directly from this definition.

The modulo function can be applied once or more times without changing the result, that is, $\overline{\overline{u}} = \overline{u}$. It follows that for all integers u and v

$$\overline{u+v} = \overline{\overline{u}+\overline{v}} \qquad \text{and} \qquad \overline{u \times v} = \overline{\overline{u} \times \overline{v}},$$

which enables the decomposition of one operation on large arguments into several operations on smaller ones. For example, for $m = 17$, let us calculate $\overline{57 \times 79} = \overline{4503}$. Most of us will find it difficult to multiply these numbers in their heads, and even more so to divide the result by 17 to find the remainder. Applying the modulo repeatedly simplifies this task to

$$\overline{57 \times 79} = \overline{6 \times 11} = \overline{66} = 15,$$

for the evaluation of which we need no paper or pencil.

Instead of performing all the necessary multiplications and applying the modulo function only once, at the end, we repeatedly apply modulo after each multiplication:

$$\overline{b^{n-1}} = \overline{\overline{\overline{b \times b} \times b} \times b \times \cdots \times b}. \qquad (5.3)$$
$$\underbrace{}_{n-1 \text{ factors}}$$

The largest integer handled can therefore not exceed m^2, for the representation of which no more than $2\log m$ bits are needed. For our example of $n \approx 2^{400}$, all numbers modulo n fit into 400 bits, $b \times b$ may require 800 bits, but is immediately reduced to $\overline{b \times b}$, which requires only 400 bits, and so on, with the necessary space alternating between 400 and 800 bits.

5.2.3 *Time complexity*

This solves the space problem, but how can we suggest performing about n multiplications in Eq. (5.3) when even the $\sqrt{n}$ divisions of ALGORITHM 5.1 were deemed to be prohibitive? Here is an elegant way to reduce this number considerably. Note that instead of calculating

$$b^8 = \underbrace{b \times b \times b \times \cdots \times b}_{8 \text{ factors}},$$

the number of multiplications can be reduced by repeatedly squaring the results:

$$b^8 = \left(\left(b^2\right)^2\right)^2.$$

If the exponent $n-1$ is not a power of 2, it can be expressed as a sum of such powers, giving, for example, $b^{12} = b^8 \times b^4$. For the general case, consider the standard binary representation of $n-1$ as a sum of powers of 2, that is, $n-1 = \sum_{i=0}^{\lfloor \log_2 m \rfloor} a_i 2^i$, where each $a_i \in \{0,1\}$. Then

$$b^{n-1} = b^{a_0} \times b^{2a_1} \times b^{4a_2} \times \cdots \times b^{2^i a_i} \times \cdots .$$

All these equations hold as well modulo n. The procedure is thus as follows: prepare a list of basis items $\overline{b}, \overline{b^2}, \ldots, \overline{b^{2^i}}, \ldots, \overline{b^{2^{\lfloor \log(n-1) \rfloor}}}$, where each element is obtained by taking its predecessor in the list, squaring it, and reducing the result modulo n; then take the subset of this list corresponding to the 1-bits in the binary representation of $n-1$ and multiply the elements of this subset, applying modulo n after each multiplication. The total number of multiplications needed to evaluate $\overline{b^{n-1}}$ is thus at most $2\log n$. The technique of decomposing the exponent into a sum of powers of 2 and applying repeatedly the modulo function is known as *Modular exponentiation*.

5.2.4 *Euclid's GCD algorithm*

It is therefore possible to check the first condition of Definition 5.1 in logarithmic time, which permits us to handle even the huge input numbers we intend to use. The second condition, however, involves the calculation of a GCD, and we argued that n might be too large to find one of its divisors, so how can we expect to find not just any divisor, but the largest one n has in common with some other integer x?

The answer to this question has in fact be given already in about 300 BCE by the Greek mathematician Euclid, known as the inventor of the probably oldest algorithm. Though it might seem surprising, it is indeed much easier to find a common divisor of two given integers a and b than detecting even the existence of any non-trivial divisor of either a or b.

Euclid's GCD algorithm is based on the recursive identity stating that for $a \geq b > 0$,

$$\mathrm{GCD}(a, b) = \mathrm{GCD}(b, a \bmod b).$$

This allows us to derive a chain of repeated GCD calls, with progressively decreasing parameters, as shown in ALGORITHM 5.3. When the second parameter reaches the value 0, the first parameter is the GCD value we were looking for, and it holds for all the input number pairs.

As an example, consider 81 and 57 to get

$$\mathrm{GCD}(81, 57) = \mathrm{GCD}(57, 24) = \mathrm{GCD}(24, 9) = \mathrm{GCD}(9, 6)$$
$$= \mathrm{GCD}(6, 3) = \mathrm{GCD}(3, 0) = 3.$$

GCD(a, b)
    if $b = 0$ then   return a
    else   return GCD(b, a mod b)

ALGORITHM 5.3: *Euclid's greatest common divisor.*

To get a bound on the number of necessary steps in Euclid's algorithm, consider the input parameters after two consecutive iterations, if they are possible:

$$\text{GCD}(a,\ b) = \text{GCD}(b,\ a \bmod b)$$
$$= \text{GCD}(a \bmod b,\ b \bmod (a \bmod b)).$$

We see that the effect of two iterations is a change in the value of the parameter from x to $x \bmod y$, with $y < x$, for both parameters a and b. This change implies a reduction of the parameter x to at most half of its initial value:

Claim: *For all integers $a \geq b > 0$*

$$a \bmod b < \frac{a}{2}.$$

Proof: Consider the two options of the size of b relative to $\frac{a}{2}$, as depicted in Figure 5.1. In both cases, the size of the zone appearing in gray is an upper bound for $a \bmod b$.

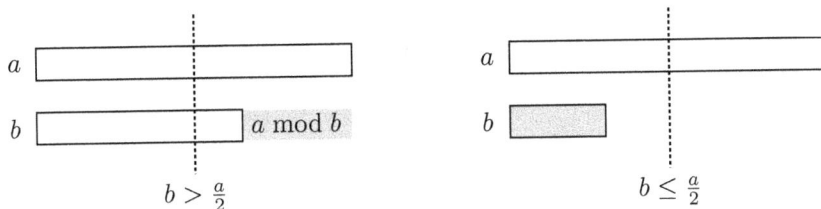

Figure 5.1: *Proof of $a \bmod b < \frac{a}{2}$.*

Formally, if $b > \frac{a}{2}$ then b fits only once into a and therefore

$$a \bmod b = a - b < a - \frac{a}{2} = \frac{a}{2}.$$

If $b \leq \frac{a}{2}$ then $a \bmod b < b \leq \frac{a}{2}$.   □

For instance, this decrease can be observed in the above example: $81 \to 24 \to 6$, and also $57 \to 9 \to 3$. Returning to the evaluation of the complexity of GCD, after two iterations the input is reduced to at most half of its size, after four iterations to a quarter, and generally, after $2i$ iterations, the input size has been cut by a factor of at least 2^i. This is true for both parameters, so the number of iterations cannot exceed $2\log\left(\min(a,b)\right)$. As one of the parameters of GCD in test (2) of definition 5.1 is n, and the number of pairs (i,m) for which this test is performed is at most $\log n$, we conclude that the time complexity for verifying whether a single candidate b is a witness is $O(\log^2 n)$.

A last issue to be settled concerns again the space complexity. We saw that $\overline{b^{n-1}}$ could be evaluated in limited space by repeatedly applying the modulo function after each multiplication, and not just once at the end. But for test (2), it is $\mathsf{GCD}(b^m - 1, n)$ that has to be evaluated, without a modulo function, and the first relevant value is $m = \frac{n-1}{2}$. For our example, we would thus be interested in $b^{2^{399}}$, and there is no way to store such a huge number! A single step of Euclid's algorithm reveals that

$$\mathsf{GCD}(b^m - 1, \, n) = \mathsf{GCD}(n, \, \overline{b^m - 1}),$$

so one can proceed as for test (1).

On modern computers, testing the primality of a given input integer n using Rabin's algorithm takes merely a tiny fraction of a second. To find the largest prime number that can be written with 400 bits, one may start by testing $2^{400} - 3$, $2^{400} - 5$, $2^{400} - 7$, all of which are composite, and keep decrementing by 2, until one gets to

$$n = 2^{400} - 593,$$

for which no witness has been found so far. This means that this number is probably prime, though we cannot really be sure. If we use $k = 20$ iterations and none of the randomly selected k integers b turn out to be witnesses, does it mean that n is prime with probability $1 - 2^{-40}$? Such a statement makes no sense, because obviously, any integer is either prime or not, and this property is an intrinsic attribute of the number itself, and not of the way it may be processed. The statement about an error probability of 2^{-40} should thus be understood in the sense that if a similar test will be performed for 2^{40} different input integers, the expected number of wrong answers will be about 1.

For the specific example of finding the largest prime below 2^{400}, the answer has been found in a linear scan within a few trials. Can one assume that this will always be the case? What if the largest prime smaller than 2^k is even smaller than, say, $2^k - 2^{80}$? This would be impossible to detect by checking the numbers in intervals of 2.

The reason justifying the sequential scan is that we know that there are many prime numbers. More precisely, we know that there are infinitely many, but even if we look only at integers up to some given upper limit N, the fact known as the *prime number theorem* states that the primes must be quite dense:

Prime number theorem: *The number of prime numbers among the integers smaller than N is asymptotically $\frac{N}{\ln N}$.*

This means that in particular, the average gap between consecutive prime numbers smaller than N is close to $\ln N$, so one can expect the size of this gap to be of the order of $\ln 2^{400} \approx 277$ for our example. The prime number theorem has many important implications, as we shall see in the next section.

5.3 Probabilistic string matching

String matching or more generally pattern matching is an important algorithmic problem with many useful applications in diverse areas of computer science, and will be devoted an entire chapter later in this book. The reason for preponing a particular string matching algorithm to our current discussion of primality is that, precisely, the surprising introduction of prime numbers into a context with which they a priori have nothing in common, will serve as an additional example of the powerful advantages of the probabilistic approach.

The string matching problem in its simplest form is the search for a character string Y within another string X. Possible interpretations range from finding a word in a text to locating an expression in the internet. We shall deal here with the general case in which both X and Y are bit-vectors, so we assume that our alphabet consists of two characters only, which we denote as 0 and 1.

Formally, we are given a text $X = x_0 x_1 \cdots x_{n-1}$ of length n and a string $Y = y_0 y_1 \cdots y_{m-1}$ of length m, and we wish to locate Y in X, that is, find an index i, $0 \le i \le n - m$, such that

$$Y = y_0 y_1 \cdots y_{m-1} = x_i x_{i+1} \cdots x_{i+m-1}.$$

If we know how to find the lowest such index i, we can as well locate *all* the occurrences of Y in X by repeating the process.

The straightforward, often called *naïve*, solution of this matching problem can be implemented by means of two nested loops as shown in ALGORITHM 5.4. The Boolean variables *found* and *match* permit exiting the loops once the first occurrence of Y in X is located.

Naïve-Matching(X, Y)

    found ← false
    for $i \leftarrow 0$ to $n - m$ while not *found* do
        match ← true
        for $j \leftarrow 0$ to $m - 1$ while *match* do
            if $y_j \ne x_{i+j}$ then
                match ← false
        if *match* = true then
            found ← true
            print 'string found at ' i
    if not *found* then
        print 'string not found'

ALGORITHM 5.4: *Naïve string matching: locating string Y in text X.*

The worst-case complexity of this algorithm is $\theta(mn)$, as can be seen for $X = 00\cdots01$ and $Y = 00\cdots01$, where the lengths of the runs of zeros in X and Y are $n-1$ and $m-1$, respectively. This is acceptable as long as the given parameters are small enough. However, for the applications we have in mind, both n and m may be large, say n could be a million and m could be 1000. The naïve approach could then require a billion comparisons, which justifies looking for alternatives.

5.3.1 *Karp and Rabin's algorithm*

The following algorithm is a probabilistic one and has been invented by R. Karp and M. Rabin [Karp and Rabin (1987)]. One may wonder how a probabilistic approach can be useful for a seemingly simple problem of locating a string. While we have no means of verifying that a number like

$2^{400} - 593$, declared by Rabin's algorithm to be a prime is indeed such, this is not the case for the current problem. If the matching algorithm returns an index i, one can easily check whether or not $Y = x_i x_{i+1} \cdots x_{i+m-1}$. Nevertheless, the algorithm we shall see is much faster than the naïve one, and hence used to *improve* the answer, whereas for other instances like primality testing, a probabilistic algorithm may be necessary to even *enable* a solution.

The idea leading to the efficiency of Karp and Rabin's algorithm is to replace the comparisons of long bit-strings by comparisons of corresponding numbers modulo p, for some well-chosen number p. It exploits the fact that numbers, bit-strings and even character strings are all stored as binary data and might therefore be considered as equivalent and used interchangeably. For example, consider the word example, its binary ASCII representation and the decimal value of this bit-string of length $7 \times 8 = 56$ bits.

characters	e x a m p l e
binary	01100101 01111000 01100001 01101101 01110000 01101100 01100101
number	28,561,332,491,021,413

The challenge is to fill in the details that result in significant savings on the one hand, yet do not deteriorate the quality of the provided answers.

Karp–Rabin-1(X, Y, k)

1 $p \leftarrow$ randomly chosen prime with k bits
2 $\overline{Y} \leftarrow Y \bmod p$
3 for $i \leftarrow 0$ to $n - m$
4      $\overline{Z_i} \leftarrow Z_i \bmod p$
5   if $\overline{Z_i} = \overline{Y}$ then
6   print 'string found at ' i

ALGORITHM 5.5: *Karp–Rabin string matching: first version.*

The algorithm first sets a parameter k which will be chosen as a function of the available hardware and of the error probability we might tolerate. The next step is the random choice of a prime number p that can be written with k bits. This is the only probabilistic command in the entire algorithm.

Define Z_i as the substring of X starting at index i and being of the same length m as the string Y we are looking for, that is

$$Z_i = x_i x_{i+1} \cdots x_{i+m-1} \qquad \text{for } 0 \le i \le n - m.$$

Using this notation, our aim is to find an index i for which $Z_i = Y$. We shall, however, instead consider both Z_i and Y as if they were huge numbers and check whether $\overline{Z_i} = \overline{Y}$, where we use the notation

$$\overline{A} = A \bmod p,$$

introduced in the previous section. Note that if k is chosen according to the word size of the given computer, say 32 or 64 bits, then all the numbers modulo p fit into a single computer word, so that the test $\overline{Z_i} = \overline{Y}$ is a single operation, whereas checking whether $Z_i = Y$ may involve $\theta(m)$ bit operations. A first attempt for a formal algorithm printing the indices of all occurrences of Y, and not just the first one, is shown in ALGORITHM 5.5.

What seems to be a single operation in line 2 has in fact to be implemented as a loop, because the bit-string Y is of length m and does, usually, not fit into a single computer word. Applying a similar technique as that used for modular exponentiation in the evaluation of $b^{n-1} \bmod n$, we first identify the bit-string $Y = y_0 y_1 \cdots y_{m-1}$ with its value when considered as the binary representation of a number:

$$Y = y_0\, 2^{m-1} + y_1\, 2^{m-2} + \cdots + y_{m-3}\, 2^2 + y_{m-2}\, 2^1 + y_{m-1}. \qquad (5.4)$$

This can be evaluated efficiently if it is rewritten into a form known as *Horner's rule*:

$$Y = (\cdots ((y_0 \cdot 2 + y_1) \cdot 2 + y_2) \cdot 2 + \cdots + y_{m-2}) \cdot 2 + y_{m-1}.$$

To get the corresponding equality modulo p, we proceed as in Eq. (5.3) to get

$$\overline{Y} = \overline{(\cdots \overline{(\overline{(y_0 \cdot 2 + y_1)} \cdot 2 + y_2)} \cdot 2 + \cdots + y_{m-2}) \cdot 2 + y_{m-1}},$$

that is, after each multiplication by 2 and addition of the current bit value y_i, the accumulated result is reduced modulo p, so it never needs more than $k+1$ bits for its representation. If we define a variable Y to hold ultimately the value of $Y \bmod p$, then line 2 can be replaced by the loop

2a Y $\leftarrow$ 0
2b for $i \leftarrow 0$ to $m - 1$
2c   Y $\leftarrow (2 \times$ Y $+ y_i) \bmod p$

The value of Y is constant throughout the algorithm and therefore can be pre-computed before entering the loop. There is, however, a similar operation for Z_i in line 4 which depends on i. If the assignment to Z_i is replaced by a similar loop as for Y, the complexity of the algorithm will again be $\theta(nm)$ as for the naïve variant. Moreover, that would mean that for each i, instead of comparing y_j with x_{i+j} for $0 \le j < m$ in m simple comparisons, we perform a much more involved loop of m steps to evaluate $\overline{Z_i}$. The algorithm can nevertheless be salvaged because the different Z_i are not independent. Actually, the consecutive strings Z_i and Z_{i+1} are overlapping in $m - 1$ bits, as can be seen in Figure 5.2. We can therefore easily express Z_{i+1} as a function of Z_i and deduce from it the sought equality modulo p. Equation (5.4) can be rewritten for Z_i as

$$Z_i = x_i\, 2^{m-1} + x_{i+1}\, 2^{m-2} + \cdots + x_{i+m-2}\, 2^1 + x_{i+m-1}.$$

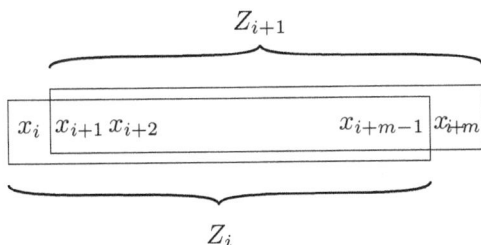

Figure 5.2: *Overlapping parts of Z_i and Z_{i+1}.*

Starting with Z_i, the following steps are needed to get to Z_{i+1}:

(1) delete the leftmost bit (by subtracting $x_i\, 2^{m-1}$);
(2) shift the $m - 1$ bits one position to the left (multiplying by 2);
(3) add the rightmost bit.

This yields

$$Z_{i+1} = 2\,(Z_i - x_i\, 2^{m-1}) + x_{i+m},$$

from which we get the following by applying modulo p:

$$\overline{Z_{i+1}} = 2\,(\overline{Z_i} - x_i\, \overline{2^{m-1}}) + x_{i+m}. \qquad (5.5)$$

Note that $\overline{2^{m-1}}$ is a constant that can be evaluated before entering the loop, so the necessary operations to derive $\overline{Z_{i+1}}$ are subtracting, or not, a constant, depending on the bit x_i being 0 or 1, shifting by one bit, adding the bit x_{i+m} and applying modulo p. In particular, the number of operations and their sizes k are independent of m.

ALGORITHM 5.6 summarizes all the details of the Karp–Rabin algorithm. It uses a variable Z holding the current value of the bit-string Z_i, which is initialized in line 2c. The constant P2 is initialized in line 2e as $\overline{2^{m-1}}$. The complexity of the algorithm is $\theta(m+n)$, which is much better than the $\theta(mn)$ we saw before.

Karp–Rabin(X, Y, k)

1   $p \leftarrow$ randomly chosen prime with k bits
2a Z $\leftarrow x_0$ Y $\leftarrow y_0$ P2 $\leftarrow 1$
2b for $i \leftarrow 1$ to $m-1$
2c   Z $\leftarrow (2 \times \text{Z} + x_i) \bmod p$
2d   Y $\leftarrow (2 \times \text{Y} + y_i) \bmod p$
2e   P2 $\leftarrow (2 \times \text{P2}) \bmod p$
3   for $i \leftarrow 0$ to $n-m$
4   if Z = Y then
5   print 'string found at ' i
6   if $i < n-m$ then Z $\leftarrow \left(2 \times (\text{Z} - x_i\, \text{P2}) + x_{i+m}\right) \bmod p$

ALGORITHM 5.6: *Karp–Rabin string matching.*

5.3.2   *Error probability*

What remains to be seen is how frequently one may get a wrong answer. The possibility of failing is only one-sided, as obviously, if $Z_i = Y$, then there will also be equality between $\overline{Z_i}$ and $\overline{Y}$ regardless of which modulus p has been used. The algorithm does therefore not miss any occurrences of Y in X, and in particular, if no index is printed, then there is not a single match.

On the other hand, it could well be that $\overline{Z_i} = \overline{Y}$ whereas $Z_i \neq Y$. For example, if $p = 7$, then $\overline{12} = \overline{19} = 5$, and yet $12 \neq 19$. This means that when the algorithm declares, in line 5, that there is a match at index i, this might be erroneous.

We would be interested in getting an estimate of the probability of such an error, similarly to what we saw for Rabin's primality test, in which the probability of falsely declaring a composite number as being prime is not more than 2^{-2k}, where k is the number of checked potential witnesses b.

Background concept: Probability

We have to clarify what is meant by the term *probability* in this and similar contexts. Only if the process under consideration involves a *stochastic* aspect, that is, one determined by some random activity, does it make sense to speak about probabilities.

Consider for example the statement $x = 7$. Whether or not this statement can be assigned a probability depends on how the value of x has been assigned. If we are told that x is chosen arbitrarily from the set $\{3, 7\}$, then we can say that the probability for $x = 7$ to hold is $\frac{1}{2}$, the number of possibilities to satisfy the request divided by the total number of possibilities; if the set is instead $\{3, 5, 6, 8\}$, the probability would be 0. However, if x is given, say x is 5, or even when x is 7, then there is no stochastic process; one can then say that the statement $x = 7$ is false in the former case and true in the latter, but not that the probability is 0 or 1, respectively. When there is nothing to choose from, the notion of probability is not defined.

Returning to Karp and Rabin's algorithm, note that the statement in line 1 is the only one in which there exists a random choice. All the other statements are deterministic, that is, well defined once p is given. Indeed, a part of the indices, at which the algorithm claims that a match has been found, might be wrong, and the fact that they are nevertheless printed depends solely on the choice of the random prime number p.

To evaluate the error probability, we proceed as follows. Define an integer $\mathcal{P}$ by

$$\mathcal{P} = \prod_{\substack{i=0 \\ Z_i \neq Y}}^{n-m} (Z_i - Y), \qquad (5.6)$$

that is, we again identify the bit-vectors Z_i and Y of length m with the numbers they stand for in a standard binary representation, and multiply all the differences $Z_i - Y$, but only for indices for which $Z_i \neq Y$, i.e., for indices i for which there is no match. $\mathcal{P}$ may be a huge number, as it is the product of almost n factors of m bits each, which might reach $\theta(nm)$ bits, about a billion for the example we gave of n and m being a million and a thousand, respectively. The size of $\mathcal{P}$ does not really bother us, because there is no intention, and no need, to actually calculate $\mathcal{P}$: it is only used as a tool to derive the error probability.

The only information we have about $\mathcal{P}$ is that it cannot be zero, since its definition skipped over all the indices i for which $Z_i = Y$. Let us then look at the equation one can derive from Eq. (5.6) by applying modulo p to both sides:

$$\overline{\mathcal{P}} = \overline{\prod_{\substack{i=0 \\ Z_i \neq Y}}^{n-m} (\overline{Z_i} - \overline{Y})}, \tag{5.7}$$

where we use the fact that the modulo function can be applied repeatedly and thus be pushed into each of the factors. Note, however, that in the enumeration of the indices i under the product sign $\prod$, it is indeed still $Z_i \neq Y$ which has to be written, and not $\overline{Z_i} \neq \overline{Y}$, because this is not a part of the formula, and rather a shortcut describing it; the number of factors has of course to be the same in both Eqs. (5.6) and (5.7), as it is not affected by the modulo function.

What can be said about $\overline{\mathcal{P}}$? That if there is even a single false alarm, that is, at least one index i for which $Z_i \neq Y$ and yet $\overline{Z_i} = \overline{Y}$, then $\overline{\mathcal{P}}$ will be zero. Recall that this is a property of the chosen prime number p. Had another prime number p' be selected, it is possible that

$$Z_i \bmod p' \neq Y \bmod p' \qquad \text{even though} \qquad Z_i \bmod p = Y \bmod p.$$

Let us refer to primes like p that may give rise to at least one wrong answer in the Karp–Rabin algorithm as *lying* primes, or *L*-primes for short. So for *L*-primes p, we know that $\overline{\mathcal{P}} = \mathcal{P} \bmod p = 0$, which means that the prime number p divides $\mathcal{P}$. Since we consider only primes written with k bits, the set of *L*-primes is finite. Let ℓ be the size of this set and denote the different *L*-primes by $p_1, p_2, \ldots, p_\ell$. For all of them we know that p_i divides $\mathcal{P}$ and therefore

$$p_1 \times p_2 \times \cdots \times p_\ell \quad \text{divides} \quad \mathcal{P}.$$

Here we have used the fact that the p_i are prime numbers. Actually, all we need is a much weaker result, namely that

$$p_1 \times p_2 \times \cdots \times p_\ell \quad \leq \quad \mathcal{P},$$

which implies a similar inequality concerning the number of bits used: the product of ℓ primes of k bits needs $k\ell$ bits, the number $\mathcal{P}$ needs up to nm bits, so one may conclude that

$$k\ell \leq nm.$$

By definition, the probability of getting (at least) a single erroneous index as output of Karp and Rabin's algorithm is the number of L-primes divided by the total number of primes with k bits. The latter is known from the prime number theorem. Summarizing,

$$\mathsf{Prob(error)} = \frac{\ell}{\frac{2^k}{k \ln 2}} = \frac{k \ell \ln 2}{2^k} \leq \frac{n m \ln 2}{2^k}.$$

For the running example, $n \approx 2^{20}$ and $m \approx 2^{10}$, hence choosing k only slightly above 30 might not yield a satisfactory behavior. But for $k = 64$, the error probability is below one billionth, so that one can probably rely on the list of declared matches and there is even no need to verify them.

The remarkable benefits of the probabilistic method for string matching are due to the facts that on the one hand, L-primes have so distinctive properties that their number cannot be too large, and on the other hand, there are many prime numbers with k bits.

How do we choose a random prime with k bits? One could toss a coin k times independently and write the heads and tails as 0s and 1s. This gives just a random k-bit number R_1, which can be tested for primality as we saw in the preceding section. If R_1 is not prime, one can start over from scratch and generate new numbers R_2, R_3, etc. The value of each R_i is of the order of 2^k, thus the expected number of trials until a prime will be detected is about $\ln 2^k = k \ln 2$. Alternatively, one may save coin tosses by searching linearly for the largest prime smaller than R_1, as we did to find $2^{400} - 593$; this will again involve $O(\log R_1)$ tests.

Since the computer does not know whether we have really picked our prime number p as just described, one could ask why we should bother through this cumbersome procedure. What would be wrong if we were to use, once and for all, our favorite personal prime number, like $2^{31} - 1$ or $2^{73} - 273$, because they are easy to remember? As a matter of fact, why do we need a prime number at all? The remainder function works just as well if the modulus is not prime!

The truth is that if p is chosen as a constant or even as a composite number, it is not the algorithm that would be hurt, but its analysis. One would still get a list of indices of presumed matches, but nothing could be said about the chances to get some false results. It could be possible that the specific constant prime we chose is an L-prime for many possible texts and strings.

We shall see more applications of Karp and Rabin's string matching technique in the next chapter dealing with data compression.

5.4   Probabilistic Quicksort

As a last example, we consider a probabilistic version of a sorting algorithm
known as Quicksort, in spite of its very bad worst-case behavior. It belongs
to the family of Divide and Conquer algorithms treated in Chapter 1. We
are given a set of n numbers stored in an array A; for the ease of description,
we shall assume that there are no duplicates.

The first step of Quicksort is to pick one of the elements of the array to
serve as *pivot*: the chosen element, denoted by P, is used to split the set A
into three disjoint parts:

the elements of A that are smaller than P $\qquad S = \{x \in A \mid x < P\}$,

the singleton containing only the pivot $\qquad \{P\}$,

the elements of A that are larger than P $\qquad L = \{x \in A \mid x > P\}$.

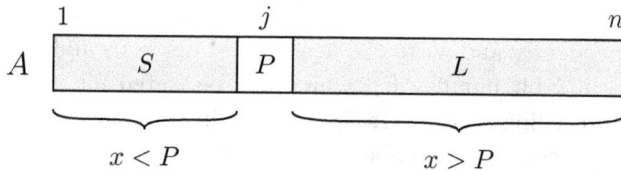

Figure 5.3: *Partition of the array into three subsets.*

The elements of these three sets are then rearranged as shown in Figure 5.3, with S on the left of P and L on the right of P. After this rear-
rangement, the pivot element P is already in the position indexed j it would
hold, had we sorted the array completely, but S and L are not sorted internally, which is why they appear in gray in Figure 5.3. As example, consider
the 20 first digits of π as a sequence of 10 integers

$$31 \quad 41 \quad 59 \quad 26 \quad 53 \quad 58 \quad 97 \quad 93 \quad 23 \quad 84;$$

if the pivot P is chosen, as is often done, as the first element 31, the
rearranged array corresponding to Figure 5.3 would be

$$26 \quad 23 \quad \mathbf{31} \quad 84 \quad 93 \quad 97 \quad 58 \quad 53 \quad 59 \quad 41, \tag{5.8}$$

where P has been emphasized. To complete the sorting task, all that needs
to be done is continuing recursively on the subsets S and L. ALGORITHM 5.7
brings the formal algorithm. We use the notation $A[\ell : u]$ to denote a sub-
array, that is, the elements of A with indices between, and including, lower

Quicksort(A)

> if $|A| > 1$ then
>> $j \leftarrow$ Split(A)
>> $S \leftarrow A[1 : j - 1]$ Quicksort(S)
>> $L \leftarrow A[j + 1 : n]$ Quicksort(L)

ALGORITHM 5.7: Quicksort.

and upper bounds ℓ and u. By convention, if $\ell > u$, then $A[\ell : u]$ is the empty set.

The decision about which element P to choose as pivot is not a part of the Quicksort specifications. To simplify the coding, programmers often use the first element $A[1]$ of the input array. The easiest way to implement the split procedure is by using an auxiliary array B, and at the end, retransfer the data back to A. In a linear scan of the input, the elements smaller than P are added starting at the left end of B, and the larger ones at the right end, using the variables s and ℓ to build the sets S and L. At the end of the loop $s = \ell$, and their common value is the index j of P, which is returned. The array in the displayed line numbered (5.8) has been built according to ALGORITHM 5.8.

Split(A)

> $P \leftarrow A[1]$
> $s \leftarrow 1$ $\ell \leftarrow n$
> for $i \leftarrow 2$ to n do
>> if $A[i] < P$ then
>>> $B[s] \leftarrow A[i]$ $s \leftarrow s + 1$
>> else
>>> $B[\ell] \leftarrow A[i]$ $\ell \leftarrow \ell - 1$
> $B[s] \leftarrow P$
> $A \leftarrow B$
> return s

ALGORITHM 5.8: Split *procedure used in* Quicksort.

In the worst case, the partition element is either the first or the last element of the array in every recursive call. One of the sets S or L is then empty, but the other is of size $n - 1$, which yields $\theta(n^2)$ comparisons. In particular, if the input is already sorted, which is not always unrealistic, the worst case is reached, whereas one would then rather expect a faster execution.

The probabilistic approach tries to avoid the worst-case behavior by choosing the pivot randomly. The intuition is that the probability for choosing then one of the extreme elements as pivot in every iteration should be so small, that the ensuing quadratic function of $\theta(n^2)$ will not be the dominant term of the complexity. More precisely, define $T(n)$ as the average number of comparisons required by Quicksort to sort n elements, assuming that the pivot is chosen uniformly among the possible choices in each iteration. The boundary condition is $T(1) = 0$.

The number of comparisons in Split part is $n - 1$, no matter which specific element is chosen as pivot P. As to the recursive steps, the relative position of P determines the sizes of the subsets S and L, as can be seen in Figure 5.3. A uniform choice of P means that the probability of P being in position j is $\frac{1}{n}$, independently of j, implying that the corresponding sizes of S and L are $j - 1$ and $n - j$, respectively. This yields the following recurrence relation for $T(n)$:

$$T(n) = n - 1 + \frac{1}{n} \sum_{j=1}^{n} (T(j-1) + T(n-j)) = n - 1 + \frac{2}{n} \sum_{i=0}^{n-1} T(i). \quad (5.9)$$

Let us first get rid of the fractional part by multiplying both sides of (5.9) by n, leaving only integers to be manipulated:

$$nT(n) = n(n-1) + 2 \sum_{i=0}^{n-1} T(i). \quad (5.10)$$

This recursion is different from the ones we have seen in Chapter 1 in that $T(n)$ does depend on all of its previous values $T(j)$ for $j < n$, and not just on some selected ones. The way to handle such recursions is to introduce an additional equation, which is similar to, but different from, the one at hand. One can then subtract the corresponding sides of the equations, and because of the similarity, many terms may cancel, often yielding another formula from which new insights may be gained.

For our application, let us rewrite Eq. (5.10) for $n - 1$:

$$(n-1)T(n-1) = (n-1)(n-2) + 2 \sum_{i=0}^{n-2} T(i). \quad (5.11)$$

By subtracting (5.11) from (5.10), one gets

$$nT(n) - (n-1)T(n-1) = 2(n-1) + 2T(n-1), \quad (5.12)$$

in which all the elements of the summation, except two extreme ones, cancel out. We thus get

$$nT(n) = 2(n-1) + (n+1)T(n-1). \tag{5.13}$$

The term on the left-hand side resembles the last term on the right-hand side, but not enough. To enable a substitution, we divide the sides of (5.13) by both n and $n+1$ to get

$$\frac{T(n)}{n+1} = \frac{2}{n+1} - \frac{2}{n(n+1)} + \frac{T(n-1)}{n}, \tag{5.14}$$

which suggests to define $G(n) = \frac{T(n)}{n+1}$. Substituting in (5.14), one gets

$$G(n) = \frac{2}{n+1} - \frac{2}{n(n+1)} + G(n-1)$$

$$= \frac{2}{n+1} - \frac{2}{n(n+1)} + \frac{2}{n} - \frac{2}{n(n-1)} + G(n-2)$$

$$\vdots$$

$$= \left[\frac{2}{n+1} + \frac{2}{n} + \cdots + \frac{2}{i+2}\right] - \left[\frac{2}{n(n+1)} + \frac{2}{n(n-1)} + \cdots + \frac{2}{(i+1)(i+2)}\right] + G(i)$$

for all $i < n$. Since $T(1) = 0$, this implies $G(1) = 0$, and one finally gets

$$G(n) = 2\sum_{i=3}^{n+1} \frac{1}{i} - 2\sum_{i=2}^{n} \frac{1}{i(i+1)}.$$

The rightmost term is a telescoping sum

$$2\sum_{i=2}^{n} \frac{1}{i(i+1)} = 2\sum_{i=2}^{n} \left(\frac{1}{i} - \frac{1}{i+1}\right) = 2\left[\frac{1}{2} - \frac{1}{3} + \frac{1}{3} - \frac{1}{4} + \cdots\right] = 1 - \frac{2}{n+1},$$

and what remains to be evaluated is a part of the *harmonic series* $\sum_{i=1}^{\infty} \frac{1}{i}$, known to be divergent. The interesting question here is, at what rate it goes to infinity. Since this and similar sums often appear in the analysis of algorithms, this is an opportunity to present a useful tool for the approximation of similar discrete sums.

Background concept: Approximating a sum by an integral

Let f be a function defined on the integers, which we assume to be monotonic. Let ℓ and u be two integers serving as lower and upper bounds for some evaluation, with $\ell \le u$. A sum of the form

$$S = f(\ell) + f(\ell+1) + \cdots + f(u) = \sum_{i=\ell}^{u} f(i) \tag{5.15}$$

does frequently occur in the evaluation of the complexity of an algorithm. To approximate S, we add the assumption that f can be extended to be defined on all the reals in the range $(\ell - 1, u + 1)$, not just on the integers. This is often indeed the case for typical complexity functions like n^2, $n \log n$, etc. Suppose also that the function f is integrable on the given range.

Refer to the graphical interpretation of the sum S in Figure 5.4, in which the function f is shown as monotonically decreasing, and the values $f(\ell), f(\ell + 1), \ldots, f(u)$ are represented by the lengths of the bars connecting the x-axis with bullets on the graph of the function. The problem is that we do not know how to evaluate the sums of lengths.

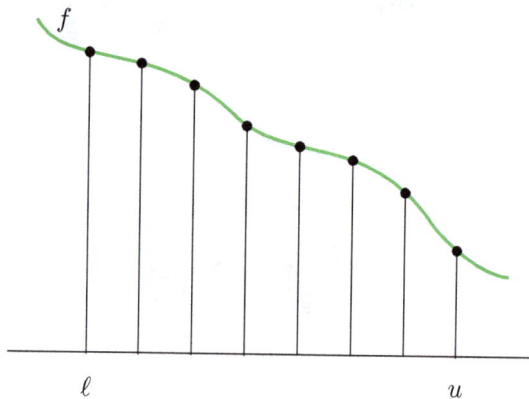

Figure 5.4: *Graphical interpretation of*
$S = \sum_{i=\ell}^{u} f(i) =$ *sum of the lengths of the bars.*

Nonetheless, we do know how to evaluate areas, which suggests to proceed as depicted in Figure 5.5. Extend the bar of length $f(\ell)$ to its left, forming a rectangle of width 1 and height $f(\ell)$, as shown in red in the figure. The area of this rectangle is thus $f(\ell) \times 1 = f(\ell)$, which is also the length of the bar. Even though lengths and areas are measured in different units, equality means here that the number measuring the length is equal to the number quantifying the area. We now repeat this process for all i from ℓ to u to get that the sum S defined in Eq. (5.15) is equal to the area in form of a staircase, delimited by the red lines in Figure 5.5. Because the function f is decreasing, this staircase is completely below the graph f, shown in green, so the area of the staircase is smaller than the area below f, with appropriate limits. This can be expressed by means of a definite integral:

$$S = \text{area of red staircase} \leq \int_{\ell-1}^{u} f(x)dx. \qquad (5.16)$$

So far, we have dealt with an upper limit for the sum S. To get a double-sided approximation, repeat the procedure, extending this time the bars to the right. This results in the blue staircase of Figure 5.5. It should be emphasized that the areas of both staircases are the same, as you can convince yourself by virtually sliding the blue staircase to the left by one unit to get an overlap with the red staircase. By switching the direction of the extension, it is now the area *below* the function that is bounded by the area below the blue staircase, and also the limits have slightly changed, to get

$$\int_{\ell}^{u+1} f(x)dx \leq S = \text{area of blue staircase}. \qquad (5.17)$$

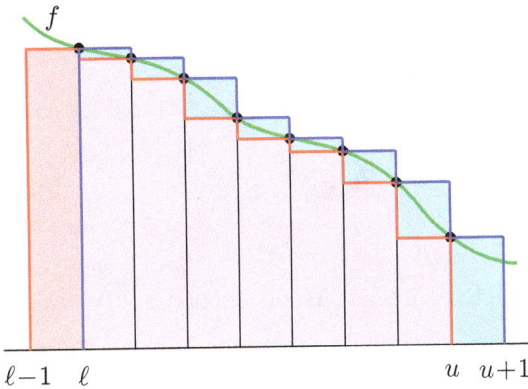

Figure 5.5: *Replacing a sum of lengths by sums of areas.*

Moreover, this approximation uses the integral of the same function twice, so only one primitive function has to be calculated. Denote by $F(x)$ the primitive function of $f(x)$, then

$$F(u+1) - F(\ell) \leq S \leq F(u) - F(\ell-1).$$

For an increasing function $f(x)$, the inequalities in (5.16) and (5.17) have to be reversed. What about non-monotonic functions? If the range can be partitioned into sub-intervals on each of which f is either increasing or decreasing, then one can apply this technique separately on each sub-range and collect the terms accordingly.

Recall that we are interested in a part of the harmonic series. The primitive function of $f(x) = \frac{1}{x}$ is $F(x) = \ln x$, the natural logarithm, which yields

$$\ln(n+1) - \ln 2 \ \leq \ \sum_{i=2}^{n} \frac{1}{i} \ \leq \ \ln(n).$$

One may thus approximate the requested sum quite precisely as belonging to a very small interval around $\ln n$.

Since G was defined as $G(n) = \frac{T(n)}{n+1}$, the conclusion is that

$$T(n) = (n+1)G(n) \in \theta(n \log n).$$

This means that the probabilistic approach succeeded to reduce a $\theta(n^2)$ worst-case complexity to an expected $\theta(n \log n)$, which is the best one could expect.

5.5 Exercises

5.1 To find the largest prime below 2^{400}, we suggested testing linearly all the odd candidates $2^{400} - 3$, $2^{400} - 5$, etc. Why did the list not start with $2^{400} - 1$?

5.2 A sequence of integers $R(i)$ is defined by the recurrence relation

$$R(i) = R(i-1) - 2\,R(i-2) + 3\,R(i-3), \qquad \text{for } i \geq 3,$$

and the boundary conditions $R(i) = 1$ for $i = 0, 1, 2$. It is easy to evaluate $R(n)$ in time $O(n)$. Note that the recurrence relation can be rewritten as an equation involving matrices:

$$\begin{pmatrix} R(i) \\ R(i-1) \\ R(i-2) \end{pmatrix} = \begin{pmatrix} 1 & -2 & 3 \\ 1 & 0 & 0 \\ 0 & 1 & 0 \end{pmatrix} \begin{pmatrix} R(i-1) \\ R(i-2) \\ R(i-3) \end{pmatrix},$$

to derive from it a way to evaluate $R(n)$ in time $O(\log n)$. *Hint:* Use the fact that the multiplication of matrices is associative and a lesson learned from modular exponentiation.

5.3 One of the conjectures of Pierre de Fermat was that all integers of the form $A_n = 2^{2^n} + 1$ are prime. Indeed, A_0 to A_4, which are 3, 5, 17, 257 and 65537, respectively, are prime numbers, but the conjecture breaks already down for $n = 5$: $A_5 = 4,294,967,297$ is divisible by 641, and $A_6 = 18,446,744,073,709,551,617$ is divisible by 274,177. If Fermat would have checked the primality of A_5 by applying his own Little Theorem with $b = 3$, he could have convinced himself that the conjuncture is wrong, because

$$3^{A_5-1} \bmod A_5 \neq 1. \qquad (5.18)$$

(a) Show that indeed there is no equality in Eq. (5.18), nor in the corresponding one for A_6.

(b) What would be the complexity of performing similar tests (only with $b = 3$) for all A_i, $0 \leq i \leq n$?

5.4 One of the TV stations in a country with 9 million inhabitants decided to interview one of the 3000 oldest people. Suppose the reporter has a list of all the citizens, but that the list does not include age. Suggest a probabilistic algorithm, based on random choices, so that the probability of finally interviewing somebody who does not belong to the targeted group is below $0.0183 \approx e^{-4}$.

5.5 Given are two integers a and b, with $a > b$. We consider the two numbers n and m defined by

$$m = 2^a + 1 \qquad \text{and} \qquad n = 2^b + 1.$$

What will be the complexity of calculating $\mathsf{GCD}(2^m, n)$ using Euclid's algorithm?

5.6 Suppose that Karp and Rabin's pattern matching algorithm is applied with a randomly chosen prime number p which is larger than 2^m, that is, the length of the prime is greater than the length of the pattern.

(a) Will the algorithm work?

(b) What can be said about the probability of error relatively to the case of using a smaller p?

(c) What will be the complexity of the algorithm in this case?

5.7 Let us extend Karp and Rabin's algorithm to work with decimal digits instead of just with binary bits, and let us search for the pattern $Y = 3\ 4$ in the text

$X = 4\ \ 7\ \ 2\ \ 5\ \ 1\ \ 7\ \ 9\ \ 4\ \ 8\ \ 5\ \ 1\ \ 3\ \ 9\ \ 6\ \ 4\ \ 2\ \ 2\ \ 8.$

Assume that the prime number used is $p = 17$. What are the matches announced by the algorithm? Assume that the value of a string is the number it represents, not the value of the ASCII codewords of the digits.

PART 4
Text Algorithms

Chapter 6

Data Compression

Computing devices have been invented to facilitate the monotonic and quite boring task of performing many high-precision calculations. Doing it faultlessly requires advanced skills and an ability to focus on details, but humans tend to relax their concentration in the long run, which ultimately leads to errors. Machines are preferable in this regard, providing more accurate results and, furthermore, achieving them much faster.

Only several decades after the advent of computers did researchers realize that it might be profitable to mobilize its advantages to improve also the processing of essentially textual data, and not just use them for number crunching. We all know, of course, that whatever is stored in the memory of our machines is ultimately translated into binary form, so that the difference between textual and numerical information alluded here seems rather artificial. The C programming language even identifies characters with small numbers between 0 and 255. Nevertheless, we shall continue to distinguish between different kinds of basic element types, because what matters is what we imagine our data to represent, rather than how it is actually coded.

This fourth part of the book is devoted to what we shall call *text algorithms*, as opposed to the others that primarily process *numbers*, though the intersection of these targets is not necessarily empty. Chapter 6 deals with data compression, which is the art to represent information using as little as possible storage space. In Chapter 7, we then return to the problem of pattern matching, for which we have seen naïve and probabilistic solutions in the previous chapter.

A simplified schematic outline of the data compression process is shown in Figure 6.1. Some DATA is given and transformed via an encoding function f into ᴅᴀᴛᴀ, which represents exactly the same information, but is physically smaller, that is, requires fewer bits for its representation. The focus here is therefore on *lossless* methods, for which the original DATA can be perfectly restored by means of a decoding function f^{-1}, which is the inverse of the encoding function f. It should be mentioned that there are application areas, such as image, video or audio data processing, which may tolerate, or often even prefer, *lossy* compression, but we shall not treat them here.

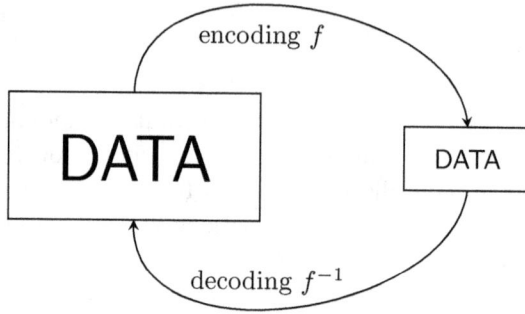

Figure 6.1: *Schematic view of lossless data compression.*

Before considering the various methods, one may ask why data compression is still an issue, given the technological advances of the past decades allowing us to store today amounts of data that were unimaginable just one generation ago. Compression methods could thus be deemed as unjustified. However, our appetite for ever-growing storage space seems to keep pace with hardware innovations, and furthermore, the reduction of the necessary space is just one of the advantages of the compression process. Here are examples of a few others:

Communication   Some data need not be stored at all, but may be transmitted over some communication channel. The bandwidth is often limited, which can force the sender to either restrict the amount of data to be sent or to accept delays. Compressing the data could alleviate the problem: the time needed for encoding on one end, and for decoding on the other, may largely be compensated for by the time saved due to the smaller amount of data to be handled.

Secondary storage Large repositories are often stored on a variety of media—tapes, disks, USB sticks, etc.—the access to which may be orders of magnitudes slower than to the main memory of our computers. Even if there is enough space to save a file in its original form, compressing it could ultimately save time by reducing the number of necessary I/O operations.

Security In certain applications, the content of our private information has to be hidden from others. This is the main concern of *cryptography*, a glimpse of which we shall get in Chapter 9. Data compression and cryptography have some common traits; they both try to reduce the *redundancy* in the given input file, albeit for different reasons. A compressed file may thus at times also be cryptographically more secure.

In addition to the above-mentioned benefits, data compression techniques are also interesting from the algorithmic point of view. While most algorithms are judged on the basis of their time and space complexities, compression methods are evaluated by yet another criterion: the achieved *compression ratio*, defined as the size of the compressed divided by the size of the original file.

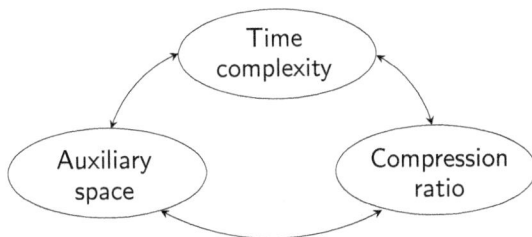

Figure 6.2: *Competing evaluation criteria for compression methods.*

These features tend to be competing, as depicted in Figure 6.2, and an improvement in the performance of one of the criteria comes generally at the expense of some deterioration for the other two. The main objective of data compression is finding an acceptable trade-off yielding good compression performance at reasonable speed and without relying on exceedingly large supplementary tables.

6.1   Statistical coding

A basic tool for achieving the compression goal is to represent the given data in a more economical way than originally given. This action is called *coding*, and the set of strings into which the elements of the original file are transformed is called a *code*. We consider here only binary codes, that is, all the components of the code are strings of 0s and 1s, though the ideas can easily be extended to larger sets than just $\{0, 1\}$. The members of a code are called *codewords*, trying to avoid the ambiguity of referring to both the set and its elements with the same term "code," as is regretfully often done.

As example, refer to some text T as the given input file. The simplest approach would be to consider T as a sequence of individual characters, belonging to a well-defined alphabet Σ, e.g., the English one. This alphabet could include lowercase a, b, c, ... and uppercase A, B, C, ..., as well as digits 0, 1, 2, ... and punctuation signs . : ; , ... We shall adhere to the convention of referring to the elements of the alphabet as *letters* and to those of the text as *characters*. One wishes to assign a code to the letters of the given alphabet. A standard such code is known by its acronym ASCII, standing for the *American Standard Code for Information Interchange*, which uses, in its extended variant, eight bits to represent a codeword, thereby providing $2^8 = 256$ possibilities.

Considering the text T as a sequence of characters is by no means the only choice; one could as well parse the text as a sequence of character pairs aa, ab, ..., zz, or triplets, or even entire words the, of, which, etc. Ultimately, any set S of strings, even one including word fragments like comput, or phrases like once⊔upon⊔a, can be chosen as the set of basic elements, as long as there is an unambiguous way to break the given text T into a sequence of elements of S. We shall still refer to such a set S as an alphabet, and to its members as letters, but these terms should be understood in the broader sense.

In its simplest setting, a code consists of codewords, all of which are of the same length, as in the ASCII example mentioned above. The encoding and decoding processes for such fixed-length codes are straightforward: to encode, the codewords corresponding to the characters of the message have just to be concatenated, which would yield the string

$$0100001101101111011001000110100101101110011001111$$

for the ASCII encoding of the word Coding. For decoding, the encoded string is parsed into blocks of the given size, eight bits in our case, and a decoding table can be used to translate the codewords back into the letters they represent. For our example string, we would get

01000011 | 01101111 | 01100100 | 01101001 | 01101110 | 01100111

C   o   d   i   n   g.

It is, however, not efficient to allot the same number of bits to each of the letters, and one may renounce the convenience of working with a fixed-length code to get a representation that is more economical in space, though harder to process. By assigning shorter codewords to the more frequent letters and encoding the rare letters by longer strings, the *average* codeword length may be reduced. The encoding of such *variable-length* codes is just as simple as for their fixed-length alternatives and still consists in concatenating the codeword strings. Decoding, on the other hand, may not be so simple, and a few technical problems may arise.

A first observation is that not every set of binary strings is necessarily a useful code. As example, consider the following four codewords:

A	B	C	D
00	010	0010	1001

for the alphabet {A, B, C, D}. The string 0001000 can only be parsed as ABA, but there are strings with more than one possible interpretation, e.g., 0010010010, whose origin could be the concatenation of both

00 | 1001 | 0010 = ADC   and   0010 | 010 | 010 = CBB.

The existence of even a single such ambiguous string should not be permitted, because it violates the reversibility of the code, without which the ability of decoding is not guaranteed for every possible message. We shall therefore restrict attention to codes called *uniquely decipherable* (UD), for which any binary string obtained by the concatenation of a finite sequence of codewords can be parsed only into the original sequence.

For a code to be UD, it is necessary that its codewords should not be too short. A precise requirement has been formulated by McMillan (1956): a necessary condition for a binary code with codeword lengths $\{\ell_1, \ldots, \ell_n\}$ to be uniquely decipherable is for the set of lengths to satisfy

$$\sum_{i=1}^{n} 2^{-\ell_i} \leq 1. \tag{6.1}$$

For example, no binary UD code could possibly exist with two codewords v_0 and v_1 of length 1 and a third codeword v_2 of any length k; clearly, v_0 and v_1 must be 0 and 1, and the k bits of v_2 could be parsed as k occurrences of either v_0 or v_1. Indeed, in this case, $2^{-1}+2^{-1}+2^{-k} > 1$, so that McMillan's inequality does not hold.

The fact that for the example of the non-UD code on a four-letter alphabet given above one gets

$$2^{-2} + 2^{-3} + 2^{-4} + 2^{-4} = \frac{1}{2} < 1$$

shows that the condition is not sufficient for the code to be UD.

6.1.1 *Universal codes*

Before tackling the problem of devising an efficient UD code for a given alphabet with specific characteristics, let us consider some simple solutions that often serve as building blocks in more involved algorithms. In many applications, the set of elements to be encoded is the set $\mathbb{N}$ of the natural numbers. Of course, the specific set we deal with will always be finite, but we may wish to prepare a code that is fixed in advance, so the set of possible elements for which codewords have to be generated may not be bounded. Just using then the standard binary representations of the integers $\{1, 2, 3, \ldots\}$ without leading zeros is not a feasible solution, because their set $\{1, 10, 11, 100, 101, \ldots\}$ is not UD. If the strings are padded with zeros on the left to turn them into fixed length, the set would be UD, but this requires knowing the maximum value in advance.

We seek a solution for encoding the integers in some systematic way, still exploiting the benefits of variable-length codes by assigning shorter codewords to the smaller integers. In many settings, the smaller the number, the more frequently it occurs. A simple way could be a *unary* encoding, using the codewords

1	2	3	4	5	6	
0	01	001	0001	00001	000001	$\cdots$

that is, the length of the n-th codeword, representing the integer n, will be n bits. Such a biased representation will be efficient only if the probability of using larger integers decreases exponentially, which is not the case in most applications. It seems preferable to let the lengths of the codewords be bounded by $O(\log n)$ to represent n. Several infinite codeword sets with such a property have been defined by P. Elias (1975) as being *universal* codes.

6.1.1.1 *Elias codes*

One of the simplest codes devised by Elias has been called the γ-code or γ for short. The codeword for an integer $n \geq 1$ relies on its standard binary representation without leading zeros, $B(n)$, of length $\lfloor \log_2 n \rfloor + 1$ bits. Since $B(n)$ by itself cannot be decoded, it is preceded by an encoding of its length, using the unary encoding seen above. Actually, since $B(n)$ always has a leading 1, it suffices to encode the length of $B(n)$ minus 1. For example, $B(314) = 100111010$, so for its encoding, the string will be preceded by eight zeros: 00000000100111010. $B(1) = 1$; hence the string to be pre-pended is empty, and thus the γ codeword for 1 is 1 itself.

Table 6.1: *Sample codewords of the γ, δ and Fibonacci codes.*

Index	Elias γ	Elias δ	Fibonacci
1	1	1	11
2	0 10	010 0	011
3	0 11	010 1	0011
4	00 100	011 00	1011
5	00 101	011 01	00011
6	00 110	011 10	10011
7	00 111	011 11	01011
8	000 1000	00100 000	000011
9	000 1001	00100 001	100011
10	000 1010	00100 010	010011
20	0000 10100	00101 0100	0101011
30	0000 11110	00101 1110	10001011
40	00000 101000	00110 01000	100100011
50	00000 110010	00110 10010	001001011
100	000000 1100100	00111 100100	00101000011
200	0000000 11001000	0001000 1001000	100000001011
500	00000000 111110100	0001001 11110100	00000001010011
1000	000000000 1111101000	0001010 111101000	0000010000000011

Note that the leading 1-bit of $B(n)$ serves a double purpose: it is both a delimiter of the prefixed sequence of zeros, thereby determining its length, and part of the binary representation of the given integer. The decoding process therefore scans the codeword from its beginning until the first 1-bit is detected. If k bits have been read, including the limiting 1-bit, the length of the codeword is $2k - 1$, and the rightmost k bits are the standard binary representation of the integer represented by the given γ codeword.

The γ code, a sample of which can be seen in Table 6.1, has the useful property that no codeword is the prefix of any other. Indeed, if two γ codewords x and y are not of the same length, then the positions of their leftmost 1-bits cannot be the same, and if $|x| = |y|$, where $|a|$ denotes the length of the string a, then one can be the prefix of the other only if $x = y$. This is known as the *prefix property*, and a code satisfying it is called a *prefix code*.[1] Being a prefix code is a sufficient condition for a code to be UD. An encoded string can be scanned from its beginning until a codeword is detected; no ambiguity is possible, because the codeword is not the beginning of any other one.

For very large integers, the γ code may be wasteful because the unary encoding of the length grows linearly. This led to the definition of an alternative, called the δ code. As for γ, the definition relies on $B(n)$, but for the encoding of its length, the unary code is replaced by the γ code, which increases only logarithmically as a function of its parameter. Moreover, the leading 1-bit of $B(n)$ may now be omitted: no delimiter is needed, as the length part is encoded by a prefix code.

Returning to the example above, $B(314) = 100111010$, without its leading 1-bit, will be preceded by the γ codeword for $|B(314)| = 9$, which is 0001001, thus the δ codeword for 314 is 000100100111010.

For the decoding of a δ codeword x, we first detect a γ codeword at the beginning of x. This γ codeword represents some integer m, so $m - 1$ more bits $b_1 b_2 \cdots b_{m-1}$ have to be read, and x is the integer whose standard binary representation is $B(x) = 1 b_1 b_2 \cdots b_{m-1}$. For example, if the given codeword is 000100111110100, its only prefix representing a γ codeword is 0001001, which is the γ encoding of the integer 9. The following 8 bits are thus read and preceded by a leading 1-bit to get 111110100, which is the binary representation of 500. Table 6.1 shows also the Elias δ codewords for the same sample as above, and includes examples of the Fibonacci code to be seen below. For better readability, small spaces have been added between the length and $B(n)$ part for both γ and δ.

Table 6.2 shows the number of bits needed to encode the integer n using unary, γ, δ and Fibonacci codes. A δ codeword is asymptotically shorter than the corresponding γ codeword, but as mentioned, it is often the case that the smaller values appear more frequently. Since it is for some of these smaller values that γ codewords are shorter than their δ counterparts, the *average* codeword length may be shorter for γ than for δ.

---

[1]One might object that it should rather be called a *non-prefix* code, but we shall keep the globally accepted nomenclature.

Table 6.2: *Codeword lengths of universal codes.*

Code	Codeword length for value $n = 1, 2, 3, \ldots$
unary	n
γ	$2\lfloor \log_2 n \rfloor + 1$
δ	$\lfloor \log_2 n \rfloor + 1 + 2\lfloor \log_2(\lfloor \log_2 n \rfloor + 1) \rfloor$
Fibonacci	$\lfloor \log_\phi (\sqrt{5}(n + \frac{1}{2})) \rfloor$

6.1.1.2 *Fibonacci codes*

The binary representation of an integer $B(n)$ has been referred to as being *standard*, because there are in fact infinitely many possible binary representations. An interesting alternative to Elias codes is obtained by basing the code on the Fibonacci sequence.

Background concept: Fibonacci sequence

The Fibonacci sequence 0, 1, 1, 2, 3, 5, 8, 13, 21,... is defined by the boundary conditions $F(0) = 0$ and $F(1) = 1$ and the recurrence relation

$$F(i) = F(i-1) + F(i-2) \qquad \text{for } i \geq 2.$$

It is named after Leonardo of Pisa, born in the 12th century, and the properties of this sequence have been investigated ever since. The numbers can be represented as linear combinations with fixed coefficients of the powers of the roots of the polynomial $x^2 - x - 1$. These roots are

$$\phi = \frac{1 + \sqrt{5}}{2} = 1.618 \qquad \text{and} \qquad \hat{\phi} = \frac{1 - \sqrt{5}}{2} = -0.618,$$

the first of which is known as the *golden ratio*. The entire sequence can be described compactly by the formula

$$F(i) = \frac{1}{\sqrt{5}} \left(\phi^i - \hat{\phi}^i \right) \qquad \text{for } i \geq 0. \tag{6.2}$$

As the absolute value of $\hat{\phi}$ is smaller than 1, $\hat{\phi}^i$ becomes negligible for large enough i; thus a good approximation for $F(i)$ is achieved by taking just the first term of the right-hand side of Eq. (6.2): $F(i) \simeq \frac{1}{\sqrt{5}}\phi^i$. Actually, $F(i)$ is $\frac{1}{\sqrt{5}}\phi^i$, rounded to the nearest integer, for all elements of the sequence, that is, $F(i) = \lfloor \frac{1}{\sqrt{5}}\phi^i + \frac{1}{2} \rfloor$ for $i \geq 0$.

Just as the powers of 2 serve as basis elements to represent the integer $n = \sum_{i=0}^{k-1} b_i 2^i$ by the binary string $B(n) = b_{k-1} b_{k-2} \cdots b_1 b_0$, one can use any other increasing sequence of integers as basis elements, and in particular, the Fibonacci sequence. Since the smallest element of the sequence should be 1 and should appear only once, we consider the elements of the Fibonacci sequence starting with index 2.

Any integer n can be represented by a binary string of length $r - 1$, $c_2 c_3 \cdots c_{r-1} c_r$, such that

$$n = \sum_{i=2}^{r} c_i F(i),$$

where it should be noted that, for reasons that will soon become clear, we have reversed the order of the indices relative to the standard notation, and they increase now from left to right rather than from right to left as usual. A priori, a number may have more than one such possible representation, as both 110 and 001 could be encodings of the number $n = 3$. To ensure uniqueness, apply the following procedure to get the representation of n:

1. find the largest Fibonacci number $F(r)$, smaller or equal to n;
2. if $n > F(r)$ then continue recursively with $n - F(r)$.

For example, $41 = 34 + 5 + 2$, so its Fibonacci representation would be 01010001. As a result of this encoding procedure, consecutive Fibonacci numbers F_k and F_{k+1} do not appear in any of these sums, because F_{k+2} will be chosen instead. This implies, for the corresponding binary representation, that there are no adjacent 1s.

This property can be exploited to devise an infinite code consisting of the Fibonacci representations of the integers, by appending a single 1-bit to each of the codewords. This 1-bit acts like a comma or blank and permits one to identify the boundaries between the codewords when several of them are concatenated. A sample of this code appears in the last column of Table 6.1. The reason for the reversal of the indices is that thereby, one gets a prefix code after appending the 1-bit, since all codewords are then terminated by 11 and this substring does not appear anywhere in any codeword, except at its suffix. This code is known as *Fibonacci code* [Fraenkel and Klein (1996)].

Equation (6.2) implies that the basis elements in the Fibonacci representation grow like ϕ^r rather than 2^r if the standard binary representation had been used. The number of bits needed to represent an integer n using the Fibonacci code is thus $\log_\phi n = 1.4404 \log_2 n$, which is about 44%

longer than the minimal $\log_2 n$ bits needed for the standard, not UD, binary representation using only the significant bits, but it is shorter than the $2 \log_2 n$ bits needed for the corresponding codeword in the Elias γ code, which *is* UD. On the other hand, in spite of the increased total number of bits, the number of 1-bits is smaller on the average: in a standard binary representation, about half of the bits are 1s, whereas for the Fibonacci codewords it can be shown that the probability of a 1-bit is only about $\frac{1}{2}\left(1 - \frac{1}{\sqrt{5}}\right) = 0.2764$, so even when multiplied by 1.44, this gives an expected number of only $0.398 \log_2 n$ 1-bits, still less than about $0.5 \log_2 n$. This property has many applications.

Another advantage of a Fibonacci code is its robustness against errors. If a bit is lost, or an extraneous bit is added, or some bit value is complemented, the resulting error will not propagate as it might do for other variable-length codes. For fixed-length codes, the addition of an irrelevant bit may even be disastrous, as all subsequent codewords are shifted. For Fibonacci codes, no more than three codewords can be affected, for example, if the boxed 1-bit is turned into a zero-bit, the three codewords 010$\boxed{1}$1-11-0011 will be interpreted as if they were only two codewords 010$\boxed{0}$11-10011. But this is an extreme case, and mostly only a single codeword is lost, or two, if the error touches the suffix 11.

6.1.2 *Huffman coding*

Universal codes are a good solution for encoding an a priori unbounded set of elements by a set of codewords that is fixed in advance. For many applications, however, it is the set of elements to be encoded that is fixed and finite. There is thus no need for an infinite code, and, moreover, one would prefer to devise a code having its codeword lengths custom-tailored to the known distribution of its elements. We shall still restrict our attention to codes with the prefix property, which is a sufficient, though not necessary, condition for a code to be UD.

There is a natural one-to-one correspondence between binary prefix codes and *binary trees*, given by the following construction. Assign labels to both edges and vertices of a binary tree according to the following rules:

(1) label each edge pointing to a left or right child with 0 or 1, respectively;
(2) the label of any vertex v of the tree is obtained by concatenating the labels on the edges on the path from the root to vertex v.

In particular, the label of the root of the tree is the empty string. The string labeling a vertex v is thus a prefix of the string labeling a vertex w if and only if v is one of the vertices on the unique path from the root to w. Therefore, the set of the labels of the *leaves* of any binary tree satisfies the prefix property and thereby forms a prefix code. Conversely, given any prefix code, one can easily construct the corresponding binary tree according to these rules. For example, the tree corresponding to the code $\{11, 011, 0011, 1011\}$, which are the first four codewords of the Fibonacci code in Table 6.1, appears in Figure 6.3, in which the leaves, labeled by the codewords, are colored gray.

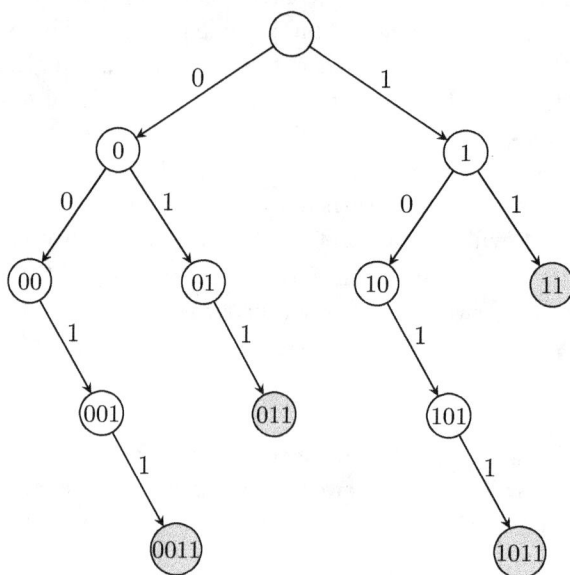

Figure 6.3: *Binary tree corresponding to the code* $\{11, 011, 0011, 1011\}$.

We shall henceforth consider a general binary tree and the corresponding prefix code as equivalent and use them interchangeably in our discussion. Recall that we have defined complete and full trees in Section 1.1. The tree in Figure 6.3 is not *complete* and thus cannot possibly be optimal from the compression point of view, because if not all the internal nodes of the tree have two children, then one could replace certain codewords by shorter ones, without violating the prefix property. It is thus possible to build another UD code with strictly smaller average codeword length. For example, the nodes labeled 01 and 101 have only a right child, so the codewords 011 and 1011 could be replaced by 01 and 101, respectively. Alternatively, one could add more codewords to the code, such as 000 or 100.

A code corresponding to a complete tree is accordingly called a *complete code*. A code is complete if and only if $\sum_{i=1}^{n} 2^{-\ell_i} = 1$, where the $\{\ell_i\}$ are the lengths of the codewords, that is, the McMillan inequality (6.1) is in fact an equality. An equivalent definition is that a complete code is a set C of binary codewords which is a binary prefix code, but adjoining any binary string x to the set turns it into a code $C \cup \{x\}$ which is not UD.

In our quest for good compression performance, we have gradually restricted our attention from general UD codes to prefix codes and finally to complete codes. Among those, we now wish to choose one which is optimal in the following sense. We are given a set of n non-negative weights $\{w_1, \ldots, w_n\}$; these could be the probabilities or (integer) frequencies of the occurrences of the letters in some extended alphabet, as discussed at the beginning of Section 6.1. The challenge is to generate a complete binary prefix code consisting of codewords with lengths ℓ_i bits, $1 \le i \le n$, with optimal compression capabilities, i.e., such that the *total length* of the encoded text $\sum_{i=1}^{n} w_i \ell_i$ is minimized. For the equivalent case where the weights w_i represent probabilities, the sum $\sum_{i=1}^{n} w_i \ell_i$ represents the *average length* of a codeword, which we wish to minimize.

If we forget for a moment that the ℓ_i represent codeword lengths and should therefore be integers, we could consider the mathematical optimization problem of minimizing the function $\sum_{i=1}^{n} w_i \ell_i$ of the n variables $\ell_1, \ldots, \ell_n$, under the constraint that the variables still have to obey the McMillan equality $\sum_{i=1}^{n} 2^{-\ell_i} = 1$, but letting the ℓ_i be positive real numbers, and not necessarily integers. This would yield $\ell_i = -\log_2 w_i$; if the w_i are probabilities, or more generally, when the weights w_i are frequencies, the minimum is achieved for

$$\ell_i = -\log_2 \frac{w_i}{\sum_{j=1}^{n} w_j} = -\log_2 p_i,$$

where the sum $W = \sum_{j=1}^{n} w_j$ of the frequencies in the denominator is in fact the total number of characters in the text, and $p_i = \frac{w_i}{W}$ is the probability of the element indexed i. The quantity $-\log_2 p_i$ is called the *information content* of the element with probability p_i and is measured in bits. For example, when $n = 3$ and the distribution is uniform, each of the three letters should ideally be encoded by $-\log_2 \frac{1}{3} = 1.585$ bits.

We can therefore derive a lower bound on the size of the compressed file as the weighted average of the information contents of its characters, that is,

$$-\sum_{i=1}^{n} w_i \log_2 p_i = W\left(-\sum_{i=1}^{n} p_i \log_2 p_i\right). \qquad (6.3)$$

The sum in parentheses, $H = -\sum_{i=1}^{n} p_i \log_2 p_i$, is known as the *entropy* of the probability distribution $p_1, \dots, p_n$, as defined by Claude Shannon [Shannon (1948)].

As the option of using fractional bits does not seem to lead to a practical implementation, the optimization problem of minimizing $\sum_{i=1}^{n} w_i \ell_i$ has been considered with the additional constraint that all the ℓ_i have to be integers. This problem has been solved in an optimal way in 1952 by D. Huffman (1952) with ALGORITHM 6.1, which maintains a forest $\mathcal{F}$ of complete binary trees. The nodes of the trees are assigned values in a field called val: the weights w_i are assigned as values to the leaves, the value of an internal node is the sum of the values of its two children, and the value of the root of a tree is considered as the value of the tree.

Huffman$(w_1, \dots, w_n)$

    $\mathcal{F} \leftarrow \{$Singleton trees with val(root) $\leftarrow w_i$, $i = 1, \dots, n\}$
    while $|\mathcal{F}| > 1$ do
        $T_1, T_2 \leftarrow$ trees in $\mathcal{F}$ with minimal val
        $T \leftarrow$ new binary tree having T_1 and T_2 as left and right subtrees
        val(root) $\leftarrow$ val(T_1) + val(T_2)
        $\mathcal{F} \leftarrow \mathcal{F} \setminus \{T_1, T_2\} \cup \{T\}$

ALGORITHM 6.1: *Construction of a Huffman tree.*

For initialization, each of the n weights w_i is assigned as val in its own small tree consisting only of the root. The main loop repeatedly finds two trees T_1 and T_2 with smallest values in $\mathcal{F}$, creates a new tree T having T_1 and T_2 as children, and replaces T_1 and T_2 by T in the forest $\mathcal{F}$. The value of T is the sum of the values of its two subtrees. At the end, the set $\mathcal{F}$ consists of a single tree, which is the sought Huffman tree.

To implement the construction, the initial weights are used to build a *min-heap* in time $O(n)$ (see the Background concept on Heaps in Section 4.2.2). In each iteration, two minima are extracted and their sum is inserted, which requires $O(\log n)$ comparisons. The complexity of constructing a Huffman tree is thus $O(n \log n)$.

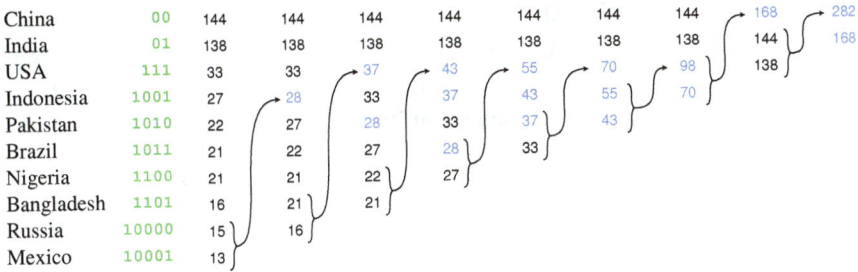

China	00	144	144	144	144	144	144	144	168	282
India	01	138	138	138	138	138	138	138	144	168
USA	111	33	33	37	43	55	70	98	138	
Indonesia	1001	27	28	33	37	43	55	70		
Pakistan	1010	22	27	28	33	37	43			
Brazil	1011	21	22	27	28	37	33			
Nigeria	1100	21	21	22	27	33				
Bangladesh	1101	16	21	21						
Russia	10000	15	16							
Mexico	10001	13								

Figure 6.4: *Huffman code for countries according to population.*

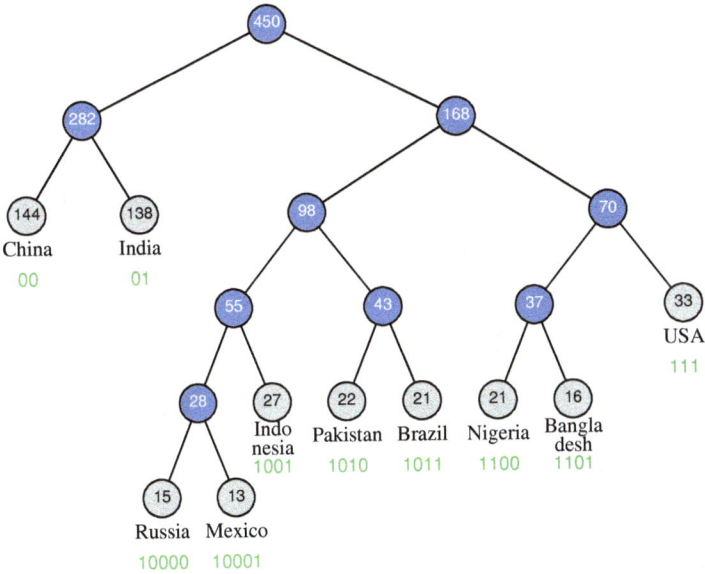

Figure 6.5: *Huffman tree of the code in Figure 6.4.*

For example, suppose we wish to build a Huffman code for the most populated countries according to the UN estimates of the number of their inhabitants in 2020. The "letters" of our alphabet are thus country names. The ordered frequencies, in tens of millions, are given in Figure 6.4 in the third column, next to the Huffman codewords, which appear in green.

Each iteration corresponds to one of the columns, removing the two smallest numbers from the list, and inserting their sum in the proper position, until only two frequencies remain. Each column represents the set $\mathcal{F}$ at the end of an iteration; the value of a tree that is not a singleton appears in blue: their roots are the internal nodes of the Huffman tree. The last iteration is omitted from the figure. It would form the final tree which appears in Figure 6.5 and shows the frequencies in its nodes. The code in green of the second column of Figure 6.4 corresponds to this tree.

Huff-decoding(S, T)

$\quad p \leftarrow root(T)$
$\quad \text{for } i \leftarrow 1 \text{ to } |S|$
$\qquad \text{if } S[i] = 0 \text{ then}$
$\qquad\qquad p \leftarrow left(p)$
$\qquad \text{else} \quad p \leftarrow right(p)$
$\qquad \text{if } p \text{ is a leaf then}$
$\qquad\qquad \text{output } lttr(p)$
$\qquad\qquad p \leftarrow root(T)$

ALGORITHM 6.2: *Decoding a string S using the Huffman tree T.*

Once the tree or, equivalently, the code, is given, the actual *compression* of a message M is obtained by concatenating the corresponding codewords. The Huffman tree T is then also a useful tool for the *decompression* process. Consider a binary string S, which is the encoding of M according to the given Huffman code, though the same procedure may be applied for any prefix code. Starting at the root, the bits of S are used to guide our navigation through the tree, where 0 and 1 indicate left and right turns, respectively. When a leaf is reached, the corresponding codeword is output, and p is reset to the root. ALGORITHM 6.2 brings the formal decoding steps. We assume that the leaves of the tree store the corresponding letters in their *lttr* fields.

In our case, the encoded elements, which are the letters of the alphabet, are country names, and they are written below their respective leaves in Figure 6.5. If S is the binary string 1011010110001, it will be decoded as Brazil, India, India, Mexico.

We now turn to the proof that Huffman's construction indeed yields a code with minimum average codeword lengths. This is by no means self-evident, because for similar optimization problems, as those we shall see in Chapters 10 and 11, no one has so far found an optimal solution in less than exponential time, so the existence of an $O(n \log n)$ algorithm may be surprising.

Theorem 6.1. *Given a set of $n \geq 2$ weights $\{w_1, \ldots, w_n\}$, Huffman's algorithm constructs an optimal code in the sense that it minimizes the average codeword length $\sum_{i=1}^{n} w_i \ell_i$, where ℓ_i is the length of the codeword assigned to weight w_i.*

Proof: The proof is by induction on the number of elements n. For $n = 2$, there exists only one complete binary prefix code $\{0, 1\}$, which therefore is optimal; this is also the only possible Huffman code for 2 elements, independently of the weights w_1 and w_2.

For general n, suppose that the claim is true for $n-1$. The proof consists of considering an optimal tree T_1 for n elements and showing that a Huffman tree must have the same average codeword length $M_1 = \sum_{i=1}^{n} w_i \ell_i$ as T_1, which implies its optimality.

It will be convenient to refer to Figure 6.6, which gives a schematical representation of the four trees we shall define. The tree T_i corresponds to the sequence of weights appearing above it, and its average codeword length M_i is written underneath.

Starting with an optimal tree T_1 for the n weights $\{w_1, \ldots, w_n\}$, pick two sibling leaves on its lowest level, corresponding to the smallest weights w_1 and w_2. There are at least two leaves on the lowest level of T_1 because the tree is complete. The fact that the lowest weights have been assigned to leaves on the lowest level follows from the optimality of T_1. Indeed, if the element with weight w_2 is not the lowest level ℓ, but on some level $m < \ell$, then there exists another element on level ℓ with weight $w_x > w_2$. If the weights w_x and w_2 are interchanged without altering the shape of the tree, one would get a code with an average codeword length of

$$M_1 - w_x \ell - w_2 m + w_x m + w_2 \ell = M_1 - (\ell - m)(w_x - w_2) < M_1,$$

that is, better than that of the optimal T_1, which is impossible.

If it is not true that the leaves assigned to the two smallest weights are siblings, then there exists another tree T_1' for which it is true; T_1' is obtained from T_1 by rearranging the weights of the leaves at the lowest level, which does not affect any of the codeword lengths, so that T_1' is also an optimal tree. There is thus no loss of generality if we assume that the two leaves we have picked are sibling nodes in T_1.

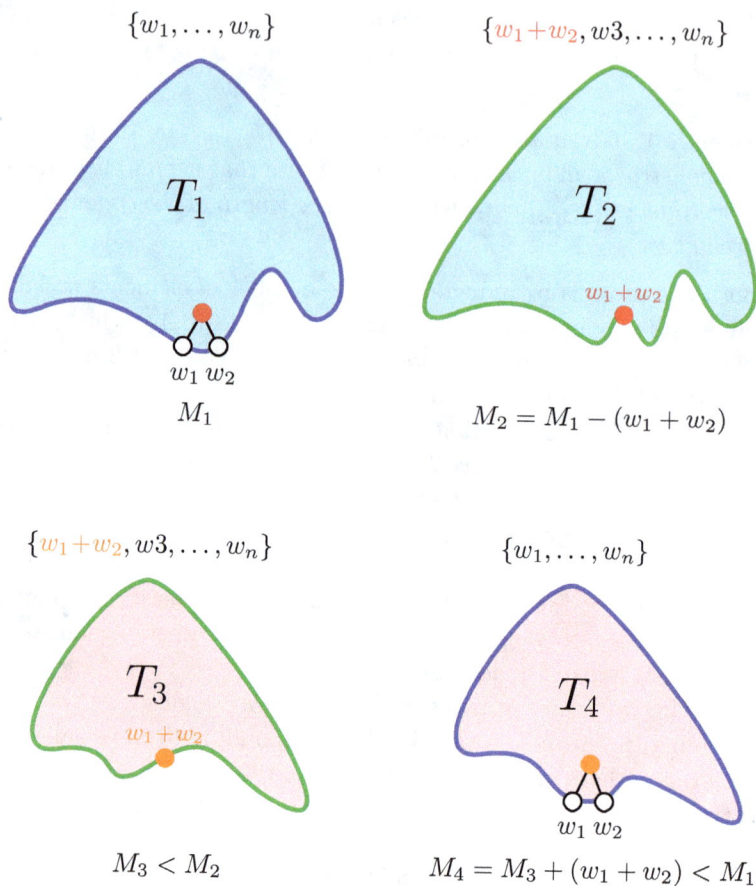

Figure 6.6: *Proof of the optimality of Huffman codes.*

The next step is to construct a new tree T_2 by a small local transformation of T_1. The subtree rooted at the red node in T_1, whose children are the leaves assigned to the weights w_1 and w_2, is replaced by a singleton tree consisting of the red node alone in T_2. The number of leaves is thereby reduced by 1, as two leaves have been removed, but a new (red) one has been added. The weight assigned to the red leaf in T_2 will be $w_1 + w_2$, whereas all the other leaves remain in place and are still assigned to the same weights w_i for $i \geq 3$ as in T_1. The average codeword length for T_2 is therefore $M_2 = M_1 - (w_1 + w_2)$.

Remember that T_1 is an optimal tree for the weights $\{w_1, \ldots, w_n\}$. Is T_2 also an optimal tree? A priori, we have no good reason to believe this. There was just a local perturbation to the structure of the tree, and the assignment of $w_1 + w_2$ as weight to the newly created red leaf seems rather arbitrary. In spite of this skepticism, we show that

Claim 6.1: T_2 is an optimal tree for the weights $(w_1 + w_2), w_3, \ldots, w_n$.

Proof: Suppose that the claim is not true. Then there exists a tree T_3 for the same set of weights, which is better in the sense that its average codeword length M_3 is strictly smaller than M_2. Note that while T_2 is almost identical to T_1, which is materialized in Figure 6.6 by giving them the same light blue color and a similar shape, the optimal T_3 may have a completely different shape than that of T_2. Accordingly, T_3 appears in the figure in pink and with a clearly different form. One of the leaves of T_3, shown in orange, corresponds to the weight $(w_1 + w_2)$.

The last step of the proof of the claim is now to perform the inverse transformation which led from T_1 to T_2. Define the tree T_4 obtained from T_3 by replacing the orange leaf by a small subtree consisting of the orange node as a root and its two children, which are assigned the weights w_1 and w_2. The number of leaves in T_4 is thereby increased to n again, and they correspond to the same weights $\{w_1, \ldots, w_n\}$ as the leaves of T_1. T_4 is similar to T_3 and is therefore also colored pink. The color of the outline of the shapes of the trees corresponds to the number of their leaves: blue for n elements and green for $n - 1$. The average codeword length for T_4 is thus

$$M_4 = M_3 + (w_1 + w_2) < M_2 + (w_1 + w_2) = M_1,$$

but this contradicts the optimality of T_1 for the given weights. □

Returning to the proof of the theorem, we know that the tree T_2, which has been shown to be optimal for $n - 1$ elements with the given weights, must have the same average codeword length as the Huffman tree because of the inductive assumption. However, the way to get from the Huffman tree for $(w_1 + w_2), w_3, \ldots, w_n$ to the Huffman tree for $w_1, \ldots, w_n$ is the same as to get from T_2 to T_1. Thus, the Huffman tree for the n elements has the same average codeword length as T_1, which shows that it is optimal. □

6.1.3 *Arithmetic coding*

We have just seen that Huffman coding provides an optimal code. It will therefore come as a surprise to find in this section a method that performs even better! The truth is that there are some implicit assumptions in the Theorem of Huffman's optimality. These conditions are so natural that they are often not deemed worth of being mentioned. Huffman codes are optimal only if each character is encoded by a fixed bit pattern made up of an integral number of bits. As it is not at all obvious how to implement an encoding of elements in fractional bits, the constraint of requiring an integral number of bits had probably been considered as self evident.

Arithmetic codes, popularized by Witten et al. (1987), overcome these limitations, by deviating from the paradigm that each element has to be encoded on its own. The output of an arithmetic encoder is a single real number x between 0 and 1, that encodes the entire text.

The binary expansion of x is calculated with increasingly higher precision in each iteration processing a single character of the text to be encoded. After having processed the last character, the number x can be stored or transmitted.

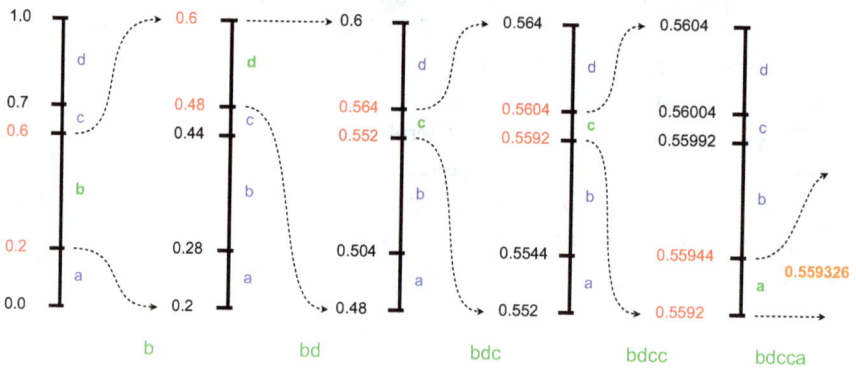

Figure 6.7: *Example of arithmetic coding.*

To derive the requested real number x, the encoding process starts with an initial interval $[0, 1)$, which will be progressively narrowed. This initial interval, as well as each of the subsequent ones to be defined, is partitioned into sub-intervals, the sizes of which are proportional to the occurrence probabilities of the letters of the given alphabet. Processing of a certain character σ of the text is then performed by replacing the current interval $[low, high)$ by its sub-interval corresponding to σ.

Figure 6.7 shows a simple example with an alphabet of four letters {a, b, c, d}, appearing with probabilities 0.2, 0.4, 0.1 and 0.3, respectively. The initial interval $[0,1)$ is partitioned, in our example, into $[0, 0.2)$ for a,

[0.2, 0.6) for b, [0.6, 0.7) for c and finally [0.7, 1) for d, as depicted on the leftmost bar in Figure 6.7. It should, however, be noted that this partition may be chosen arbitrarily, and that neither the sequence of letters nor the sequence of sub-interval sizes need obey any order rules.

Suppose that the text we wish to encode is bdcca. The first character is b, so the new interval after the encoding of b is [0.2, 0.6). This interval is now partitioned similarly to the original one, that is, the first 20% are assigned to a, the next 40% to b, etc. The new sub-division can be seen next to the second bar from the left. The second character to be encoded is d, so the corresponding interval is [0.48, 0.6). Repeating now the process, we see that the next character, c, narrows the chosen sub-interval further to [0.552, 0.564), and the next c to [0.5592, 0.5604), and the final a to [0.5592, 0.55944). The formal encoding procedure appears in ALGORITHM 6.3, getting as input the text $T = T[1]T[2]\cdots$, the alphabet $\Sigma = \{\sigma_1, \sigma_2, \ldots\}$, and the corresponding boundaries of the sub-intervals $\mathcal{P} = \{P_0, P_1, P_2, \ldots\}$, where the sub-interval $[P_{i-1}, P_i)$ is assigned to the letter σ_i, for $i = 1, 2, \ldots$. Note that the P_i are *cumulative* probabilities: if the probability of occurrence of σ_i is p_i, then $P_j = \sum_{i=1}^{j} p_i$ for $j > 0$ and $P_0 = 0$.

Arith-encoding$(T, \Sigma, \mathcal{P})$

---

    high ← 1   *low* ← 0
    for j ← 1 to $|T|$ do
        size ← *high* − *low*
        i ← index of letter in Σ such that $\sigma_i = T[j]$
        low ← *low* + *size* × P_{i-1}
        high ← *low* + *size* × P_i
    return any number x in [*low*, *high*)

ALGORITHM 6.3: *Arithmetic encoding.*

To allow unambiguous decoding, it is this last interval [0.5592, 0.55944) that should in fact be transmitted. For a larger text, this would, however, be rather wasteful: as more characters are encoded, the interval will get narrower, and many of the leftmost digits of its upper bound *high* will tend to overlap with those of its lower bound *low*. In our example, both bounds start with 0.559. The way to overcome this inefficiency is by adding some additional information, serving as an independent indicator of when the decoding process should be stopped.

One of the possibilities is to transmit the number of characters in the text. It then suffices to send only a single number within the final interval instead of its two bounds. But in certain scenarios, this size of the text is not known in advance. As an alternative, it has become customary to add a special end-of-file character # at the end of the message. The extended alphabet $\Sigma \cup \{\#\}$ and a single additional character replacing the text T by $T\#$ in ALGORITHM 6.3 will hardly have any effect on the compression, and their influence on large enough texts will be negligible. On the other hand, since the end of the decoding process will be indicated by the appearance of #, we again need to transmit only a single number in the final interval. For our example interval $[0.5592, 0.55944)$, the best choice would be $x = 0.559326$, because its binary representation 0.100011110011 is the shortest among the numbers of the interval.

Arith-decoding$(x, \Sigma, \mathcal{P})$

1    $high \leftarrow 1 \qquad low \leftarrow 0$

2   repeat forever

3       $size \leftarrow high - low$

4       $k \leftarrow$ index of letter σ_k in Σ for which
   $$low + size \times P_{k-1} \leq x < low + size \times P_k$$

5   if $\sigma_k =$ end-of-file then STOP

6   else output σ_k

7       $low \leftarrow low + size \times P_{k-1}$

8       $high \leftarrow low + size \times P_k$

ALGORITHM 6.4: *Arithmetic decoding.*

Decoding is then just the inverse of the above process. Since x is between 0.2 and 0.6, we know that the first character must be b. If so, the interval has been narrowed to $[0.2, 0.6)$. We thus seek the next sub-interval which contains x, and find it to be $[0.48, 0.6)$, which corresponds to d, etc. Once we get to $[0.5592, 0.55944)$, the process has to be stopped by some external condition, otherwise we could continue this decoding process indefinitely, for example by noting that x belongs to $[0.559248, 0.559344)$, which could be interpreted as if the following character were b, etc. This is formally summarized in ALGORITHM 6.4, whose parameters are the alphabet Σ and its partition $\mathcal{P}$ as for the encoding, in addition to the encoded number x, which replaces the text T.

One may ask how this arithmetic coding scheme is related at all to compression. The answer is that a frequently occurring letter only slightly narrows the current interval and will tend to require fewer bits to represent a number in it. Indeed, the number of bits needed to represent a number depends on the required precision: the smaller the given interval, the higher the precision necessary to specify a number within it. Just as for a number Q larger than 1, one needs $\lceil \log_2 Q \rceil$ bits to represent its standard binary form, it is similarly true that to represent a number within an interval of size is $q < 1$, $\lceil -\log_2 q \rceil$ bits may be needed.

It turns out that not only does arithmetic coding achieve some compression, it does it, furthermore, optimally, and in fact reaches the theoretical lower bound given by the entropy, as shown in the following:

Theorem 6.2. *Given is a text $T = x_1 x_2 \cdots x_W$ consisting of characters x_i, each of which belongs to an alphabet $\{\sigma_1, \ldots, \sigma_n\}$. Let w_i be the number of occurrences of letter σ_i, so that $W = \sum_{i=1}^{n} w_i$ is the total length of the text, and let $p_i = w_i/W$ be the probability of occurrence of letter σ_i, $1 \le i \le n$. The average codeword length after applying arithmetic encoding is then optimal, that is, equal to the entropy $H = -\sum_{i=1}^{n} p_i \log_2 p_i$.*

Proof. Denote by p_{x_j} the probability associated with the j-th character of the text. After having processed the first character, x_1, the interval has been reduced to size p_{x_1}, after the second character, the interval is further narrowed to size $p_{x_1} p_{x_2}$, etc. The size of the final interval after the whole text has been processed is $p_{x_1} p_{x_2} \cdots p_{x_W}$. Therefore the number of bits needed to encode the full text is

$$-\log_2 \left(\prod_{j=1}^{W} p_{x_j} \right) = -\sum_{j=1}^{W} \log_2 p_{x_j} = -\sum_{i=1}^{n} w_i \log_2 p_i = W H,$$

where we get the second equality by a change of variables: instead of summing over the sequential appearance of the characters x_j of the text, we sum over the different letters σ_i of the alphabet, multiplied by their frequencies. The last equality is the same we have seen in Eq. (6.3) where the entropy H is defined. We can amortize the $W H$ bits needed for the entire text to get that the average number of bits per single character is just H. $\square$

Note that our presentation and its analysis are oversimplified as they do not take into account the overhead incurred by the end-of-file symbol. There are also other technical problems that need to be dealt with for a practical implementation. Our computers have only limited precision, which does not allow the computation of a single number encoding the entire, a priori unbounded, input file. We thus need a mechanism of transmitting bits as soon as they are unambiguously determined, after which the intervals have to be rescaled. The technical details can be found in Witten et al. (1987).

6.2   Dictionary methods

The statistical methods we have seen in the previous section build the encodings of their alphabets as a function of the probabilities of the different elements. An alternative approach is to target the alphabet itself. We have already mentioned that the term *alphabet* should be understood in a broader sense. When the basic building blocks to be encoded include variable-length substrings of the text, one often refers to their set as a *dictionary* D. The encoding of a text $T = x_1 x_2 \cdots x_n$, whose characters x_i belong to some fixed alphabet Σ, by means of a dictionary D is then done by replacing substrings of T that belong to D by pointers to D. In their simplest form, the pointers are represented by some fixed-length code, so that each requires $\ell = \lceil \log_2 D \rceil$ bits. Compression is obtained by the fact that ℓ, the number of bits needed for a pointer q, is less than the length of the string represented by q. Even if this is true only at the average, there will still be a gain.

The construction of a good dictionary D for a given input text T is not an easy task, because the strings in D are not restricted and may be overlapping. Let us then first assume that the dictionary is given. Even then, there could be several possibilities to decompose the text T into a sequence of elements of D, an action called *parsing*. Formally,

$$T = w_1 w_2 \cdots w_t \qquad \text{such that} \qquad w_i \in D \quad \text{for } 1 \leq i \leq t.$$

Consider, for example, the dictionary $D = \{$a, b, c, ab, bb, bc, cbb, bbbb, caac, caacb$\}$, and the text

$$T = \text{a b b b b b c a a c b b b b.}$$

Table 6.3 shows some of the parsing possibilities and their costs. A *greedy* parsing tries, in a left-to-right scan, to match at each stage the longest possible element from the dictionary. This is fast, but not necessarily optimal.

A seemingly better heuristic targets first the longest possible match of an element $d \in D$ within T. Once this is located, it partitions the string T into three parts, $T = T_1 d T_2$, where T_1 is a prefix and T_2 is a suffix of T, each possibly empty, and then continues recursively. The result of applying this *longest fragment* approach to our running example is shown in the second line of Table 6.3.

Table 6.3: *Different text parsing methods.*

Method	Parsing	Codewords	Cost
greedy	ab - bb - bc - a - a - cbb - bb	7	26
longest fragment	a - bbbb - caacb - bb - b	5	24
min # codewords	a - bbbb - caac - bbbb	4	27
min encoding	a - bb - bb - caac - bb - bb	6	21

Neither of the two methods yields a minimum number of codewords, as can be seen by the counter-example shown in the third row. Moreover, only for our simplistic approach of encoding the pointers by a fixed-length code, using 4 bits for each pointer on our example, is it the number of elements that should be minimized. In a more general scenario, the dictionary D will be encoded by some variable-length code, for example a Huffman code based on the frequencies of occurrence of the different dictionary items in the parsing. Note that this may lead to a chicken-and-egg problem: we need the frequencies to derive the Huffman code, whose codeword lengths influence the optimal parsing, which in turn determines the frequencies.

Suppose then that in addition to the text and the dictionary, we are also given an encoding function $\mathcal{E}(d)$ defined for all $d \in D$, producing a binary variable-length code. The lengths in bits of the elements in our example could be, e.g.,

a	b	c	ab	bb	bc	cbb	bbbb	caac	caacb
1	2	4	7	3	5	6	9	8	9

The generalized problem is then to find a parsing $T = w_1 w_2 \cdots w_t$ as above, minimizing the total length $\sum_{i=1}^{t} |\mathcal{E}(w_i)|$ of the encoded text, where $|x|$ stands for the length of a string x. As can be seen in Table 6.3, the three solutions above would yield, respectively, encodings of lengths 26, 24 and 27 bits for the given codeword lengths, none of which is optimal.

We perform a reduction to a graph theoretic problem we have seen in Chapter 4 to build the optimal solution. Define a weighted directed graph $G = (V, E)$ in which the vertices are $V = \{0, 1, \ldots, n\}$, and vertex i corresponds to the character x_i of the string; in addition, there is also a vertex numbered 0. The edges correspond to substrings of the text that are elements of the dictionary D, and are defined, for all $0 \le i < j \le n$, by

$$(i, j) \in E \qquad \text{if and only if} \qquad x_{i+1} x_{i+2} \cdots x_j \in D,$$

with a weight of $|\mathcal{E}(x_{i+1} \cdots x_j)|$ assigned to the edge (i, j).

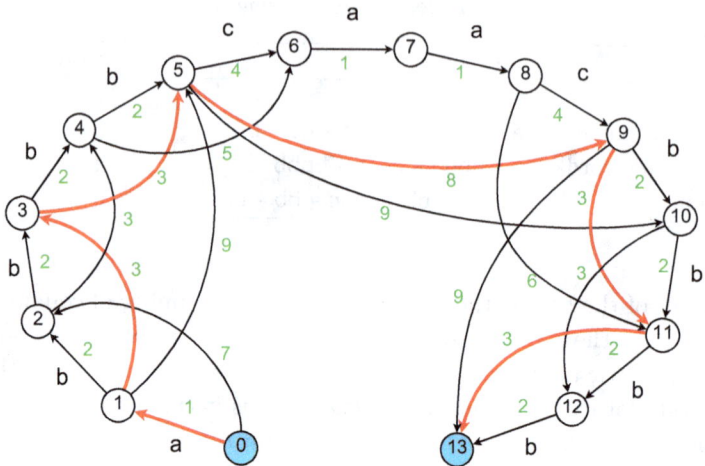

Figure 6.8: *Reduction of optimal parsing to shortest path.*

Figure 6.8 shows the graph corresponding to our running example with $n = 13$, in which the weights on the edges appear in green. The edge $(i-1, i)$ is labeled by the character x_i. The solution we are looking for corresponds to the *shortest path* from vertex 0 to vertex n, which are emphasized as light blue nodes in the figure. The edges of such a path are shown in red, and the weight of this path, which is the cost in bits of the optimal parsing of the text, is 21.

6.2.1 *Ziv–Lempel 77*

Remember that finding an optimal parsing is only useful if the dictionary is known to both the encoding and decoding sides. However, in the standard setting of the problem, it is only the text T that is given, and the dictionary $D(T)$ can be chosen by the encoder as a function of the given text, and exploiting its peculiarities. In addition, even if we could agree on which strings should be included in D, the dictionary has also to be transmitted to the decoder, which puts an additional burden on the process and severely limits the size of D.

An ingenious solution, addressing both the problems of how many and exactly which elements to include in D, has been found by J. Ziv and A. Lempel (1977), who suggested that the text T itself could be used as the dictionary! Since the text is being built incrementally by the decoder, one can use the already known prefix at every stage and consider it as a dynamically growing dictionary, without having to transmit, besides it, anything else. To distinguish it from a different procedure published by the same scientists a year later, the method is often referred to as ZL77.

So what exactly are the elements of D and how can they be addressed? In fact, every substring that has occurred earlier can be considered as a dictionary element. The encoding is implemented by replacing a substring s, starting at the current position in T, by a pointer to an earlier occurrence of s in T, if there is one and its length is at least two characters. The form of these pointers will be pairs (off, len), where off gives the *offset* of the occurrence, that is, how many characters do we have to search backwards, and len ≥ 2 is the *length* of s, the number of characters to be copied. If no such substring can be copied, the current (single) character is sent. The output of one of the variants of ZL77 thus consists of a sequence of elements, each of which is either a single character, or an (off, len) pointer, and a flag bit is added for differentiation.

The following three examples should clarify these notions. The first, in French, is due to Paul Verlaine, in which

```
les-sanglots-longs-des-
violons-de-l'automne-
```

could be replaced by

```
les-sanglot(9,2)(5,2)(7,2)(6,2)d(19,3)
        vio(13,3)(12,4)(21,2)'automn(10,2);
```

the second, in German, is taken from a poem by Johann Wolfgang von Goethe:

```
Besen-Besen-seids-gewesen-
```

to be replaced by

```
Besen-(6,6)(4,2)ids-gew(14,5);
```

the third is from the Vulgate Latin translation of the Bible:

```
vanitas-vanitatum-dixit-Ecclesiastes-
vanitas-vanitatum-omnia-vanitas
```

yielding

```
vanitas-(8,6)tum-dix(10,2)-Ecclesi(26,2)t(6,2)-
            (36,18)om(10,2)a(24,8).
```

Actually, transforming the text into a sequence of characters or (off, len) pointers is only an intermediate phase: this sequence needs to be encoded by means of some encoding function $\mathcal{E}$. The numbers off and len could, theoretically, be as large as the length of the text up to that point. One usually sets some upper bound on off, which effectively means that we are looking for previous occurrences only within some fixed-sized window preceding the current position. The window size WS could be 4K or 16K, so that $\log_2 4K = 12$ or $\log_2 16K = 14$ bits would suffice to encode any offset. Similarly, one could impose an upper limit on len, of, say, 16 or 256 characters, thus the value of len can be encoded in 4 or 8 bits, respectively.

However, most of the copied items tend to be shorter; therefore it will be wasteful to use always the maximum number of bits. On the other hand, it will not be unrealistic to find some longer copy items, in particular for the interesting special case in which len $>$ off, that is, the current string and its earlier copy are overlapping. For instance, a string of m identical characters ccc$\cdots$c (think of a sequence of consecutive zeros or blanks) can be encoded by c$(1, m-1)$.

ZL77-encoding(T)

    $i \leftarrow 1$
    while $i \leq n$ do
        find largest $j \geq 0$ for which $\exists\, k$ such that $i - WS \leq k < i$ and
            $T[i, \ldots, i+j] = T[k, \ldots, k+j]$
        if no such k exists or $j = 0$ then
            output $0 \parallel \mathcal{E}(T[i])$ // single character
        else
            output $1 \parallel \mathcal{E}(i - k, j + 1)$ // (off,len) pair
    $i \leftarrow i + j + 1$

ALGORITHM 6.5: *Ziv–Lempel 77 encoding.*

The generic encoding procedure appears in ALGORITHM 6.5, in which $\parallel$ is the concatenation operator. Ziv and Lempel suggested to locate, at each stage, the *longest* matching substring, but this may be time consuming. A faster variant has been proposed in [Williams (1991)]: a small hash table is addressed by the current pair of characters. The hash table stores, for each letter pair, a pointer to its last encountered occurrence in the text.

Figure 6.9 shows an example in which the text at the current position d starts with XYZAC$\cdots$. The hash table is accessed at the entry $h(\text{XY})$, which points to the previous occurrence of XY. If this previous occurrence falls outside of the current history window, as in our example for FR, indicated by the broken line arrow, it would simply be ignored. Finally, the two occurrences are extended as far as possible, in our case by one more character. The resulting (off, len) pair in this case is $(d - h(\text{XY}), 3)$. The encoding then shifts d by 3, to continue with AC$\cdots$.

This heuristic does obviously not guarantee that the longest match has been found, but trading a possible deterioration of compression performance for increased speed seems often to be a reasonable choice. Williams (1991) also suggests an encoding function $\mathcal{E}$ representing single letters by their 8-bit ASCII codewords, and off and len values by 12 and 4 bits, respectively. The fact that the compressed file consists of items of one or two bytes, after having packed the flag bits in groups of 16, again helps in reducing the processing time.

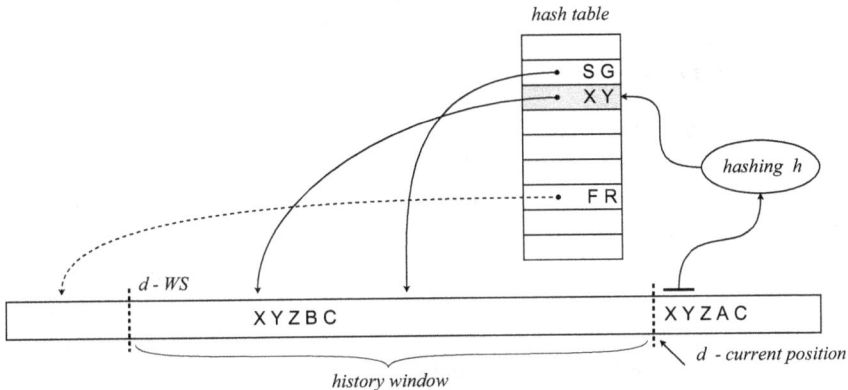

Figure 6.9: *Finding a previous occurrence of a string using a hash table.*

The corresponding decoding in ALGORITHM 6.6 gets the encoded bit-vector B as input and builds the reconstructed text in T. Note that decoding is very fast, regardless of whether the encoding has located the longest matches or only some approximation.

Many variants of ZL77 have emerged over the years, like the popular gzip, which adds a layer of Huffman coding. Instead of using a simple fixed encoding function $\mathcal{E}$ as suggested above, two Huffman codes are devised according to the statistics of the given input file, the first encoding the

ZL77-decoding(B)
_____
> $i \leftarrow 1$
>   while $i \leq n$ do
>> $b \leftarrow$ next bit of B
>>   if $b = 0$ then
>>>   next 8 bits are encoding of $T[i]$
>>> $i \leftarrow i + 1$
>>   else
>>>   *off* $\leftarrow$ next 12 bits of B
>>>   *len* $\leftarrow$ next 4 bits of B
>>> $T[i, \ldots, i + len - 1] \leftarrow T[i - off, \ldots, i - off + len - 1]$
>>> $i \leftarrow i + len$

ALGORITHM 6.6: *Ziv–Lempel 77 decoding.*

individual letters and the different values of the len parameters, and the second built as a function of the distribution of the various off offset values.

6.2.2   *Ziv–Lempel 78*

The idea of using the text T itself as an implicit dictionary may be easy to implement, but can, on the other hand, also be wasteful: many of the substrings of T may never be addressed by (off, len) pointers, and even if the strings are recurring, they may be too far from their duplicates to be included in the sliding window. This led Ziv and Lempel to develop a different adaptive compression procedure [Ziv and Lempel (1978)] using an *explicit* dictionary D. The problem of having to transfer D from the encoder to the decoder is circumvented by letting both sides dynamically generate the same dictionary in synchronization. At the end of the encoding and decoding processes, the dictionary is not needed anymore and may be discarded.

ALGORITHM 6.7 shows the encoding procedure, using an implementation known as LZW [Welch (1984)]. The dictionary D is initialized in line 1 to contain all the single letters of the alphabet Σ, often the 256 possible ASCII letters, and the string *str* is initialized in line 2 as the (empty) NULL string Λ. A main property of the dictionary D is that if it contains some string *str*, then also all the prefixes of *str* belong to D.

The main loop 3–11 processes the input text in several iterations, each producing a single codeword. The current string is extended in 4–6 as far

ZL78-encoding(T)

1    $D[1, \ldots, |\Sigma|] \leftarrow \Sigma \qquad q \leftarrow |\Sigma| + 1$

2    $str \leftarrow \Lambda \qquad i \leftarrow 1$

3   while $i \le n$ do

4   while $i \le n$ and $(str \parallel T[i]) \in D$ do

5            $str \leftarrow str \parallel T[i]$

6            $i \leftarrow i + 1$

7   output $index(str)$

8   if $i \le n$ then

9            $D[q] \leftarrow str \parallel T[i]$

10          $str \leftarrow T[i]$

11          $i \leftarrow i + 1 \qquad q \leftarrow q + 1$

12   if $|str| = 1$ then   // special case: last element is single character

13   output $index(str)$

ALGORITHM 6.7: *Ziv–Lempel 78 encoding.*

as possible, until a hitherto unknown string $R = str \parallel T[i]$ is detected. The index in D of the prefix str of R is added as the next element in the compressed file in line 7, R itself is adjoined as a new element to the dictionary, and the current string is updated to hold just the last character of R in lines 9–10. The special case when the last encoded element consists of a single character is dealt with in lines 12–13.

Consider the example in Figure 6.10. The processed strings R are underlined, the prefix str appearing in black and its single character extension in gray. Note that consecutive strings R are overlapping: the last character at each stage becomes the first character in the following iteration.

The list at the right side of the figure is the dictionary D built by the encoder, and independently, and using an altogether different algorithm, also by the decoder. It is initialized by the alphabet of four letters $\{A, H, M, N\}$ and adds, in order, a new entry for each index in the compressed file, except for the last one. When the end of the input is reached, the encoder may discard the dictionary D and only transmits the sequence of pointers as output. For our example, this sequence is

$$4 \quad 1 \quad 5 \quad 2 \quad 7 \quad 3 \quad 1 \quad 10 \quad 12 \quad 6,$$

as shown in the middle of Figure 6.10.

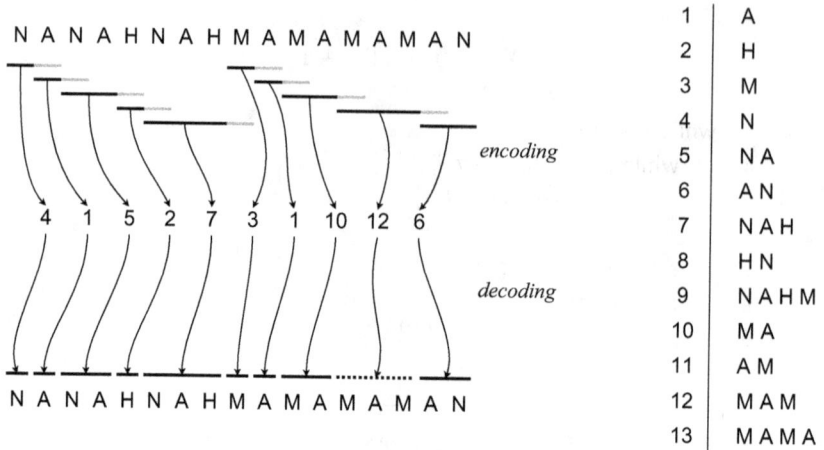

Figure 6.10: *Example of Ziv–Lempel 78 encoding and decoding.*

Once an encoding function is given, it is often straightforward to derive the corresponding decoding, but for ZL78 this is a bit more tricky. The reason for this complication is that the compression process and its inverse are not completely symmetric, and there is a delay in the construction of the dictionary D. The decoding also adds a new entry to D for each index in the compressed file, except for the first one, but it lags behind the encoding by one step.

Let us inspect, for instance, the fourth iteration of the encoding in our running example of Figure 6.10. After having sent the index 2 of the string $str = $ H to the output in line 7, the element $R = $ HN is adjoined as entry 8 to D in line 9. However, when the decoding will get to the fourth element to be decoded, the codeword 2 will be correctly interpreted as standing for H, but the element HN at entry 8 in D is not yet known to the decoder! The same situation arises also for any other decoded element. In this case, the delay does not cause any trouble, because the next requested element is 7, which is already known. But there are cases which require a special treatment, as we shall see below.

Actually, the decoding process, formally given in ALGORITHM 6.8, alternates between two tasks: it has to translate the sequence of input indices into character strings, which is done in lines 3–11, and it needs to reconstruct the dictionary D, in lines 12–18. The dynamic nature of the algorithm implies that this has to be performed almost in parallel—each task depends on the other, so the one cannot be started until the other has been completed.

For the first part, p stores the current value from the list of pointers to be decoded, and i and j are, respectively, the indices of the first and last characters of the current codeword in the reconstructed text T. For the second part, q is the index of the next string S to be stored in the dictionary D, str is the first character of S and k is the index of the last character of S.

ZL78-decoding(L)

1    $i \leftarrow 1$ $q \leftarrow |\Sigma| + 1$ $k \leftarrow 2$

2   while list L is not exhausted do

3       $p \leftarrow$ next index from L

4   if $p = q$ then   // special case: value in D not yet defined

5           $D[p] \leftarrow D[last] \parallel$ first char of $D[last]$

6           $q \leftarrow q + 1$ $k \leftarrow k + |D[p]| - 1$

7       $j \leftarrow i + |D[p]| - 1$

8       $T[i, \ldots, j] \leftarrow D[p]$

9   if $i = 1$ then   // special case: initialize str

10         $str \leftarrow T[1]$

11       $i \leftarrow j + 1$ $last \leftarrow p$

12   while $k \le j$ and $(str \parallel T[k]) \in D$

13         $str \leftarrow str \parallel T[k]$

14         $k \leftarrow k + 1$

15   if $k \le j$ then

16         $D[q] \leftarrow str \parallel T[k]$

17         $str \leftarrow T[k]$

18         $k \leftarrow k + 1$ $q \leftarrow q + 1$

ALGORITHM 6.8: *Ziv–Lempel 78 decoding.*

Because of the mentioned delay, it may so happen that the element that should be decoded is not yet defined in the decoder's copy of D, as in the case of decoding 12 in our example of Figure 6.10; the corresponding string MAM is indicated by the dotted line. Consider the example from the decoder's point of view after having decoded 10: the text consists currently of 12 characters NANAHNAHMAMA and D has 11 entries, the two last being MA and AM. The attempt to add another element W to D fails: W has to start with the last character of the last element adjoined to D, that is, the character $T[11] = $ M, but both M and its extension MA are already in D, which concludes this iteration. Returning to line 3, the next value of p is 12, but $D[12]$ is not yet known!

Decoding is still possible if the alternation between its two sub-tasks is done in smaller steps. Even though, in our example, we ignore the value of $D[12]$ when it is needed, we do know that this string must start with MA, so suppose it is MAx for some string x of characters of Σ. Accordingly, the text reconstructed so far is

$$T = \text{NANAHNAHM}\overline{\text{A}}\underline{\textbf{MA}}\text{MA}x, \qquad (6.4)$$

where the last inserted element $D[11] = $ AM is overlined, and the just added characters MAx appear in gray. At this stage, we can already extend the known prefix MA of $D[12]$ and conclude that it is MAM, which appears underlined.

In fact, this example is representative of all occurrences of this special case. It will only occur when the current pointer p is just the index q of the following element to be added to D and can thus be detected as in line 4. Moreover, the missing string will then be an extension of the last processed element V, in our case $V = D[10] = $ MA, which is boldfaced in Eq. (6.4), and the extension is in fact a single character, which must be the first character of V, in our case M. This may be implemented by remembering the last processed index in a variable *last*.

6.3   Burrows–Wheeler transform

The following technique differs from those seen so far by the fact that it is actually not a compression method by itself. It is a transformation, named after its inventors Burrows and Wheeler (1994). It gets a string or text T of length n as input, and returns a specific permutation of its characters, that is, another string BWT(T) of the same length. The reason for including this method in the current chapter is that the transformed string BWT(T) is generally much easier to compress than the string T it originated from. To understand why, consider the output obtained by applying BWT to the three example strings we saw in Section 6.2.1.

T	BWT(T)
sanglots-...	lsseessss'--dnldnnv---ogomaooitlllngtee-uoa-
Besen-...	nnsn--isssswBBg-eeeed-eeee
vanitas-...	tmmsass-ivvvvvttitttEc-tlnsxnnnnndcuuomaaaaa-eaaeaaiiiiiiisaatts----i

As can be seen in these examples which are quite representative, the BWT has a tendency to regroup identical characters together. The characters in BWT(T) may not appear in strict alphabetical order, yet the strings could be judged as "close" to being sorted, if the criterion of closeness is a higher average length of runs of identical characters. The appearance of such runs, like ssss or iiiiii, is the key to applying afterwards simple compression techniques, such as *move-to-front* or *run-length coding*, which exploit this repetitiveness.

This leads us to the obvious question, why not simply arrange the characters of T in order, since such a string would be even more compressible than BWT(T). The last of the above strings would thus produce

-------Eaaaaaaaaaaaaaccdeeiiiiiiiiiimmmnnnnnnnossssssttttttttttuuvvvvvx

which could be run-length-compressed by a sequence of (character, repetition count) pairs, where the parentheses are only added for better readability, as

$$(-, 7)(\mathsf{E}, 1)(\mathsf{a}, 12)(\mathsf{c}, 2)(\mathsf{d}, 1)(\mathsf{e}, 2)(\mathsf{i}, 9)(\mathsf{m}, 3)(\mathsf{n}, 6)(\mathsf{o}, 1)(\mathsf{s}, 6)(\mathsf{t}, 9)(\mathsf{u}, 2)(\mathsf{v}, 5)(\mathsf{x}, 1).$$

The size of the compressed text would then be $O(|\Sigma| \log n)$, where $n = |T|$ is the length of the text, an order of magnitude smaller than $|T|$ itself, since the size of the alphabet Σ is deemed to be constant relative to n.

The reason for not preferring this alternative is that sorting the characters of a text T is not a reversible operation, since any permutation of T would yield the same sorted order. The surprising feature of BWT is that it is a bijective mapping, so that an inverse transform BWT^{-1} does exist. The schematic diagram in Figure 6.11 depicts the layout of the compression system in broad terms.

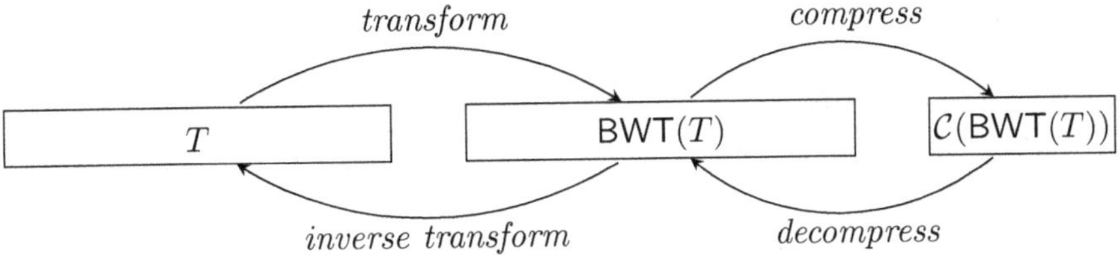

FIGURE 6.11: *Compression using the Burrows–Wheeler transform.*

The upper part of the figure, referring to the encoding, shows that a text T is first transformed into a more compressible form BWT(T) of the same size, and only then compressed by means of some compression function $\mathcal{C}$.

To recover the original text, as shown in the lower part of Figure 6.11, the compressed file R has first to be decompressed, after which the inverse transform can be applied, yielding $\mathsf{BWT}^{-1}(\mathcal{C}^{-1}(R)) = T$.

Burrows and Wheeler called their method *block sorting*, and we shall explain the details using the small input text $T = \mathsf{NANAHNAHMAN}$, as shown in Figure 6.12. As a first step, the left-hand side matrix is produced by rotating the input string i characters to the left, for $0 \le i < |T|$, and storing the i-th rotated string in the row indexed i. We thus get the original string both in the first row (indexed zero) and in the first column, where it is emphasized. A small bar | indicates in each row the end of the original text, i characters to the left of the end of the string. Note that in this version of the algorithm, this bar is not considered as a letter by itself, and it is only added as a visual help for the explanation.

The next step sorts the rows of the matrix in lexicographic order, and results, in our example, in the right-hand side matrix of Figure 6.12, known as the BW matrix. The numbers on both sides of the matrix are explained below. Clearly, every column of both matrices contains some permutation of the characters in T. The first column of the BW matrix, shown in blue, is the sequence of sorted characters, from which the original sequence cannot be deduced. It is, however, the *last* column of this matrix, shown in red, that has been chosen as $\mathsf{BWT}(T)$. The surprising fact is that this sequence, which for our example is NNNMAAHHANA, *does* allow the retrieval of T, or, more precisely:

Rotations		BW matrix	
N A N A H N A H M A N |	0	A H M A N | N A N A H N	0
A N A H N A H M A N | N	1	A H N A H M A N | N A N	1
N A H N A H M A N | N A	2	A N A H N A H M A N | N	2
A H N A H M A N | N A N	3	A N | N A N A H N A H M	0
H N A H M A N | N A N A	0	H M A N | N A N A H N A	0
N A H M A N | N A N A H	1	H N A H M A N | N A N A	1
A H M A N | N A N A H N	0	M A N | N A N A H N A H	0
H M A N | N A N A H N A	0	N A H M A N | N A N A H	1
M A N | N A N A H N A H	1	N A H N A H M A N | N A	2
A N | N A N A H N A H M	2	N A N A H N A H M A N |	3 ←—
N | N A N A H N A H M A	3	N | N A N A H N A H M A	3

FIGURE 6.12: *Example of the Burrows–Wheeler transform.*

Theorem 6.3. *Given any string T, BWT(T), which is the last column of the BW matrix, suffices to reconstruct the entire matrix.*

Proof: Since the last column is a permutation of the characters of the text, all one needs to do to get the **first** column of the matrix is to sort the last column. But for each row, the element in the first column is in fact adjacent in the text to the character in the last column, because the rows are rotations. If one now concatenates, element by element, the first column *after* the last one, one gets a list of all the character pairs in T. Sorting this list of pairs lexicographically yields the first **two** columns of the BW matrix.

This sequence of concatenating and sorting steps can be continued: concatenating the sorted pairs after the last column yields the set of all the character triplets in T, and sorting this set gives the first **three** columns of the matrix. After $|T|$ such iterations, the whole BW matrix has been restored. □

BWT	sort	cat	sort	cat	sort	cat	sort
N	A	NA	AH	NAH	AHM	NAHM	AHMA
N	A	NA	AH	NAH	AHN	NAHN	AHNA
N	A	NA	AN	NAN	ANA	NANA	ANAH
M	A	MA	AN	MAN	ANN	MANN	ANNA
A	H	AH	HM	AHM	HMA	AHMA	HMAN
A	H	AH	HN	AHN	HNA	AHNA	HNAH
H	M	HM	MA	HMA	MAN	HMAN	MANN
H	N	HN	NA	HNA	NAH	HNAH	NAHM
A	N	AN	NA	ANA	NAH	ANAH	NAHN
N	N	NN	NA	NNA	NAN	NNAN	NANA
A	N	AN	NN	ANN	NNA	ANNA	NNAN

FIGURE 6.13: *Progressively restoring the columns of the BW matrix.*

Figure 6.13 shows the first few steps of this process on our example. As for the previous figure, the BWT and the sorted sequence appear, respectively, in red and blue.

Actually, the result of the transform alone is not sufficient to deduce the original string T, since any rotation of T, $\mathcal{R}(T)$ will produce the same BW matrix as T itself, and therefore BWT$(\mathcal{R}(T)) = $ BWT(T). We therefore need, in addition to the permuted string BWT(T) itself, also a pointer to the row of the BW matrix holding the original string T. This pointer is shown in Figure 6.12 as a small red arrow in the 10th row.

The proof of Theorem 6.3 provides a decoding algorithm for the BWT, but not an efficient one, as it includes n applications of sorting on n elements each, which already requires $\Omega(n^2 \log n)$ comparisons. A much faster procedure, working in linear time, can be based on the following observation.

Refer again to the first column of the BW matrix in Figure 6.12, that contains all the characters of T in alphabetical order. There are four As, and accordingly, there are also four corresponding As in the last column, in which the characters appear in a different order. However, the relative order of the different occurrences of a given letter is preserved. In other words, the order of the four As in the blue column is the same as for the corresponding four As in the red column, as indicated by the green numbers 0,1,2,3 next to them. The same is true for all the other characters, the two Hs, the single M and the four Ns, as shown by the matching colors.

To see why this is true for A, remember that the first four rows of the matrix are ordered, but since they start with the same letter A, they are in fact ordered according to their suffixes of length $n-1$. On the other hand, the four rows ending in A are ordered by their prefixes of length $n-1$. But these are in fact the same four strings, because each of these prefixes is obtained from the corresponding suffix by wrapping around the first character to get another rotation.

row	cum	sort	BWT	cnt		index	j
0	0	A	N	0	7	7	5
1		A	N	1	7	8	2
2		A	N	2	7	9	0
3		A	M	0	6	6	8
4	4	H	A	0	0	0	6
5		H	A	1	0	1	3
6	6	M	H	0	4	4	7
7	7	N	H	1	4	5	4
8		N	A	2	0	2	1
9		N	(N)	3	7	10	10
10		N	A	3	0	3	9

FIGURE 6.14: *Linear time evaluation of the inverse* BWT.

Therefore, to reconstruct T when $\mathsf{BWT}(T)$ is given, one needs to locate the position of each character belonging to the red (last) column in the blue (first) column of the BW matrix, as shown by the solid arrows in

Figure 6.14. For our example, we got a pointer to the line indexed 9, so we know that N, in the circle, is the last character in T. But this N is the fourth in BWT, and thereby corresponds to the fourth N in the sorted (blue) column, which appears in the line indexed 10. Thus the character just preceding the N in T is the one in line 10 in BWT, which is A. This is the fourth A, corresponding in the blue column to the fourth A, in line 3. Following the dotted line, we then get the next character M, and so on.

How do we know that a given character is the fourth A and what the index of the corresponding fourth A in the blue column will be? All we need is to keep the relative rank of the occurrence of each character within BWT(T), shown in green in Figure 6.14, as well as the cumulative number cum[σ] of characters in BWT(T) that are alphabetically smaller than the current letter σ; this is shown in the violet column in the figure. Given the current index p, the following index is then calculated as

$$p = \mathsf{rank}[p] + \mathsf{cum}[\mathsf{BWT}[p]].$$

These values appear in the column headed "index" in Figure 6.14.

BWT-inverse($S, n, start$)

    for $i \leftarrow 1$ to $|\Sigma|$ do
        cnt[σ_i] $\leftarrow 0$
    for $i \leftarrow 1$ to n do
        rank[i] $\leftarrow$ cnt[$S[i]$]
        cnt[$S[i]$] $\leftarrow$ cnt[$S[i]$] $+ 1$
    cum[σ_1] $\leftarrow 0$
    for $i \leftarrow 2$ to $|\Sigma|$ do
        cum[σ_i] $\leftarrow$ cum[σ_{i-1}] $+$ cnt[σ_{i-1}]
    $p \leftarrow start$
    for $j \leftarrow n$ to 1 by -1 do
        $T[j] \leftarrow S[p]$
        $p \leftarrow$ rank[p] $+$ cum[$S[p]$]

ALGORITHM 6.9: *Inverse* BWT.

The text T is generated backwards, starting at its last character, which is given by the additional pointer. The last column of Figure 6.14, headed j, gives, for each row, the index at which the corresponding character is stored. The decreasing sequence of these indices from $n - 1$ to 0 can be

obtained starting at the circled character and following the arrows. ALGO-
RITHM 6.9 is the formal inverse BWT, and runs in time $O(|T| + |\Sigma|)$. Its
input parameters are the following: the array S holding the BWT transform
and its size n, and the index *start* of the row indicated by the red arrow in
Figure 6.12.

The BWT is the basis of the open source compression program bzip2,
which often performs better than gzip and LZW. It is also connected to the
suffix array, a data structure we shall see in the next chapter.

6.4 Deduplication

Deduplication is a lossless data compression technique that is somewhat
similar to LZ77, but extends its basic idea a step further to a larger scale.
The elements to be replaced by pointers are not just short substrings but
entire blocks of data, referred to as *chunks*. The application area of deduplication is quite different from that of standard lossless compression. As we
saw in the previous section, the compressibility of some input text varies,
and purely random data cannot be compressed at all. There are, however,
applications in which even such incompressible files, if they appear more
than once, may yield some savings.

6.4.1 *Content addressable storage*

An example could be a large backup system, in which, to prevent the loss
of data, the entire electronic data of some corporation has to be saved
weekly or sometimes even daily. Such backup data are characterized by the
property that only a small fraction of it differs from the already stored copy
produced a day or a week ago. Deduplication handles such files by storing
duplicate chunks only once. The challenge is, of course, how to locate as
much of the duplicated data as possible, and doing it efficiently.

One of the approaches to deduplication systems is known as *Content
Addressable Storage* (CAS). The input database, often called a *repository*,
is partitioned into chunks and a hash function h is applied to each of these
input chunks. Hash functions are generally not one-to-one, but here, a
special type of hashing is used, which is *cryptographically strong*. That is,
for chunks C_i and C_j and some small predetermined constant ε,

$$C_i \neq C_j \quad \rightarrow \quad \Pr\big(h(C_i) = h(C_j)\big) < \varepsilon.$$

In other words, the probability of getting identical hash values in spite of
applying h to different chunks is so low that one can safely ignore it, so

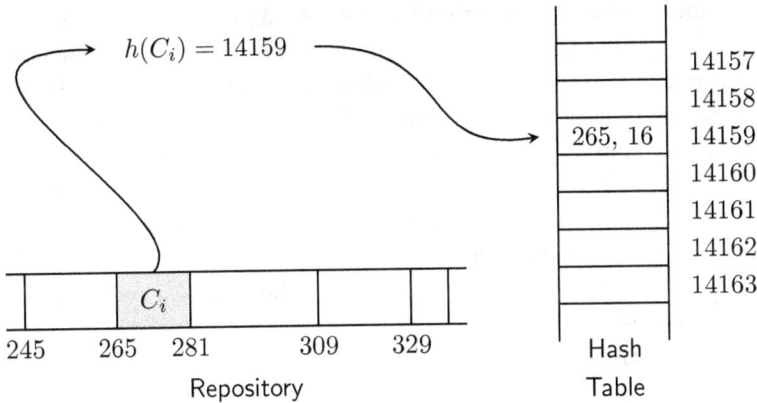

FIGURE 6.15: *CAS deduplication system.*

that equal hash values may be assumed to imply identical chunks. The address of a chunk C_i and its size are stored in a hash table at index $h(C_i)$, as depicted in the example in Figure 6.15. A B-tree could also be used as an alternative to the hash table. For our example, the chunk C_i starts at address 265 and applying the hash h to C_i yields the value 14159. The address 265 and the size 16 of the chunk C_i are therefore stored in the hash table at address 14159.

For our weekly backup, the new copy of the data, which is often called a *version*, is also partitioned into similar chunks. The hash value of each of these new chunks is searched for in the hash table, and if it is found, one may conclude that the new chunk is an exact copy of a previous one, so all one needs to store is a pointer to the earlier occurrence. For our example, if for a new version chunk C_j we also get $h(C_j) = 14159$, the pair $(265, 16)$ is retrieved from the hash table, and we conclude that C_j is identical to the chunk of length 16 starting at 265, which is C_i.

How should the chunk size be chosen? The easiest solution would be to define all the chunks to be of the same size. This may, however, cause problems in the presence of small insertions and deletions in the version, relative to the original chunks of the repository. All the chunk boundaries after the inserted or deleted bytes in the version would be shifted, so the chunks of version and repository would not perfectly overlap anymore and therefore produce completely different hash values. To avoid this problem, one can define the boundary of the chunk to be dependent on the content itself, which implies variable-length chunks.

A simple method for cutting the data string consisting of a sequence of bytes $s_1 s_2 \cdots$ into pieces is to use an idea we saw in the implementation of the Karp–Rabin pattern matching algorithm in Section 5.3.1. Recall that the method used the hash function $h(Z_i) = Z_i \bmod P$, where P is some large prime number, and relied on $h(Z_i)$ to derive $h(Z_{i+1})$ efficiently, see Eq. (5.5).

We now apply a similar hash function to any consecutive sequence of ℓ bytes. We shall refer to such a sequence as a *seed*. Starting with the byte indexed ℓ, each byte can be considered as the last of a seed. If, for some arbitrarily chosen constant Y chosen from the set of the possible values of $h(X)$, we get that

$$h(s_{j-\ell+1} s_{j-\ell+2} \cdots s_j) = Y, \tag{6.5}$$

the current last byte s_j of the seed is chosen to be the last byte of the current chunk. A good hash function returns values that are uniformly distributed; thus if the size of the set of possible hash values is N, then the probability of getting the equality in Eq. (6.5) is $\frac{1}{N}$, independently of the specific value Y chosen. This yields an expected size of N bytes for the chunks. In practice, one can control the lower and upper limits of the chunk sizes by not even checking at the beginning, and by choosing a maximal cutoff point.

6.4.2 *Similarity-based deduplication*

For very large repositories, say of about one petabyte (1 PB $= 2^{50}$ bytes), a chunk size of a few kilobytes may be too small: the number of chunks which is the number of entries in the hash table would then be too large to fit in the main memory. However, increasing the chunk size comes at the price of reducing the number of duplicates. This led to the development of deduplication systems relying on the *similarity* rather than the *identity* on chunks [Aronovich et al. (2009)]. A chunk C_i may be deduplicated by means of a similar chunk C_j by storing a pointer to C_j and a list of differences between C_i and C_j. As a consequence of this shift from identity to similarity, much larger chunk sizes can be considered without jeopardizing the hashing approach.

The algorithm of similarity-based deduplication relies on the assumption that similar chunks are in fact almost identical, with minor locally restricted perturbations. An example of these could be a pair of files which differ only in a few words, such as a date or name that has been changed. We therefore expect the chunks we define as being similar to share long stretches of identical characters and we aim at detecting some of these. The problem is, of course, that we ignore the relative locations of the matching substrings within a new version chunk and a corresponding repository chunk. To enable the use of a hash table as for the CAS approach in Figure 6.15, we need to represent each repository chunk by a small *signature* or *fingerprint*, as follows.

Suppose we are given a repository of about 1 PB and partition it into large chunks of $N = 16$ MB each. Choose a prime number P of length about 60 bits, so that all calculations modulo P can be performed efficiently on a 64-bit machine. We shall use the hash function $h(X) = X \bmod P$, where we identify the character string X with its binary representation and the number it stands for. Given a chunk $C = c_0 c_1 \cdots c_{N-1}$, where each c_i is a character, define a block B_i as a substring of length $\ell = 500$ characters of C, starting at index i, that is

$$B_i = c_i c_{i+1} \cdots c_{i+\ell-1} \qquad \text{for } 0 \le i \le N - \ell,$$

and consider the hash values $h(B_i)$ produced by all the $N - \ell + 1$ (overlapping) blocks in the chunk C. As a first attempt, let us choose the four largest so-obtained hash values as the signature of C. The idea behind this choice is that these maximal values are robust against small inserts and deletes, and even if one of these perturbations happens to fall within the range of some block B_i that yielded the maximum hash, there are still three other values left, which are unlikely to be simultaneously invalidated.

However, while hash values are uniformly distributed over their range, this property is not true for the maximum of several hash values. The suggested signatures will thus all belong to some very narrow interval which is likely to cause collisions. This difficulty is overcome by the following amendment. Each of the chosen blocks $B_{m_1}, B_{m_2}, B_{m_3}, B_{m_4}$ yielding the maximal values serves a double purpose:

(1) its location m_i in the repository chunk is well defined and thus reconstructable in a new version chunk;
(2) its value $h(B_{m_i})$ is used as a signature to identify it.

We can maintain the full functionality of the deduplication system if, instead of storing the values $h(B_{m_i})$, which are not uniformly distributed, we shift our attention a few bytes further and store instead the values $h(B_{m_i+k})$, for $i = 1, 2, 3, 4$ and k being some small constant, say $k = 8$. One of the properties of hash functions is that small changes in the argument imply large changes in the values. Contrarily to $h(B_{m_i})$, the new values $h(B_{m_i+k})$ are likely to behave uniformly.

Summarizing, we apply a Karp–Rabin hash to all the blocks of the chunk to locate the maximal values. This defines the locations m_i in the chunk, but the actually stored values are obtained by applying the hash h not to B_{m_i}, but rather to a block B_{m_i+k} shifted by a few bytes to keep a uniform distribution of the values. The values kept for each chunk are the four hash values (8 bytes each), along with the indices of the corresponding blocks and offsets within the blocks (additional 8 bytes each), that is, 64 bytes for each of the 1PB/16MB $= 2^{50}/2^{14} = 2^{26}$ chunks. The total overhead is therefore $2^{26} \times 64 = 2^{32} = 4$GB, which is reasonable for a 1PB repository.

6.5   Structured Data

All the methods we have seen so far are designed as general-purpose compression schemes and do not assume any knowledge about the nature of the file to be compressed. There are, however, situations in which some inherent structure of the input is known, as for various lists, inventories, databases and the like. This knowledge can be exploited to devise custom-tailored compression techniques.

6.5.1   *Dictionaries and concordances*

A *dictionary* is a list of different words, generally stored in alphabetical order. There are standard dictionaries in each language, like Webster for American English, Duden for German or Petit Larousse for French, but there are also more artificial dictionaries as those compiled for information retrieval systems. The latter contain *all* the different words in the underlying textual database, and would thus store the terms go, goes and went as different items.

An easy way to deal with such lists is based on the observation that consecutive terms in a sorted dictionary generally share a common prefix. One may thus omit such a prefix if one remembers its length, to be copied from the preceding item. This is known as the *prefix omission method*, an example of which is shown in Figure 6.16, based on some consecutive terms of *Webster's* dictionary.

The left column gives the original list, each element x of which is replaced by a (*copy, suffix*) pair, where *copy* is the number of characters to be copied from the preceding entry, and *suffix* is what remains of x after having removed its prefix of length *copy*. To maximize the savings, one usually selects the longest possible common prefix, though the algorithms

Term	Copy	Suffix
nation	0	nation
national	6	al
nationalism	8	ism
nationalist	10	t
nationalistic	11	ic
nationalistically	13	ally
nationality	9	ty
nationalization	9	zation
nationalize	10	e
nationally	8	ly
native	4	ve
nativism	5	ism
nativity	6	ty

FIGURE 6.16: *Compression of a dictionary by prefix omission.*

work just as well if one imposes an upper limit ℓ to the *copy* value, for example to facilitate its encoding. In our case, if $\ell = 7$, nationally would be compressed to $(7, \text{lly})$ instead of $(8, \text{ly})$.

Prefix omission can also be applied to the compression of concordances. Large information retrieval systems and search engines do not locate the key words of a query by a linear search in the entire database — that could be done in linear time, but for texts of the order of TB or even PB, this would still be prohibitive. Instead, the search is based on *inverted files*: in a pre-processing stage, the list of different terms in the text is extracted and stored in a dictionary, which contains also pointers to the *concordance*. The concordance is a sorted list, giving for each term the list of its exact locations in the text.

Famous concordances are those of the Bible, in various languages, for which each location can be described hierarchically as a quadruple (b, c, v, w), where b is some index of the *book*, c is the *chapter* number within the book, v is the *verse* number within the chapter and w is the index of the current *word* within the verse. There are, of course, other alternatives. Figure 6.17 is an excerpt of a concordance of the collected works by Shakespeare. It shows the first 14 of the 145 occurrences of the term question in his plays. Each occurrence is represented by the quintuple (p, a, s, ℓ, w), giving, respectively, the indices of the *Play*, *Act*, *Scene*, *Line* and *Word*; the line number is absolute within the given Play, while a, s and w are relative within the next level of the hierarchy.

Original					Compressed	
Play	*Act*	*Scene*	*Line*	*Word*	*Copy*	*Suffix*
ALL'S-WELL	I	1	115	4	0	A1-1-1-115-4
ALL'S-WELL	II	1	817	4	1	2-1-702-4
ALL'S-WELL	II	2	854	4	2	1-37-4
ALL'S-WELL	II	2	862	1	3	8-1
ALL'S-WELL	II	5	1301	1	2	5-439-1
ALL'S-WELL	III	5	1640	1	1	3-5-339-1
ANTONY-CLEO	II	2	737	4	0	A2-2-2-737-4
ANTONY-CLEO	II	2	783	4	3	46-4
ANTONY-CLEO	III	13	2254	3	1	3-13-1471-3
AS-YOU-LIKE	II	4	778	6	0	A3-2-4-778-6
AS-YOU-LIKE	II	7	1072	4	2	7-294-4
AS-YOU-LIKE	III	4	1624	9	1	3-4-552-9
AS-YOU-LIKE	V	2	2253	8	1	5-2-629-8
AS-YOU-LIKE	V	4	2555	3	2	4-302-3

FIGURE 6.17: *Compression of a concordance by prefix omission.*

As the sequence of concordance elements corresponding to a certain term is usually sorted lexicographically, prefix omission may be applied as for dictionaries, using, for consecutive entries, the number of identical fields in the quintuple instead of the length of the common prefix. For example, the last two entries of the occurrences of the term question in Figure 6.17 have their first two fields (Play and Act) in common, so only s, ℓ and w are kept. In addition, in this particular case in which the values in the ℓ field form a non-decreasing sequence within each play, one may store the increment from the previous entry, $2555 - 2253 = 302$ in our example, instead of the current ℓ value 2555 itself. The compressed entry will thus be $(2, 4\text{-}302\text{-}3)$.

The concordance elements are stored in blocks of up to k entries, and the first entry in each block is given explicitly, that is, with $copy = 0$. The size k of the block is chosen not too large, to allow decoding from the beginning of the block in reasonable time, and on the other hand not too small, to reduce the effect of not compressing the first entry. ALGORITHM 6.10 presents the formal decoding of such a concordance block; the fields p, a, s, ℓ and w have been renamed as $f[1], \ldots, f[5]$, respectively.

Concordance-decode

while list is not exhausted do
    $(copy, suff)$ ← read next comprerssed entry
    break $suff$ into fields $f[copy + 1], \ldots, f[5]$
    if $copy > 0$ then
        $(f[1], \ldots, f[copy]) \leftarrow (oldf[1], \ldots, oldf[copy])$
        if $copy < 4$ then   // ℓ field holds difference
            $f[4] \leftarrow f[4] + oldf[4]$
    output $(f[1], \ldots, f[5])$
    $(oldf[1], \ldots, oldf[5]) \leftarrow (f[1], \ldots, f[5])$

ALGORITHM 6.10: *Decoding of a concordance block.*

6.5.2 *Sparse bitmaps*

In certain applications, a hierarchical structure like (b, c, v, w) for each entry is more than is needed, and the concordance consists, for each term x, only of the list of documents in which x occurs. Since this list can be sorted, it is a list of increasing integers, and there are many other scenarios in which such lists are needed.

As an alternative equivalent form of these lists, one may use a bit-vector or *bitmap*, in which the bit indexed i is set to 1 if and only if the integer i appears in the list. For example, the sequence 3, 9, 10, 19, 26 could be represented by

index	1	5	10	15	20	25	30

bitmap 0 0 1 0 0 0 0 0 1 1 0 0 0 0 0 0 0 0 1 0 0 0 0 0 0 1 0 0 0 0 0 $\cdots$

Such bitmaps appear in a myriad of different applications: occurrence maps for the different terms in an information retrieval system, in which each bit position stands for a document number; black and white pictures, where each bit represents a pixel; annotations in the internal nodes of wavelet trees, etc. When the number of 1-bits is close to half the size of the map, and there is no obvious pattern in their occurrences, the bitmap may not be compressible at all. For many applications, however, the average bitmap will be *sparse*, that is, with a low probability p_1 for a 1-bit (the symmetric case of low probability p_0 of a 0-bit can be dealt with by complementing the bitmap). Here are a few ways to compress such vectors.

List of indices: When there are very few 1-bits, we can revert to the list of integers we started with. The displayed bitmap here above could thus be compressed by the sequence 3, 9, 10, 19, 26, using $\log n$ bits for each element, where n is the length of the bitmap.

List of differences: Instead of recording the absolute index of a 1-bit, use the increment from the previous one, as we did for the Line number in the compression of the concordance. For the example above, this would yield 3, 6, 1, 9, 7.

Huffman or arithmetic coding: Partition the map into blocks of k bits, collect statistics about the frequency of occurrence of the 2^k possible bit patterns and apply Huffman or arithmetic coding accordingly. Because of the sparseness, the block Z consisting of zeros only will have a high probability. Huffman coding cannot assign less than one bit to the codeword representing Z, which limits the compression gain when p_1 is very small, but arithmetic coding does not have such limitations.

Encoding the length of 0-bit runs: The following method has been suggested by Teuhola (1978) and is similar to the Elias γ-code of Section 6.1.1.1. The method has an integer parameter $k \geq 0$ whose value is chosen according to the average length of a 1-terminated 0-bit run in the entire collection of bitmaps to be compressed. Refer to Figure 6.18 for an example of compressing a run of length $n = 98$ with $k = 2$.

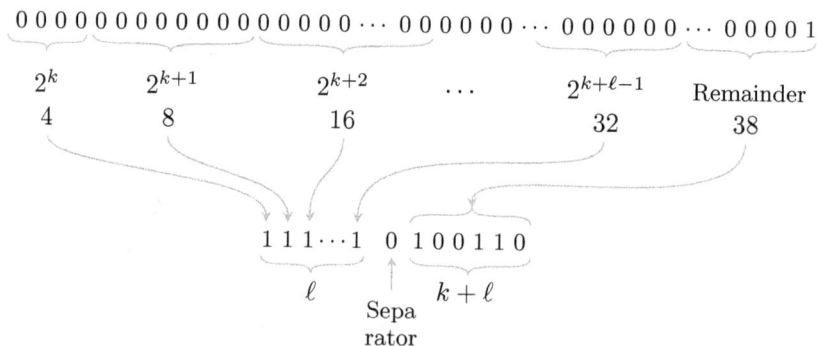

FIGURE 6.18: *Encoding the length of a run of zeros.*

We try to decompose the run of 0-bits into increasingly longer blocks of zeros. Starting with a block of length 2^k, we double the length as often as possible, that is, the largest integer $\ell \geq 1$ is sought such that

$$2^k + 2^{k+1} + \cdots + 2^{k+\ell-1} \leq n.$$

The remainder $n - \sum_{i=0}^{\ell-1} 2^i$ is therefore smaller than $2^{k+\ell}$ and can thus be represented in $k + \ell$ bits.

To encode the length n, ℓ 1-bits are used to represent each of the ℓ blocks of zeros of lengths 2^k to $2^{k+\ell-1}$; this string of 1s is followed by a 0-bit, acting as a separator, from which ℓ may be deduced by the decoder; finally, the length of the remainder, encoded in standard binary form using $k + \ell$ bits, is appended.

Hierarchical compression: Instead of processing the bitmaps by runs from left to right, the hierarchical compression, mentioned in [Wedekind and Härder (1976)], is efficient when there are clusters of small dense areas in the sparse bitmap. Rename the given bitmap of length n_0 bits as b_0 and partition it into consecutive blocks of k bits each, as shown in Figure 6.19 for $k = 4$. Generate a new bitmap b_1 of length $n_1 = \frac{n_0}{k}$ bits, and set its i-th bit to 0 if and only if the corresponding i-th block in b_0 contains only zeros, that is, each bit in b_1 is the result of ORing the bits in the corresponding block in b_0. The procedure is then repeated to generate b_2 from b_1, and so on, until a level r is reached for which the size $n_r = \frac{n_0}{k^r}$ of the bitmap b_r is just a few bytes.

By recording the non-zero blocks, shown in gray in Figure 6.19, top-down, left to right, this operation is reversible: b_{j-1} may be restored by inserting 0-blocks as indicated by the 0-bits in b_j, for $j = r, r-1, \ldots$, to finally get the original b_0.

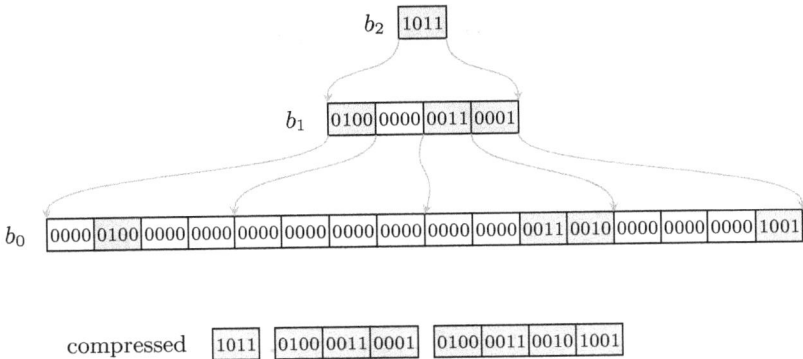

FIGURE 6.19: *Hierarchical compression of a sparse bitmap.*

It will be convenient to choose the parameter k as a multiple of the byte size. For example, with $k = 16$, bitmaps of length $n_0 = 2^{16} = 65536$ bits can be compressed with just $r = 3$ additional levels, of sizes 4096, 256 and 16 bits, respectively, to end up with a highest level b_3 of just two bytes.

6.6   Exercises

6.1 For the set of weights $\{6, 4, 2, 2, 1, 1\}$, build a Huffman tree which is neither of minimal nor of maximal depth.

6.2 Huffman's algorithm can be generalized to support a k-ary encoding, for $k \geq 2$, and not just the binary case as we have seen. At each step, k elements are combined, the tree will be k-ary and the edges emanating from an internal node will be labeled $0, 1, \ldots, k - 1$. For example, for the set of $n = 10$ weights $\{1, 1, 2, 3, 6, 7, 7, 10, 20, 30\}$, one of the possible 4-ary Huffman codes could be $\{0, 1, 3, 20, 22, 23, 210, 211, 212, 213\}$.

  (a) Note that for the given example, $n \bmod (k - 1) = 1$. What is the problem if this does not hold, for example, for $n = 12$, still with $k = 4$?
  (b) Suggest how to overcome the problem in (a).
  (c) Adjoin the weight 8 to the set given above and build a 4-ary Huffman tree for this set of $n = 11$ weights.

6.3 Given is the alphabet $\{$a, b, c, d, e, f, g, h$\}$, and the following text of 50 characters, in which the letters appear with frequency 12, 11, 8, 7, 4, 4, 3 and 1, respectively:

```
a b c f h a a a a b b b b d c d a a a a b c d e g b
c d c d f b b c d e g a b c e f a a b b c d e f g
```

We are interested in compressing this text using several techniques, and comparing their performances. For each of the following, perform the compression and calculate the size of the compressed file.

  (a) Devise a 4-ary Huffman code for the alphabet, that is, the codewords are built over the set of symbols $\{\alpha, \beta, \gamma, \delta\}$, rather than the usual $\{0, 1\}$. For the final binary encoding, the 4-ary set is converted to binary, giving, in an optimal way, one of the symbols the codeword 0, another the codeword 01 and the two others the codewords 001 and 000, or the four symbols are replaced by 00, 01, 10, and 11.

(b) Compress the text using LZW, starting with a dictionary of 8 elements. For the encoding, start with using 4 bits per pointer, which are increased as necessary.

(c) Compress the text with LZ77 in which characters and (offset, lengths) pairs can appear in any order, to be indicated by flag-bits. For the encoding assume 3 bits for a character, 2 bits for a length (what lengths can one encode then?), and offsets of 3 or 5 bits, to all of which, the necessary flags have to be added.

6.4 Given are the codewords $a = 111$, $b = 11100$, $c = 11101$ and $d = 111011$. Prove or disprove the claim that $\{a, b, c, d\}$ is a UD (uniquely decipherable) code.

6.5 Any Huffman code is a prefix code, and by reversing all the codewords, one would get a *suffix* code, in which no code-word is the proper suffix of any other. A code having both prefix and suffix properties is called an *affix* code. The code $\{0100, 0101, 0110, 0111, 1000, 1010, 1011, 1100, 1101, 1110, 1111\}$ is an affix code, as is any fixed-length code, but it is not complete, since $8 \times 2^{-4} = \frac{1}{2} < 1$.

(a) Turn the code into a complete affix code by adjoining six more codewords: one of length 3 bits, one of length 4 bits and four of length 5 bits.

(b) An affix code is also called *never-self-synchronizing*. Why?

(c) Show that there are infinitely many variable-length binary complete affix codes.

6.6 Show that the infinite Fibonacci code $\{11, 011, 0011, 1011, 00011, 10011, 01011, 000011, \ldots\}$ is complete by evaluating $\sum_{i=1}^{\infty} 2^{-\ell_i}$, where ℓ_i is the length of the i-th codeword of the sequence.

6.7 We have seen that arithmetic coding is better than Huffman coding, and in fact the ratio, for a given sequence of files, of the length of their Huffman encoding to that of their arithmetic encoding may not be bounded. Show that when we are not sure of the probabilities, just the opposite may be true. In other words, give an example of two sequences of (finite) probability distributions such that when we assume one, but the other is true, then not only are we better off by using Huffman instead of arithmetic, but also the ratio of the lengths of the arithmetically encoded files to the lengths of the Huffman encoded files may tend to infinity.

6.8 We are given a set of n numbers in the range from 0 to $2^k - 1$, and would like to store them in less than nk bits. Suggest a compression method based on an idea similar to prefix omission, by partitioning the binary representations of the numbers into two parts: the d rightmost bits, and the $k - d$ leftmost bits. The total cost of storing the set should not be more than $(d + 1)n +$ some overhead, not depending on n, which should be determined.

6.9 Give an example of a text T of length n over an alphabet of size at least 2, such that each character of the alphabet appears in T with probability $O(1)$ (which means, the number of occurrences of each character is $O(n)$), and for which LZW yields a compressed text of size $O(\sqrt{n})$.

6.10 What is the connection between Teuhola's bitmap compression method and Elias's γ code? What is the purpose of the parameter k in Teuhola's technique?

6.11 Find a necessary and sufficient condition on a string T for its Burrows–Wheeler transform $\mathsf{BWT}(T)$ to be a sorted string.

6.12 We saw that the $\mathsf{BWT}(T)$ is not enough to reconstruct the original string T, and we had to add a pointer to the row in the BW matrix holding the string T itself. This pointer can be saved by the convention of replacing the string T by $T\$$, where $\$$ stands for a character not belonging to the alphabet Σ from which the characters of T are drawn. For a long string, this addition is negligible and the starting point for decoding can be detected by locating the (single) $ in BWT($T\$$). To see how the addition of this terminal character affects the transform, compare the strings BWT($T\$$) to the strings BWT(T) given at the beginning of Section 6.3 for the three strings T starting with `sanglots-`, `Besen-` and `vanitas-`.

6.13 What are the conditions for getting not only an output of the same length, but *identical* output files by Huffman and arithmetic coding?

6.14 Given are n ordered integer sequences $A_1, \ldots, A_n$, of lengths $\ell_1, \ldots, \ell_n$, respectively. We wish to merge all the sequences into a single ordered sequence. The number of steps needed to merge a elements with b elements is $a+b$. Find an optimal way to merge the n sets, that is, such that the number of comparisons is minimized.

Chapter 7

Pattern Matching

The string matching problem, which is a special case of the more general *pattern matching* problem, is tackled differently depending on the sizes of its input parameters. It may be defined generically for a *text* $T = T[0]T[1]\cdots T[n-1]$ of length n and a *string* $S = S[0]S[1]\cdots S[m-1]$ of length m characters, all belonging to some pre-defined alphabet Σ. We are interested in finding the starting index of the first location of S in T, if it appears there at all, or to get appropriate notice, if not. In the example of Figure 7.1, the string $S = \text{OT-TO}$ is indeed found in T, starting at position 10.

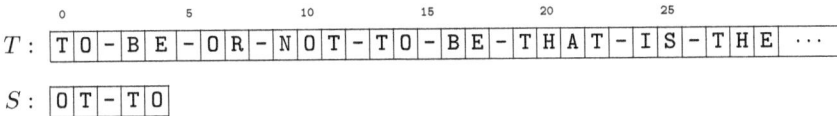

Figure 7.1: *Schematic view of the string matching problem.*

For large input texts, however, when instead of a single text, the search should be applied to an entire corpus, a large collection and ultimately, to the entire World Wide Web, our matching algorithms do not scale efficiently, forcing another approach, like *inverted files*, which have been mentioned in Section 6.5.1. In this chapter, we restrict our attention to methods solving the generic problem stated earlier. One of these methods, the probabilistic method of Karp and Rabin, has already been treated in Section 5.3, which also includes what we call the naïve direct solution in ALGORITHM 5.4.

We imagine the string to be aligned directly below the text, both left-justified, so that one can compare the corresponding characters, until either the string is exhausted, or a mismatch is found. In the latter case, the string is moved forward to a new potential matching position. It should be clear that when we say that the pattern is "moved", this is just a metaphor to facilitate the understanding. Obviously, both the text and the string remain in the same location in memory during the entire search process, and it is by changing the values of the pointers to S and T that their suggested movement is simulated.

A first step to improve the naïve approach has been published by Knuth, Morris and Pratt [Knuth et al. (1977)] and is known by the initials of its inventors as the KMP algorithm. It inspects every character of the text and the string exactly once, yielding a complexity of $n + m$ rather than $n \times m$ comparisons. The same year, an even better algorithm was found by Boyer and Moore (1977), with the surprising property that the expected number of comparisons is less than the length of the text, or in other words, the algorithm is likely not to inspect all the characters of the text! We shall see in the next section how this could be possible.

7.1   Boyer and Moore's algorithm

Boyer and Moore's algorithm is one of those highlights in which an utterly simple twist to a long established straightforward approach turns out to yield a significant improvement. In fact, the main idea of the algorithm can be encapsulated in the phrase:

Start processing the string from the end.

Applying this idea to the naïve ALGORITHM 5.4 would mean to reverse the direction of the inner loop, which scans the string, from

$$\text{for } j \leftarrow 0 \text{ to } m - 1 \text{ do}$$

to

$$\text{for } j \leftarrow m - 1 \text{ to } 0 \text{ by } -1 \text{ do.}$$

This does not seem to make any difference. Indeed, the reversal of the processing direction is useful only if, in addition, the partial information gathered from past comparisons may be exploited by means of some small simple auxiliary arrays.

Refer to Figure 7.2 for a running example of some arbitrary text T and the string $S = \text{MONOTONE}$ taken from the continuation of the poem

Chanson d'automne by Verlaine mentioned earlier. The pointers i and j show, respectively, the initial indices in text and string, with $j = m - 1$ pointing to the end of the string, but with a quite unusual initial position for i: not at one of the extremities of the text T, but rather at position $m - 1$, corresponding to the last character of the string S.

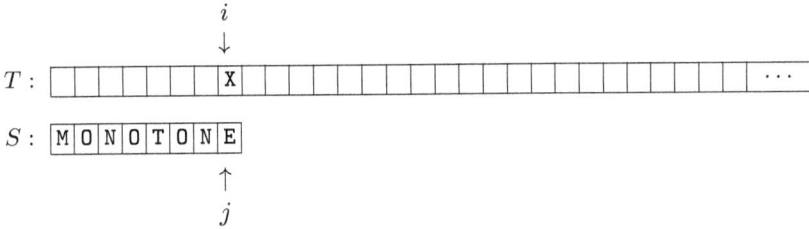

Figure 7.2: *Initial positions of the pointers to text and string.*

What is then the advantage of starting from the end? The first comparison, in the example of Figure 7.2, would be of character E in S against an X in T, which yields a mismatch. We thus know that the current position of S is not the one we seek, so the string has to be moved. Does it make sense to move it just by a single position to the right? Or even by two, three or up to seven? That would still leave the X in position 7 of T over one of the characters of S and necessarily lead to some mismatch, since X does not appear at all in S. We conclude that S can be moved at once beyond the X, that is, so that its beginning should be aligned with position 8. But recall that we have decided to scan the string S always from its end, so the position for the next comparison should be at $i = 15$ in T, as shown in Figure 7.3. Note that the first seven characters in T have not been inspected at all, and nevertheless, no match of S in T may have been missed.

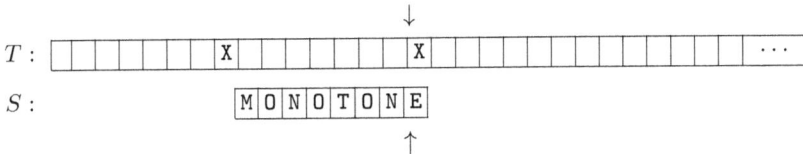

Figure 7.3: *Positions of the pointers after the first shift.*

We just saw an example in which the string could be moved by a step of size m, which depended critically on the fact that X did not appear in

S. However, it requires m comparisons to check this fact, which seems to void our alleged savings! Suppose then that in the position of the second comparison, indexed 15, there is again an X, as in Figure 7.3. There is clearly no need to verify again that X does not appear in S, if we can remember it from earlier checks.

7.1.1   *The bad character heuristic*

And what if there is a mismatch, but the character of the text *does* appear somewhere in the string, as shown in Figure 7.4?

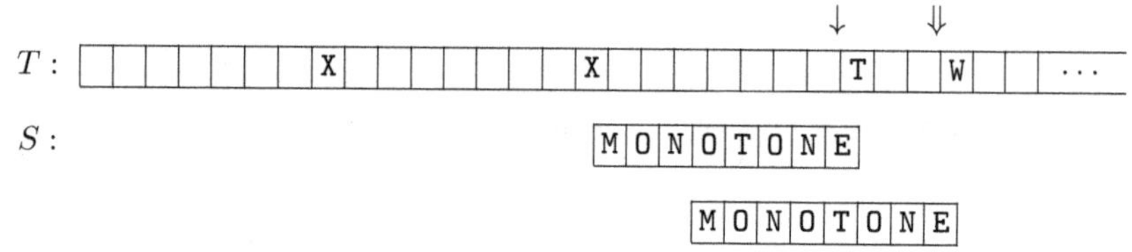

Figure 7.4:   *Mismatching character appears in S.*

One may again argue that nothing can be gained from shifting the string by just one or two positions, and that we should, for the given example, shift it by three, in order to align the two Ts, as shown in the last line of Figure 7.4. Since the next comparison should again be at the end of the string, the pointer i to T is advanced by 3 to the position indicated by the double arrow $\Downarrow$.

To unify the cases of a mismatching character that appears or not in the string, one may define an auxiliary table Δ_1 of integers to store the size of the possible jump of the pointer. Formally, Δ_1 is defined, for a given string S, as a function from the alphabet Σ to the integers by

$$\Delta_1[x] = \ell,$$

if the string S can be safely moved forward by ℓ positions in the case of a mismatch at its last character x. The main step of the scanning algorithm, which updates the current pointer i to the text, will thus be

$$i \longleftarrow i + \Delta_1[T[i]]. \tag{7.1}$$

For our example string $S =$ MONOTONE, the Δ_1 table is given in Table 7.1. The alphabet is supposed to be $\Sigma = \{$A, B, $\ldots$, Z$\}$. The entries for the characters that appear in S are emphasized in red. The table can be built as shown in ALGORITHM 7.1.

Delta$_1$-construction(S, m)

1	for $j \leftarrow 0$ to $m - 1$ do
2	$\quad \Delta_1[S[j]] \leftarrow m$
3	for $j \leftarrow 0$ to $m - 1$ do
4	$\quad \Delta_1[S[j]] \leftarrow m - 1 - j$

ALGORITHM 7.1: *Construction of the Δ_1 table for Boyer and Moore's algorithm.*

Each entry is first initialized with the length of the string m, 8 in our example. By then processing the string left to right, the index stored in the table corresponds to the rightmost appearance, should a character appear more than once in S, such as N and O in our example. The complexity of building Δ_1 is $m + |\Sigma|$.

Table 7.1: *Table Δ_1 for example string $S = $ MONOTONE.*

A	B	C	D	E	F	G	H	I	J	K	L	M
8	8	8	8	0	8	8	8	8	8	8	8	7
N	O	P	Q	R	S	T	U	V	W	X	Y	Z
1	2	8	8	8	8	3	8	8	8	8	8	8

Consider now also the possibility of a match in the last position, as in Figure 7.5, where the single arrow shows the position of the first comparison for the current location of S, as usual at the end. While we keep getting matches, both pointers i and j are decremented and the process is repeated. The loop terminates either when the entire string is matching and a success may be declared:

$$\text{if } j = -1 \qquad \text{return } i + 1;$$

or when after $k > 0$ steps backwards, we again encounter a mismatch, as indicated by the double arrow in Figure 7.5, where the mismatch occurs for $k = 2$.

In this case, all we can do is to shift the string beyond the current position, that is, by six positions forward in our example. In the general case, the shift of the string will be by $\Delta_1[T[i]] - k$, as shown in the last line of Figure 7.5.

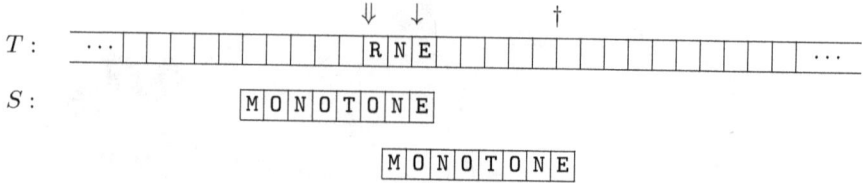

Figure 7.5: *First mismatch after $k > 0$ matches: bad character.*

It is, however, the pointer i to the current position in T which needs to be updated, not that of the string S. Since the following comparison should again be according to the final position in the string, at $j = m - 1$, we should compensate for having moved backwards k positions by adding k again, getting the correct updated value of i

$$i + (\Delta_1[T[i]] - k) + k = i + \Delta_1[T[i]].$$

This is the same update as before; hence the assignment in line (7.1) is actually valid for every value of $k \geq 0$ and not only for the case $k = 0$ of a mismatch at the end of the string. Returning to our example of Figure 7.5, the current position of i, indicated by the double-arrow $\Downarrow$, points to R, which does not appear in S; i is thus incremented by $\Delta_1[\text{R}] = 8$, and will point, after the shift, to the position indicated by the dagger sign †.

A special case forces us to amend the assignment (7.1): when the mismatching character of T appears in S to the right of the current position, then $\Delta_1[T[i]] < k$, so to get an alignment, we would actually shift the string *backward*. But this cannot be justified, as the string has only been shifted over positions for which no match could be missed. For example, this special case would occur in Figure 7.5 if there would be an N or an E instead of an R at the position indicated by the double arrow. It therefore makes no sense to increment the pointer i by k or less, so that the minimal increment should be at least $k + 1$. This yields as corrected update

$$i \longleftarrow i + \max(k + 1, \ \Delta_1[T[i]]). \tag{7.2}$$

7.1.2   *The good suffix heuristic*

Reconsider now the case in which the first mismatch occurs after k steps backward, for $k > 0$, as in Figure 7.6.

We may relate to the given situation as if it were a half-empty glass, concentrating on the fact that a mismatch occurred on our $(k + 1)$st trial. But we can also take a more optimistic look, considering the half-filled glass,

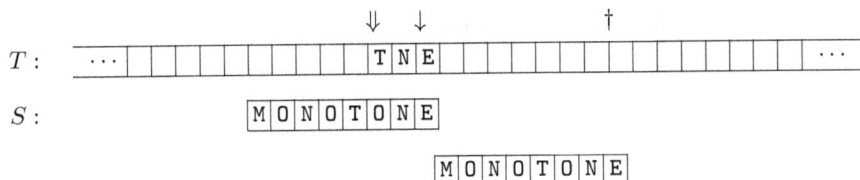

Figure 7.6: *First mismatch after $k > 0$ matches: good suffix.*

which emphasizes the fact that the k first comparisons were successful. This implies that when the mismatch occurs, we know already what characters appear in the text at positions $i+1, \ldots, i+k$: these must be the characters of the suffix of length k of S. These contradicting approaches are often referred to, respectively, as the *bad character* and *good suffix* heuristics.

According to the latter, we may check whether or not the given suffix $S[m-k] \cdots S[m-1]$ occurs again elsewhere in the string S. This suffix is NE in Figure 7.6, which does not appear again in S. The string can therefore be shifted beyond the position where the present iteration started, as shown in the last line of Figure 7.6, and the following comparison is at the position indicated by the dagger sign. The pointer i has thus been incremented by 10 from its current position, which is indicated by the double arrow, instead of increasing it only by 3, had we used $\Delta_1[\mathtt{T}]$.

Similarly to what we did for Δ_1, the search for recurring suffixes will not be performed on the fly, and will rather be prepared, independently of the text, in a pre-processing stage. As there are only m possible suffixes, a table Δ_2 with m entries will be defined, which assigns a value to each of the possible positions $j \in \{0, \ldots, m-1\}$ in the string S. $\Delta_2[j]$ is defined as the number of positions one can move the pointer i when the first mismatch is at position j in S, where we have, as usual, started the comparisons from the end, that is, from $j = m - 1$.

To prepare the Δ_2 table, note that the increments of i actually consist of two independent parts: on the one hand, the number of steps we have moved the pointer backward to get to the current position *within* the string; on the other hand, the size of the shift to reposition the string itself to the following position yielding a potential match. We have moved $k = m-1-j$ steps to the left, so that i can be increased back to point to the last position of the string again, by adding k to i. Then the string S should be shifted by r positions to align the matching suffix with its earlier occurrence in S, and the value of $\Delta_2[j]$ will be the sum of these two increments, $k + r$.

Both approaches, bad character or good suffix, are correct, so we can just choose the maximal increment at each step, which turns the main command into

$$i \longleftarrow i + \max(\Delta_1[T[i]], \Delta_2[j]). \tag{7.3}$$

Note that $k+1$ is not needed anymore as an additional parameter of the maximum in (7.3) as it did in (7.2), because

$$\Delta_2[j] = k + r \geq k + 1.$$

Table 7.2 depicts the Δ_2 table for our example string. For example, the values in columns 6, 5 and 4 correspond, respectively, to a situation in which the first mismatch between T and S, when searching from the end of S, has occurred with the characters, N, O and T, which means that there has been a match for E, NE and ONE, respectively. None of these suffixes appears again in S, so in all these cases, S may be shifted by the full length of the string, which is 8. Adding the corresponding values of k, 1, 2 and 3, finally gives Δ_2 values of 9, 10 and 11. The rightmost column 7 is a special case: it corresponds to an empty matching suffix which therefore reoccurs everywhere. We can thus only shift by 1.

Table 7.2: *Table Δ_2 for example string $S = $ MONOTONE.*

j	0	1	2	3	4	5	6	7
$S[j]$	M	O	N	O	T	O	N	E
shift	8	8	8	8	8	8	8	1
k	7	6	5	4	3	2	1	0
$\Delta_2[j]$	15	14	13	12	11	10	9	1

We may be mislead by the regular form of this table, which is due to the fact that the last character E occurs only once and thus all suffixes are unique. Let us see what happens if we change the string slightly to $S = $ MONOTONO. The value in column 6 corresponds to a mismatch with N after having matched the suffix O, which appears again, at position 5 of S. The string must thus be moved by 2 positions to align the two occurrences of O. For $j = 5$ and 4, the corresponding suffixes NO and ONO are of lengths 2 and 3, and they reoccur at positions 2 and 1, yielding each a shift of 4 positions. For columns $j < 4$, the suffixes we are looking for are TONO or its extensions, and none of them appears again in S, so the string can

Table 7.3: *Table Δ_2 for example string $S =$ MONOTONO.*

j	0	1	2	3	4	5	6	7
$S[j]$	M	O	N	O	T	O	N	O
shift	8	8	8	8	4	4	2	1
k	7	6	5	4	3	2	1	0
$\Delta_2[j]$	15	14	13	12	7	6	3	1

j	0	1	2	3	4	5	6	7
$S[j]$	M	O	N	O	T	O	N	O
shift	8	8	8	8	4	**8**	2	1
k	7	6	5	4	3	2	1	0
$\Delta_2[j]$	15	14	13	12	7	**10**	3	1

be shifted by its full length, 8. This yields the table in the upper part of Table 7.3.

A more careful inspection reveals that the value for $j = 5$ should be amended. It corresponds to a scenario in which there has been a match with the suffix NO, followed by a mismatch with the preceding character O. This means that there is an occurrence of the substring NO in the text, which is preceded by some character that is not O. Therefore, when we seek a reoccurrence of NO, the one starting at position 2 is not a plausible candidate, because it is also preceded by the same character O. We already know that this led to a mismatch with the string S at the current position; it will therefore again fail after the shift.

The definition of a reoccurring substring should therefore be refined: we seek the previous occurrence, if there is one, of a suffix S' of S, but with the additional constraint that this previous occurrence and the occurrence at the end of S should be preceded by *different* characters. For $S' =$ NO in our example, there is no such reoccurrence; hence the correct shift of the string is by the full length 8, and not just by 4. The other entries remain correct, which yields the table in the lower part of Table 7.3. For example, for $j = 4$, we search for another appearance of the suffix ONO that is not preceded by T, and indeed, the previous ONO, starting at position 1, is preceded by M, giving a shift value of 4.

In certain special cases, there is a need for yet another slight rectification. To show it, reverse the order of the first four characters of our example string to $S =$ ONOMTONO. According to what we have seen so far, this would yield the Δ_2 table in the upper part of Table 7.4. The suffix NO appears earlier, but both its occurrences are preceded by the same charac-

Table 7.4: *Table Δ_2 for example string $S = $ ONOMTONO.*

j	0	1	2	3	4	5	6	7
$S[j]$	O	N	O	M	T	O	N	O
shift	8	8	8	8	5	8	2	1
k	7	6	5	4	3	2	1	0
$\Delta_2[j]$	15	14	13	12	8	10	3	1

j	0	1	2	3	4	5	6	7
$S[j]$	O	N	O	M	T	O	N	O
shift	5	5	5	5	5	7	2	1
k	7	6	5	4	3	2	1	0
$\Delta_2[j]$	12	11	10	9	8	9	3	1

ter O, so the suffix is considered as if it did not reappear, yielding a shift of 8. The suffix ONO, on the other hand, appears again as prefix of S, but is not preceded there by T, thus we can shift only by 5.

Figure 7.7 brings an example of how this definition of the Δ_2 table may lead to an erroneous behavior of the matching algorithm. Suppose that while processing the string S from its end, the first mismatch occurs for $j = 3$, comparing the character X in the text with the character M in S, after having matched already the suffix TONO. As this suffix does not appear again in S, the upper Δ_2 table of Table 7.4 suggests that the string be shifted by 8, which results in an increment of the pointer i by 12, from the position indicated by the single arrow $\downarrow$ to that indicated by the double arrow $\Downarrow$. But if the text happens to be as indicated in the figure, this shift of 12 positions results in a missed occurrence, shown in red.

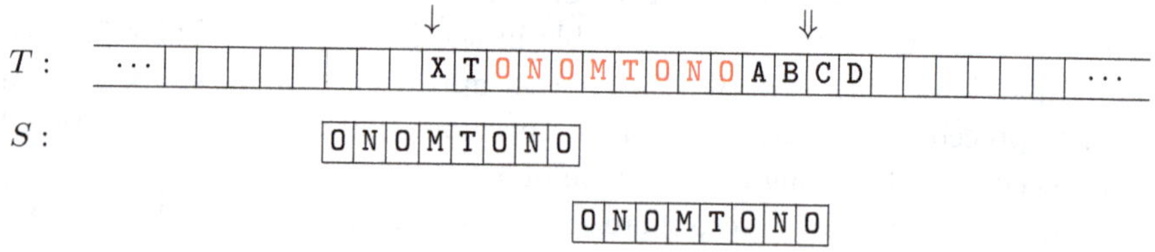

Figure 7.7: *Missing an occurrence of S because of a wrong Δ_2 value.*

So what is wrong in our definition of Δ_2? Actually, it is correct, unless the string S has the special property that it contains a proper suffix which is also a prefix. This is the case for our example string $S = $ ONOMTONO, for which O and also ONO are both prefixes and suffixes of S. This property allows different occurrences of S to overlap in the text. To remedy the flaw in such special cases, the definition of Δ_2 has to be refined.

We are looking in S for a previous occurrence of a proper suffix S' of S, not preceded by the same character; the definition of the term *previous occurrence* should, however, be taken in a relaxed sense, allowing it to extend beyond the boundary of the string S, as depicted in Figure 7.8. The string S appears with a yellow outline and the suffix S' in dark blue.

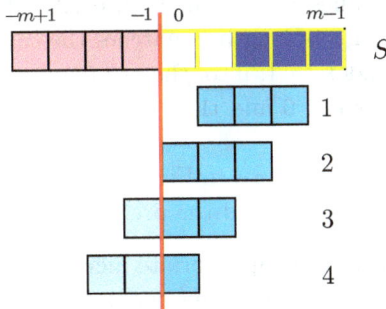

Figure 7.8: *Construction of the Δ_2 table.*

In the lines numbered 1 to 4, the suffix S' is shown shifted by 1 to 4 positions to the left, respectively, as long as there is at least an overlap of one character with the corresponding positions in S. For each of these positions, S' is compared to the corresponding substring of S, but only to the right of the red line, which indicates the left border of S. The characters of S' to the left of this demarcation line appear in lighter blue and are just ignored. An alternative way to explain the reoccurrence is to extend the string S to its left by $m-1$ *don't-care* characters, shown in pink in Figure 7.8. Matching characters have then to be identical to the right of the red line, but to its left, any character pair is declared as matching.

We are now ready to correct our example of Figure 7.7. Filling the table right to left, the first correction is at position $j = 5$, corresponding to the suffix $S' = $ NO, which reoccurs at position 1, but is preceded there by the same character O. There is no additional reoccurrence of NO, but its suffix $S'' = $ O is also a prefix of S. The correct shift is therefore only 7, not 8. For position $j = 3$ in S, there is a mismatch with X in T after having matched already the suffix $S' = \text{TONO}$, which does not reoccur in S. But the suffix $S'' = \text{ONO}$ of the suffix S' does reoccur as prefix of S, so the pattern can only be shifted by 5 positions to get an overlap of the two occurrences of ONO, and not by 8 positions as we erroneously evaluated earlier. The same will be true for mismatches at the other positions j with $j < 3$. This yields the table in the lower part of Table 7.4, in which, as before, the changes relative to the table in the upper part are emphasized.

The construction of the Δ_2 table is summarized in ALGORITHM 7.2. The main loop in line 1 can in fact be executed in any order. Line 3 sets the maximal shift size m as default for the case when no matching reoccurrence of the suffix is found. The loop starting in line 4 unifies two distinct parts:

- for $t + 1 \geq 0$ (lines 1 and 2 in Figure 7.8), it checks whether the current suffix of S of length k matches the substring of S of length k starting at index $t+1$; if so, the preceding characters are checked to differ, unless $t < 0$ and thus the reoccurrence does not have a preceding character;
- for $t + 1 < 0$ (lines 3 and 4 in Figure 7.8), we define $d = r - (t+1)$ and compare the suffix and the prefix of S, both of length $k - d$.

If a match has been found, the appropriate increment for i is stored in Δ_2. The while clause in line 4 checking whether the value in $\Delta_2[j]$ has changed ensures that the stored value corresponds to the rightmost reoccurrence, if there is one at all.

ALGORITHM 7.3 summarizes the formal description of the Boyer–Moore string matching algorithm. Its worst case is still $O(mn)$, but for many reasonable input texts and patterns, it runs in sub-linear time, that is, with less than $m + n$ comparisons. Boyer and Moore evaluated that the average time is approximately $\frac{n}{m-1}$, which yields the surprising property that for a given text T, the time spent to locate a string, or assert that it does not occur, is *inversely* proportional to its length!

Another criterion influencing the complexity is the size of the alphabet Σ. Generally, the larger the alphabet, the faster the search, because there are options for increased jump sizes. In particular, in the case of a binary alphabet working with bit-vectors instead of character strings, Δ_1 will be effective only for very particular strings with long suffixes consisting of only zeros or only ones. One can adapt the Boyer–Moore algorithm to be more efficient also for a binary alphabet, as proposed in [Klein and Ben-Nissan (2009)], or one may use in this case the Karp–Rabin approach as presented in Section 5.3.

Delta$_2$-construction(S, m)

1   for $j \leftarrow m - 1$ to 0 by -1 do
2      $k \leftarrow m - 1 - j$
3      $\Delta_2[j] \leftarrow m + k$ // set default value
4   for $t \leftarrow j - 1$ to $-k$ by -1 while $\Delta_2[j] = m + k$ do
5        $r \leftarrow \max(t + 1, 0)$
6a   if $S[r] \cdots S[t + k] = S[r + j - t] \cdots S[m - 1]$
            // compare suffix and prefix of length $k - d$
6b   and $(t < 0$ or $S[t] \neq S[j])$ then
            // check that preceding characters differ
7             $\Delta_2[j] \leftarrow (j - t) + k$

ALGORITHM 7.2: *Construction of the Δ_2 table for Boyer and Moore's algorithm.*

Boyer–Moore(T, n, S, m)

1    $i \leftarrow m - 1$
2   while $i < n - m$ do
3      $j \leftarrow m - 1$
4   while $T[i] = S[j]$ do   // compare backwards
5        $i \leftarrow i - 1$ $j \leftarrow j - 1$
6   if $j = -1$ then return $i + 1$ // match found
7      $i \leftarrow i + \max(\Delta_1[T[i]], \Delta_2[j])$ // mismatch at current i

ALGORITHM 7.3: *Boyer and Moore's string matching algorithm.*

7.2   Position and suffix trees

A common characteristic of both Knuth, Morris and Pratt's string matching algorithm and that by Boyer and Moore is that the improvement over the naïve algorithm is achieved by pre-processing the string S to be located. This string is assumed to be fixed, and it may be handled even before the text is known. In certain different search scenarios, we wish to find many different strings S, but it is the text T that is assumed to be given and fixed in advance.

If indeed the same text will be used for many searches with different strings S, it may be advantageous to pre-process the text T, rather than the string S, even though the size n of T can be orders of magnitude larger than m, the length of one of the strings. Examples for such texts, which are used time and again, are famous court decisions, literature masterpieces or the human genome.

The data structure we shall see in this section is known as *position tree* [Weiner (1973)] and its special case, the *suffix tree* [Ukkonen (1995)]. It has many practical applications and is in widespread use in text processing and computational biology.

As an example of the usefulness of position trees, consider the problem of finding the longest substring of T that appears at least twice in the text. The solution of this problem may have an impact on text compression, which has been treated in the previous chapter. As mentioned, the elements to be encoded, forming what we called the *alphabet* Σ, are not necessarily single characters or words, and may in fact be any collection of strings. Since compression results from the replacement of these alphabet elements by (shorter) pointers, it makes sense to strive for including the longest possible strings in Σ. On the other hand, shorter strings may be preferable, if they are frequent enough. Ultimately, the most promising strings are those with the highest length $\times$ frequency products.

The simplest way to find the longest reoccurring substring in a text T of length n is to iterate over all the $\binom{n+1}{2} = \theta(n^2)$ substrings s of T (each of which is defined by its limiting positions), and to check by means of, e.g., Boyer and Moore's algorithm, whether s appears again in T at some different location. The algorithm would thus require $\theta(n^3)$, which is prohibitive for large n. We shall see how a position tree provides a much faster solution.

As in the previous section, we consider a text $T = T[0]T[1]\cdots T[n-1]$ of length n characters over some alphabet Σ; the indices 0 to $n-1$ will be called *locations* or *positions*. For each $i = 0, \ldots, n-1$, we say that the string $S = S[0]\cdots S[r-1]$ *identifies* position i if S appears only once in T, starting at position i, that is,

$$S = S[0]\cdots S[r-1] = T[i]T[i+1]\cdots T[i+r-1].$$

The shortest string which identifies position i is called *the identifier* of i. For example, if $T = \texttt{MONOTONE}$, then $\texttt{ON}$ does not identify position 1, because it also reoccurs starting at position 5; NOT identifies position 2, but it is not *the* identifier, because there is a shorter string, NO, which also identifies position 2.

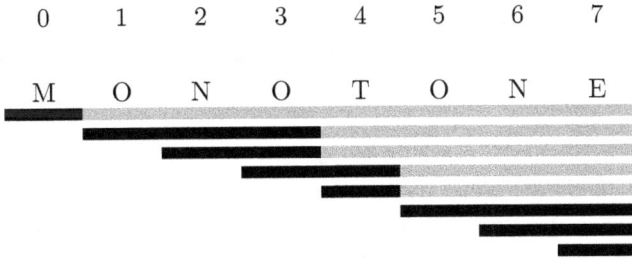

FIGURE 7.9: *Position identifiers on example text* MONOTONE.

Figure 7.9 returns to our running example MONOTONE and shows the identifiers of all its positions underlined in black. At first sight, one could think that the identifier can be found for every position, as in this example: one could simply consider all the *suffixes* and shorten them as much as possible. However, if we change the text to MONOTONO, we see that each of O, NO and ONO appears more than once, so the identifiers of positions 5, 6 and 7 would not be defined.

An elegant way to avoid the problem of not having the identifiers well defined for certain positions of certain input strings is to append an additional character, often denoted by $, which does not belong to the assumed alphabet Σ, at the end of the given text T. This is the same amendment as suggested in Exercise 6.12 of the previous chapter. We shall therefore extend the alphabet to $\Sigma \cup \{\$\}$ and replace the text $T = T[0] \cdots T[n-1]$ by $T\$$, in which the letter $ occurs only once, in the last position, indexed n. As a consequence, all the suffixes of $T\$$ are unique, and the suffix $T[i]T[i+1] \cdots T[n]$ identifies position i. For $T = $ MONOTONO, the identifier of position 6 would be NO$, and that of position 2 would be NOT.

Of course, we pay a price for the convenience of circumventing the troubles, but for large enough texts, the addition of a single character does hardly have any impact. The solution of appending a special letter $ has thus become widespread and is quite standard in order to deal with this and similar string problems.

To understand how these definitions may help in solving string matching problems, we recall here the details of a data structure known as a *trie*.

Background concept: Trie data structure

A *trie* is a tree-like data structure that extends the notion of a binary tree associated to a binary prefix code, as the one shown in Figure 6.3 in the previous chapter. We are given an alphabet $\Sigma = \{\sigma_1, \ldots, \sigma_r\}$ of size r, and a set S of k strings $S = \{s_1, \ldots, s_k\}$, with $s_i \in \Sigma^*$, that is, each string s_i is a sequence of letters of Σ.

The tree is defined as follows: any node may have between 0 and $r = |\Sigma|$ children; nodes with no children are leaves and there are k leaves in the tree, labeled 1 to k, each one being associated with another of the k given strings s_i. The edges are labeled by letters of the alphabet: if an internal node v has $\ell \geq 1$ children $v_{i_1}, v_{i_2}, \ldots, v_{i_\ell}$, where the set $\{i_1, i_2, \ldots, i_\ell\}$ is a subset of size ℓ of the possible indices $\{1, 2, \ldots, r\}$, then the edge (v, v_{i_j}) is labeled by the letter σ_{i_j}, for $1 \leq j \leq \ell$. It is customary to label the edges from an internal node to its children in lexicographic order of the labeling letters. For $1 \leq t \leq k$, the string s_t is associated with the leaf v_t such that s_t is obtained by concatenating, in order, the labels on the edges from the root to v_t. The resulting tree is called the *trie* associated with the set of strings S. Conversely, one can also consider the trie as *defining* the set S.

In particular, if S consists of the codewords of a binary prefix code, then $\Sigma = \{0, 1\}$ and the associated trie is the binary tree of this code, such as the one in Figure 6.3. For the special case of a Huffman code, the tree will also be complete, that is, there will be no internal nodes with just one child.

The *position tree* of a string $T\$$ of length $n + 1$ is defined as the trie associated with the identifiers of the positions i of the string, $0 \leq i \leq n$. Similarly, the *suffix tree* of such a string is obtained by taking the $n + 1$ suffixes $T[i]T[i + 1] \cdots T[n]$ instead of the (mostly shorter) location identifiers.

In Figure 7.10, the position identifiers are underlined in black, and their extensions to the corresponding suffixes are shown in gray, for the text NANAHNAHMAN\$. The alphabet is $\Sigma = \{\$, A, H, M, N\}$. The associated trees appear in Figure 7.11. The position tree is shown as black nodes and edges, and the leaves are labeled by the starting positions of the corresponding identifiers, so that there are $12 = |T\$|$ leaves in our example.

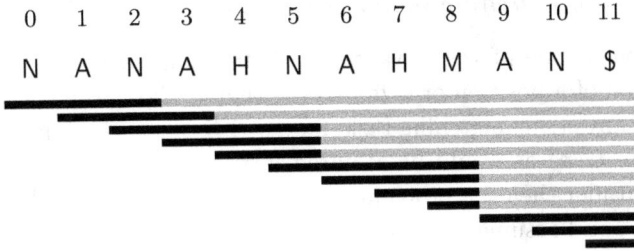

FIGURE 7.10: *Position identifiers on example text* NANAHNAHMAN$.

The suffix tree is obtained by adding also the grayed nodes and edges to the position tree (and moving, on each branch, the index of the suffix to the new leaf).

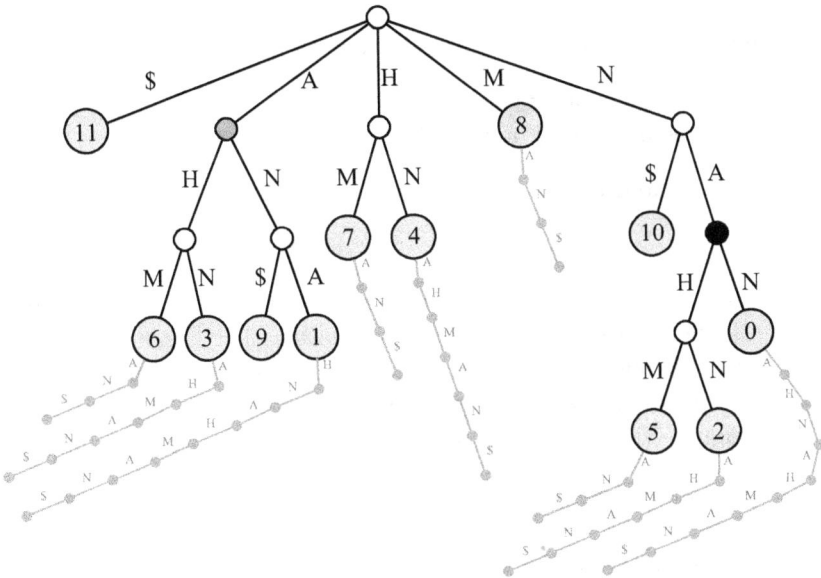

FIGURE 7.11: *Position and suffix tree.*

7.2.1 *String matching using position trees*

As mentioned, one of the advantages of investing so much effort in the construction of a position or suffix tree is that it allows the subsequent reduction of any string matching task, for a string of length m, in time $O(m)$, rather than $O(n)$ as would be required, e.g., by Boyer and Moore's algorithm. Consider then the problem of locating a string $S = S[0] \cdots S[m-1]$ in a text T, and assume that we have already constructed the position tree $\mathcal{T}$ of $T\$$.

Rather than searching within the text T, we navigate through the tree $\mathcal{T}$ as guided by the characters of the string S. Starting at t_0, the root of $\mathcal{T}$, we follow the edge labeled by $S[0]$, if it exists, to reach node t_1; from there we continue by the edge labeled $S[1]$, if it exists, to reach node t_2, etc. Repeating this iteration as often as possible, suppose that the node reached in $\mathcal{T}$ is t_r, after having processed $S[0] \cdots S[r-1]$. The possible options are

> 1. $r = m$ — the entire string S has been processed;
> 2. $r < m$ — the loop ended before reaching the end of S.

Independently of these, there are also two options concerning the node t_r:

> a) t_r is a leaf of $\mathcal{T}$;
> b) t_r is an internal node of $\mathcal{T}$.

There are thus four combinations to be handled.

1. $r = m$:

> a) t_r is a leaf of $\mathcal{T}$: this is the special case in which the string S happens to be the identifier of some location i and thus appears exactly once in T, at location i, which is stored in the leaf t_r. For example, searching with $S = $ NAHM in the position tree of Figure 7.11 leads to the leaf t_4 labeled with position 5.

> b) t_r is an internal node of $\mathcal{T}$: in this case, the string S appears more than once in T. The locations of all its occurrences are written in the leaves of the subtree of $\mathcal{T}$ rooted by t_r. For example, for $S = $ NA, t_2 will be the black internal node in the position tree of Figure 7.11; the positions written in the leaves of the corresponding subtree are 0, 2 and 5, and these are the starting positions of NA in T.

2. $r < m$:

a) t_r is a leaf of $\mathcal{T}$: this is the only case in which the position tree does supply all the necessary information. We have reached a leaf, labeled by position j, but the requested string S is not yet exhausted. For example, when looking for $S = $ HNAHM, t_2 will be the leaf labeled 4, which means that the prefix HN of S occurs once, in position $j = 4$. We thus have to check directly in the text T whether or not the missing suffix of S appears there. Note that this is not a search, but just a verification: we know exactly where this suffix should occur and check if

$$S[r] \cdots S[m-1] = T[j+r] \cdots T[j+m-1].$$

For our example, the suffix AHM is compared to $T[6]T[7]T[8]$—since this is a match, the conclusion is that S occurs once in T, at position $j = 4$. Had the string example been $S = $ HNAN, its suffix AN would have been compared to $T[6]T[7]$, yielding a mismatch, with the conclusion that S does not occur in T.

b) t_r is an internal node of $\mathcal{T}$: this means that the internal node t_r does not have an outgoing edge with the requested label $S[r]$, which means that the prefix $S[0] \cdots S[r]$, and therefore the entire string S, does not occur at all in T. For example, $S = $ AMA would bring us to t_1, the gray node in Figure 7.11, which does not have an outgoing edge labeled M, so there is no match for S in T.

The resulting formal algorithm summarizing this string matching procedure is shown in ALGORITHM 7.4, and its complexity is clearly independent of the size n of the text T.

7.2.2 *Compact position and suffix trees*

The construction of such a position or suffix tree could be done by processing the position identifiers or suffixes one by one, following already existing edges, or creating new ones when necessary. For suffixes, this requires $\sum_{i=1}^{n} i = \theta(n^2)$ steps for $n = |T|$. Position identifiers are generally shorter, but the complexity will still be $\theta(n^2)$ for certain texts. Weiner (1973) showed how to reduce the time to be linear in the number of nodes of the position tree, but even this may be quadratic in n, as can be seen for a text of the form $0^n 1^n 0^n 1^n \$$ over a ternary alphabet $\{0, 1, \$\}$. What comes to our rescue is the fact that if the size of the tree is $\Omega(n^2)$, it must include

Match($S, m, \mathcal{T}$)

1   $t \leftarrow$ root($\mathcal{T}$)
2   for $r \leftarrow 0$ to $m - 1$ do
3   if t is a leaf labeled j then   // case 2.a)
4   if $S[r] \cdots S[m-1] = T[r+j] \cdots T[j+m-1]$ then
5   return one occurrence at $j - r$
6   else return no occurrence found
7   else   // t is not a leaf
8   if t has a child x such that (t, x) is labeled $S[i]$ then
9               $t \leftarrow x$
10   else return no occurrence found   // case 2.b)
        // here when $i = m - 1$ and loop on i did not reach a return statement
11 retrieve labels $\ell_1, \ldots, \ell_s$ of all the leaves in the subtree rooted by t
        // when t is a leaf, then $s = 1$ // case 1.a)
12 return found s occurrences at $\ell_1, \ldots, \ell_s$ // case 1.b)

ALGORITHM 7.4: *String matching by means of position tree.*

many nodes with a single outgoing edge, and such trees may be *compacted* without affecting their functionality. This compaction is achieved by allowing labels to be strings in Σ^*, and not just single letters, and by replacing certain chains of edges by single edges. Specifically, we apply repeatedly the transformation shown in Figure 7.12 to selected sub-parts of the tree, as often as possible.

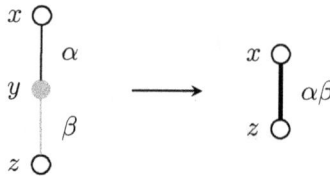

FIGURE 7.12: *Single compaction step.*

If there is a node y in the tree having only a single outgoing edge to a node z and with the edge (y, z) labeled β, then the node y and its outgoing edge (shown in gray) are removed from the tree; the incoming edge (x, y)

labeled α is replaced by an edge connecting directly x to z, and the new edge (x, z) (emphasized by the thicker line) is labeled by the concatenation $\alpha\beta$ of the labels of (x, y) and (y, z).

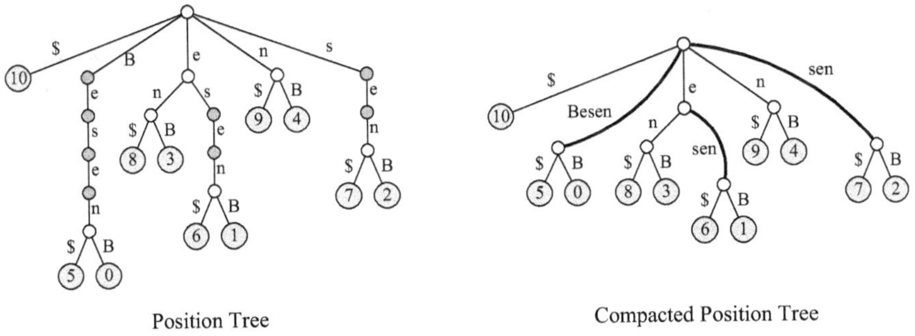

Position Tree   Compacted Position Tree

FIGURE 7.13: *Position tree and compacted position tree for* BesenBesen$.

Figure 7.13 is an example of the compaction process for the input string BesenBesen$. Nodes of the position tree that are removed are shown in gray and the new edges in the compacted tree are emphasized, as in Figure 7.12. To evaluate the number of nodes in a compacted position tree with $n + 1$ leaves, note that the worst case, from the point of view of the number of nodes, would be for a complete binary tree, in which every internal node has two children. The internal nodes of a compacted position tree have *at least* two children, so their number cannot be larger as in the binary case. It is well known that the number of internal nodes of any complete binary tree with $n + 1$ leaves is n, so we conclude that the number of nodes in a compacted position tree is at most $2n + 1$. For example, the compacted tree in Figure 7.13 is almost binary (only the root has more than two children). It has 11 leaves and only 7 internal nodes.

All the algorithms working with position trees can be adapted to work with compacted position trees. The details are more involved, as new nodes have to be inserted by splitting edges, but the overall complexity will not increase by an order of magnitude, and in fact remain linear.

The conclusion is that if our data are such that a simple approach could lead to quadratic complexities, we should prefer the compacted versions. On the other hand, if only pathological versions of our input data would not yield a linear behavior, using standard position trees is much simpler.

7.2.3 *Suffix array and connection to* BWT

A data structure closely related to suffix trees is called *suffix array* and has been suggested by Manber and Myers (1993). It is an integer array, listing the starting positions of the lexicographically sorted suffixes of a given string S. It can be obtained from the suffix tree by traversing it in depth-first order and retrieving the labels of the leaves, which are thereby scanned left to right. Applying this to the tree of Figure 7.11, one gets the suffix array in Table 7.5. Conversely, given the suffix array, the corresponding suffix tree can also be derived in linear time.

Table 7.5: *Suffix array* SA[j] *for example string* S = NANAHNAHMAN$.

j	0	1	2	3	4	5	6	7	8	9	10	11
$S[j]$	N	A	N	A	H	N	A	H	M	A	N	$
SA[j]	11	6	3	9	1	7	4	8	10	5	2	0

The suffix array also provides a noteworthy connection between suffix trees and the Burrows–Wheeler transform we have encountered in Chapter 6. Let us revisit the construction process of the BW matrix, as shown in Figure 6.12 for S = NANAHNAHMAN, and adapt it slightly to account for the added $ at the end of the string. The resulting matrices appear in Figure 7.14.

Rotations

```
N A N A H N A H M A N $    0
A N A H N A H M A N $ N    1
N A H N A H M A N $ N A    2
A H N A H M A N $ N A N    3
H N A H M A N $ N A N A    4
N A H M A N $ N A N A H    5
A H M A N $ N A N A H N    6
H M A N $ N A N A H N A    7
M A N $ N A N A H N A H    8
A N $ N A N A H N A H M    9
N $ N A N A H N A H M A    10
$ N A N A H N A H M A N    11
```

BW matrix

```
$ N A N A H N A H M A N    11
A H M A N $ N A N A H N     6
A H N A H M A N $ N A N     3
A N $ N A N A H N A H M     9
A N A H N A H M A N $ N     1
H M A N $ N A N A H N A     7
H N A H M A N $ N A N A     4
M A N $ N A N A H N A H     8
N $ N A N A H N A H M A    10
N A H M A N $ N A N A H     5
N A H N A H M A N $ N A     2
N A N A H N A H M A N $     0
```

FIGURE 7.14: *Connection between suffix array and* BWT.

The left matrix contains all the rotations of the input string, with row i being rotated by i characters, for $0 \leq i \leq 11$, as indicated by the numbers in the adjacent blue column. These numbers also represent the lengths of the suffixes after the $. The matrix on the right is obtained by sorting the rows lexicographically, maintaining the connection between a row and its index. The last column of this matrix, shown in red, is the BWT of the input string (which is different from that of Figure 6.12, because of the $). The pink numbers in the rightmost column are a permutation of the blue column corresponding to the new sorted order. As one can see, this column in fact contains the suffix array of the input string.

This is not a coincidence. To get the BW matrix, the rotated rows are sorted, but in fact, this is equivalent to sorting only the prefixes up to and including the $ in each row. The reason is that this letter appears only once in each row, and with different offsets, so if there are ties when comparing different rows by scanning them left to right, they are solved at the latest with the first appearance of $. This is the reason why the suffixes after $ have been grayed out in the right-hand matrix of Figure 7.14: they do not affect the order of the rows. We conclude that the order, and therefore the BWT, is determined by the prefixes up to $ of the rows, which are the suffixes of the input string yielding the suffix array.

7.3 Exercises

7.1 Build the Δ_2 tables for the strings NANAHNAHMAN and HEEADEEACHEE.

7.2 Give an example of a string for which all the entries of the Δ_2 table would contain the value 7.

7.3 Suppose that instead of the corrected update of Eq. (7.2) we would use the original incrementing step of Eq. (7.1), without the maximum function. Build an example of a string of length 6 over the alphabet $\Sigma = \{A, B, C\}$ and a corresponding small text for which the Boyer and Moore algorithm, using only Δ_1 and not Δ_2, would then enter an infinite loop.

7.4 Draw the compacted suffix tree corresponding to the compacted position tree in Figure 7.13. What are the differences between the two in the case of a general input string $S\$$?

7.5 What is the number of nodes in the position tree of the string $0^n 1^n 0^n 1^n \$$?

7.6 For a string S of length n,

(a) show how to find its longest substring that appears at least twice in S;

(b) show how to find its longest substring that appears at least twice in S, with the additional constraint that the occurrences may not overlap;

(c) show how to find its longest substring that is also a substring of a string S' of length n'.

PART 5
Numerical Algorithms

Chapter 8

Fast Fourier Transform

8.1 Extended Pattern Matching

The pattern matching algorithms we have dealt with so far are concerned with locating exact copies of the given input pattern within a large string. There are, however, applications for which merely *approximate* matches are required. Consider, for example, a standard problem in bio-informatics, consisting of locating a *read*, playing the role of a pattern, in a *genome*, which represents the text; both read and genome are encoded as strings over the alphabet {A, C, G, T}.

The problem is that in real-life applications, the representation is not perfect. Errors occur, the reasons for which may be either biological (a mutation changing an A into a T) or technical (the sequencing machinery receiving biological material as input and producing some reads as output is prone to erroneous interpretations). If a part of the given genome is

position	1 2 3 4 5 6 7 8 9 10 11 12 13 14 15 16 17 18 19 20 21
genome $G =$	C C G T A A T A T A G G G C G T C T T A C $\cdots$

and we are looking for $R =$ T G T A T G, there is no perfect match of R in G. Nevertheless, the substring starting at the 7th position of G, T A T A G G, differs only by two characters (those in positions 8 and 11 of G) from R, whereas all other locations trigger more mismatches. If we are willing to accept up to three differences, then positions 2, 5, and 14 would also be considered as matching.

The naïve pattern matching algorithm comparing each character of the pattern of length m to each character of the string of length n is easily extended to encompass also this generalized problem. Using $n \times m$ comparisons, it can produce a sequence of $n - m + 1$ integers $d_1, \ldots, d_{n-m+1}$, where d_i gives the number of mismatches between R and the substring of length m of G starting at position i. For the above example, this sequence would start with 6, 3, 4, 5, 3, 5, 2, 5, ...

Another, related, extension of the original problem involves the usage of *wildcards*, also called *don't-care* characters, often symbolized by an underscore _ , which is supposed to match any nonspecifically determined character. Thus the pattern A B _ D could match A B C D, but also A B A D, and A B B D, etc. The naïve algorithm works also in the presence of don't-cares by simply skipping the comparison if one of the characters to be compared is _ .

On the other hand, some faster algorithms, like those due to Knuth–Morris–Pratt, Boyer–Moore, or Karp–Rabin, can generally not be easily extended to deal with these approximate pattern matching variants. This leads to the following interesting approach to solve pattern matching problems.

In a first stage, we show how pattern matching can be considered as a special type of multiplication, more precisely, a multiplication of polynomials. Bearing that in mind, it will be easy to generalize the problem to its approximate alternatives. In a second stage, we shall then concentrate on accelerating the polynomial multiplications by means of a surprising algorithm, known as *Fast Fourier Transform*, which has also a myriad of different applications, not at all related to pattern matching.

Consider the small example of Figure 8.1. Let $T = t_0 t_1 t_2 t_3 t_4$ and $P = p_0 p_1 p_2$ be strings representing, respectively, a text and a pattern, both over an alphabet Σ. We shall apply on them an operation that we call "multiplication" because of its resemblance to the way we multiply numbers and polynomials, but the details of the operation, denoted by the operator $\cdot$, will be specified later. Let us see what happens if we try to evaluate $D = T \cdot P^R$, where x^R stands for the string obtained by reversing x, so that $P^R = p_2 p_1 p_0$.

Focusing on the rectangle in the middle, we see that the pattern P appears in each of the columns alongside the different substrings of T of length $|P|$. This suggests us to define the multiplication operator to yield 0 in the case of a match between t_i and p_j, and 1 otherwise, so that the sum of the elements in each column will be the number of mismatches in the corresponding position. Formally, we define

$$a \cdot b = \begin{cases} 0 \text{ if } a = b \\ 1 \text{ otherwise} \end{cases} \qquad (8.1)$$

		t_0	t_1	t_2	t_3	t_4
				p_2	p_1	p_0
		$t_0 \cdot p_0$	$t_1 \cdot p_0$	$t_2 \cdot p_0$	$t_3 \cdot p_0$	$t_4 \cdot p_0$
	$t_0 \cdot p_1$	$t_1 \cdot p_1$	$t_2 \cdot p_1$	$t_3 \cdot p_1$	$t_4 \cdot p_1$	
$t_0 \cdot p_2$	$t_1 \cdot p_2$	$t_2 \cdot p_2$	$t_3 \cdot p_2$	$t_4 \cdot p_2$		
$D[0]$	$D[1]$	$D[2]$	$D[3]$	$D[4]$	$D[5]$	$D[6]$

FIGURE 8.1: *The connection between pattern matching and multiplication.*

In particular, the original matching problem is solved by the equivalence

$$D[i + m - 1] = 0 \iff \text{there is a match at position } i \quad \text{for } 0 \le i \le n - m.$$

The extension of the problem to deal also with don't-care characters _ is then straightforward: all one needs to do is define the outcome of any multiplication involving _ as being a match, or formally, $a \cdot \_ = 0$. For example, if $T = \text{TAATA}$ and $P = \text{A\_T}$, one gets

		T	A	A	T	A
			T		_	A
		T $\cdot$ A	A $\cdot$ A	A $\cdot$ A	T $\cdot$ A	A $\cdot$ A
	T $\cdot$ _	A $\cdot$ _	A $\cdot$ _	T $\cdot$ _	A $\cdot$ _	
T $\cdot$ T	A $\cdot$ T	A $\cdot$ T	T $\cdot$ T	A $\cdot$ T		
		2	0	1		

The number of steps performed by this multiplication operator when applied to input strings of lengths n and m is $O(mn)$, but this is also the complexity of the naïve algorithm counting mismatches at each position, even in the presence of don't cares. The processing of the multiplication can, however, be significantly accelerated, as will now be shown.

8.2 Multiplying polynomials

A polynomial is a special type of a simple function, and we shall restrict the discussion to polynomials of a single variable x. The general form of a polynomial $P(x)$ is

$$P(x) = a_{n-1}x^{n-1} + a_{n-2}x^{n-2} + \cdots + a_2x^2 + a_1x + a_0 = \sum_{j=0}^{n-1} a_j x^j,$$

where $n-1$ is the *degree* of P and $a_{n-1}, \ldots, a_0, a_1$ are the n *coefficients* defining it.

In fact, a polynomial can be determined by the string of its coefficients $[a_{n-1} \cdots a_1 a_0]$, so that one could use the string $[4 \ \text{-}1 \ 0 \ 3]$ as a shortcut for $Q(x) = 4x^3 - x^2 + 3$. This is much like writing a number as a string of digits, with the difference that the coefficients are not confined to integers in the range 0 to 9, but can take any value, even a negative one. Accordingly, the multiplication of polynomials is quite similar to that of numbers, and it is even easier because of the lack of *carries*, that is, the operations in one column have no influence on adjacent ones. For example, multiplying the polynomial $Q(x)$ by $R(x) = 3x^2 + 2x - 1$, which can be represented by the string $[3 \ 2 \ \text{-}1]$, could be performed by executing

		4	-1	0	3	
			3	2	-1	
		-4	1	0	-3	
	8	-2	0	6		
12	-3	0	9			
12	5	-6	10	6	-3	

The resulting polynomial is thus $S(x) = 12x^5 + 5x^4 - 6x^3 + 10x^2 + 6x - 3$. Note that in the first and fourth columns from the left, the numbers added up to 10 or more, yet there was no carry to the next column. Nevertheless, the number of operations needed to multiply in this way two polynomials of degrees n and m, respectively, is clearly $\theta(nm)$.

The key to lowering this complexity is the fact that there exists also a different form to determine a polynomial, not just the one based on its coefficients. This is most easily understood for a polynomial of degree 1, whose graph is a straight line, like, e.g., the function $2x - 1$, depicted in Figure 8.2. As is well known, such a line in the plane is determined by any two points on it, so choosing for example $(1,1)$ and $(3,5)$, shown in the figure as small black bullets, the function $2x - 1$ is the only first-degree polynomial whose graph traverses these points.

A similar fact is actually true also for higher order polynomials: any three different points determine the unique parabola, corresponding to a

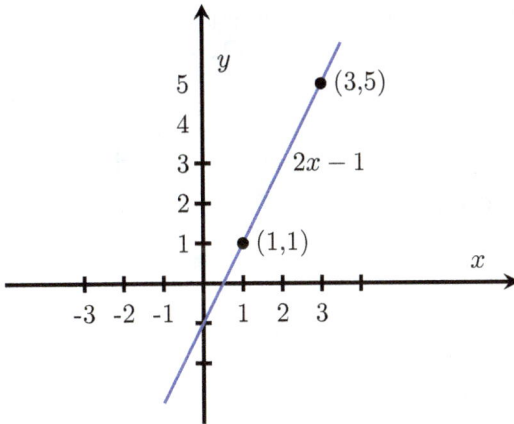

FIGURE 8.2: *Graph of the polynomial $2x - 1$.*

second degree polynomial, passing through them, and more generally, any n different points uniquely define a polynomial of degree $n - 1$. The polynomial $Q(x)$ of our running example can thus be fixed by four points, for example (-1,-2), (0,3), (1,6) and (2,31), and three points are sufficient to determine $R(x)$. The product $S(x) = Q(x)R(x)$ can then be found by multiplying the values at the same chosen points, but three or four points are not sufficient to define $S(x)$, since its degree is 5. To get the six necessary points, $Q(x)$ and $R(x)$ have to be calculated for more values of x, as shown in the example of the following table:

x	-2	-1	0	1	2	3
$Q(x)$	-33	-2	3	6	31	102
$R(x)$	7	0	-1	4	15	32
$Q(x)R(x)$	231	0	-3	24	465	3264

To facilitate the discussion, we shall henceforth assume that the polynomials to be multiplied are both of degree $n - 1$. This is no real restriction, as one can always add terms with coefficients 0 to pad the lower degree polynomial. The same argument applies also below when we restrict the values of n to be powers of 2.

Using the representation of a polynomial of degree $n - 1$ by a set of n points, it is thus possible to multiply two such polynomials Q and R by evaluating pointwise products $Q(x)R(x)$ at $2n - 1$ chosen points, which can be done in $O(n)$ time. The problem is that while these points indeed mathematically determine the resulting polynomial S of degree $2n - 2$, we might want to know the values of S also at some other points.

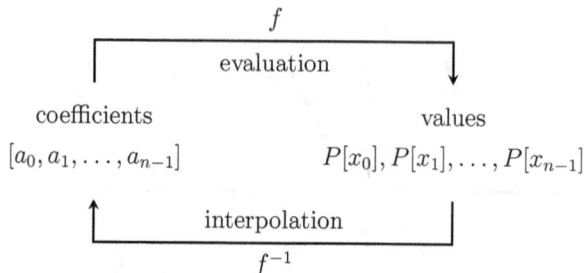

$$f$$

	evaluation

coefficients   values

$$[a_0, a_1, \ldots, a_{n-1}] \qquad\qquad P[x_0], P[x_1], \ldots, P[x_{n-1}]$$

	interpolation

$$f^{-1}$$

FIGURE 8.3: *The connection between two ways to represent a polynomial.*

Figure 8.3 schematically depicts the relation between these two possibilities to determine a polynomial $P(x)$. In the first form, P is given by the list of coefficients $[a_0, a_1, \ldots, a_{n-1}]$, and can be *evaluated* at n different points, say, $x_0, x_1, \ldots, x_{n-1}$. If, on the other side, the polynomial is given by its values at these n points, the derivation of the corresponding coefficients is called *interpolation*. Each of the representations may be convenient in other contexts; there is therefore a need to be able to switch efficiently from one representation to the other. We denote the passage from coefficients to values by the function f, and its inverse accordingly by f^{-1}.

Evaluating $P(x) = \sum_{j=0}^{n-1} a_j x^j$ at a single point x can be done in time $O(n)$ since x^j can be obtained from x^{j-1} in a single operation, for $0 < j < n$. One could also use *Horner's rule* that implements an equivalent form of $P(x)$ as

$$P(x) = \Big(\big((a_{n-1}x + a_{n-2})x + \cdots \big)x + a_1 \Big)x + a_0,$$

for example,

$$Q(x) = 4x^3 - x^2 + 3 = ((4x - 1)x + 0)x + 3.$$

Thus $\Omega(n^2)$ operations are needed to evaluate P at n points, which seems to disqualify this approach altogether, as the naïve way of calculating the coefficients of the product polynomial $S(x)$ of two polynomials P and Q, both of degree $n - 1$, also requires only $O(n^2)$ time, and there is a priori nothing to be gained from switching representation. But there is a degree of freedom we have not exploited so far: a polynomial of degree $n - 1$ can be determined by *any* set of n points; by cleverly choosing a very specific set of such points, the overall complexity of evaluating the n values may be reduced.

8.3   The Fourier transform

Our problem is now to evaluate the polynomial $P(x)$ at n points x_0, x_1, $\ldots, x_{n-1}$, where we assume, without loss of generality as explained before, that n is a power of 2. To reduce the overall work, we would like a part of the points to be as similar as possible to others. They cannot be chosen identical, since different points are required, but some points could be chosen as the *opposite* of others, say

$$x_{\frac{n}{2}+j} = -x_j \qquad \text{for} \qquad 0 \le j < \tfrac{n}{2}. \tag{8.2}$$

Thus only $\frac{n}{2}$ points $x_0, \ldots, x_{\frac{n}{2}-1}$ have to be chosen, and the $\frac{n}{2}$ other points are their opposites. To see why this is a good choice, consider the values of P for two opposing arguments x and $-x$. It will be convenient to separate the even indices from the odd ones, because $x^j = (-x)^j$ if j is even, and $x^j = -(-x)^j$ if j is odd.

$$P(x) = \sum_{j=0}^{n-1} a_j x^j = \sum_{j=0}^{\frac{n}{2}-1} a_{2j}\, x^{2j} + \sum_{j=0}^{\frac{n}{2}-1} a_{2j+1}\, x^{2j+1}$$

$$= \sum_{j=0}^{\frac{n}{2}-1} a_{2j}\, (x^2)^j + x \sum_{j=0}^{\frac{n}{2}-1} a_{2j+1}\, (x^2)^j$$

$$= PE(x^2) + x\, PO(x^2) \tag{8.3}$$

We have thus expressed the polynomial $P(x)$ of order $n-1$ as some combination of two polynomials $PE(y)$ and $PO(y)$, both of order $\frac{n}{2} - 1$, the first being built from the $\frac{n}{2}$ even indexed coefficients of P, and the second from the $\frac{n}{2}$ odd indexed coefficients, and where these polynomials are applied on $y = x^2$. A similar derivation yields

$$P(-x) = PE(x^2) - x\, PO(x^2). \tag{8.4}$$

Note that $PE(x^2)$ and $PO(x^2)$ need to be evaluated only once to give both $P(x)$ and $P(-x)$ after multiplying the latter by x and adding or subtracting, respectively. Summarizing, the polynomial $P(x)$ of degree $n-1$ can be evaluated at the n points $x_0, \ldots, x_{\frac{n}{2}-1}, -x_0, \ldots, -x_{\frac{n}{2}-1}$ by evaluating two polynomials $PE(y)$ and $PO(y)$, both of degree $\frac{n}{2} - 1$, at $\frac{n}{2}$ points $x_0^2, \ldots, x_{\frac{n}{2}-1}^2$, and then performing $O(n)$ additional operations.

This is reminiscent of the Divide and Conquer paradigm we saw in Chapter 1, and if we denote by $T(n)$ the time needed to evaluate an $n-1$-degree polynomial at n such points, this would give

$$T(n) = 2\, T(\tfrac{n}{2}) + O(n), \tag{8.5}$$

the solution of which is $T(n) = O(n \log n)$, as we have seen, e.g., for merge-sort.

There is, however, a little technical problem. The main idea of the above attempt to lower the complexity is to substitute the set of n independent points at which $P(x)$ has to be evaluated by a set of $\frac{n}{2}$ independent points and their opposites. There is a hidden assumption in Eq. (8.5) that this substitution can be continued recursively. The recursive equivalent of Eq. (8.3) is

$$PE(x^2) = PEE(x^4) + x^2 \, PEO(x^4).$$

This would mean that instead of choosing the $\frac{n}{2}$ points $y_0, \ldots, y_{\frac{n}{2}-1}$, only $\frac{n}{4}$ of them need to be selected, say $y_0, \ldots, y_{\frac{n}{4}-1}$, and the others can again be chosen as the opposites, as we did in Eq. (8.2), by setting

$$y_{\frac{n}{4}+j} = -y_j \qquad \text{for} \qquad 0 \le j < \tfrac{n}{4}. \tag{8.6}$$

But all the values of y are squares, more precisely, we defined

$$y_j = x_j^2 \qquad \text{for} \qquad 0 \le j < \tfrac{n}{2}, \tag{8.7}$$

and combining equations (8.6) with (8.7) yields

$$x_{\frac{n}{4}+j}^2 = -x_j^2 \qquad \text{for} \qquad 0 \le j < \tfrac{n}{4}. \tag{8.8}$$

This seems to be impossible, as we have been told that squares are always positive! Indeed, the last equation has no solution in $\mathbb{R}$, the set of real numbers, but it has a solution if we extend our attention to $\mathbb{C}$, the set of *complex* numbers.

Background concept: Complex numbers

Define the set $\mathbb{C}$ of complex numbers as the set of 2×2 matrices of the form $\begin{pmatrix} a & -b \\ b & a \end{pmatrix}$, where $a, b \in \mathbb{R}$ are real numbers. The set is clearly closed under addition of matrices, but also under their multiplication, since

$$\begin{pmatrix} a & -b \\ b & a \end{pmatrix} \begin{pmatrix} c & -d \\ d & c \end{pmatrix} = \begin{pmatrix} ac - bd & -ad - bc \\ bc + ad & -bd + ac \end{pmatrix} = \begin{pmatrix} e & -f \\ f & e \end{pmatrix}$$

where $e = ac - bd$ and $f = bc + ad$. In particular, when $d = 0$, the subset of matrices of the form $\begin{pmatrix} c & 0 \\ 0 & c \end{pmatrix}$ have the property

$$\begin{pmatrix} c & 0 \\ 0 & c \end{pmatrix} \begin{pmatrix} a & -b \\ b & a \end{pmatrix} = \begin{pmatrix} a & -b \\ b & a \end{pmatrix} \begin{pmatrix} c & 0 \\ 0 & c \end{pmatrix} = \begin{pmatrix} ac & -bc \\ bc & ac \end{pmatrix} = c \begin{pmatrix} a & -b \\ b & a \end{pmatrix},$$

so that, as far as multiplication is concerned, the matrix $\begin{pmatrix} c & 0 \\ 0 & c \end{pmatrix}$ can be identified with the real number c. Another particular case is $c = 0$ and $d = 1$. The result of squaring this particular matrix is

$$\begin{pmatrix} 0 & -1 \\ 1 & 0 \end{pmatrix} \begin{pmatrix} 0 & -1 \\ 1 & 0 \end{pmatrix} = \begin{pmatrix} -1 & 0 \\ 0 & -1 \end{pmatrix},$$

and the right-hand-side matrix has been identified with the number -1. We shall denote[1] the matrix $\begin{pmatrix} 0 & -1 \\ 1 & 0 \end{pmatrix}$ by i. The last equation can thus be rewritten as

$$i^2 = -1.$$

In fact, this allows us to represent any complex number as

$$\begin{pmatrix} a & -b \\ b & a \end{pmatrix} = a \begin{pmatrix} 1 & 0 \\ 0 & 1 \end{pmatrix} + b \begin{pmatrix} 0 & -1 \\ 1 & 0 \end{pmatrix} = a + b\, i.$$

Any polynomial of degree n has n roots in $\mathbb{C}$, and in particular, equations like (8.8) can be solved.

Returning to our attempt to solve Eq. (8.8), we get

$$x_{\frac{n}{4}+j} = i\, x_j \qquad \text{for} \qquad 0 \le j < \tfrac{n}{4}.$$

In particular, if $n = 4$, only a single element x_0 needs to be selected, and if we choose $x_0 = 1$, the set of four points on which the polynomial should be evaluated is $\{1, i, -1, -i\} = \{i^k \mid 0 \le k < n\}$. If $n > 4$, the next step in the recursion allows us to reduce the choice from $\frac{n}{4}$ to $\frac{n}{8}$ elements, according to

$$x_{\frac{n}{8}+j} = -x_j^4 \qquad \text{for} \qquad 0 \le j < \tfrac{n}{8}.$$

This can be obtained if $x_{\frac{n}{8}+j}^2 = i\, x_j^2$ and finally $x_{\frac{n}{8}+j} = \sqrt{i}\, x_j$ for $0 \le j < \frac{n}{8}$.

Denoting one of the square roots of i by z, for instance $z = \frac{\sqrt{2}}{2}(1 + i)$, the set of eight points on which the polynomial is evaluated for $n = 8$ could be $\{1, z, i, iz, -1, -z, -i, -iz\} = \{z^k \mid 0 \le k < n\}$.

---

[1] The widespread use of the square root operator to define i as $\sqrt{-1}$ has been deliberately avoided here. While it could be justified to use this notation as a shortcut, using it as an operator is a mathematical aberration and could have disastrous implications, as shown in Exercise 8.1. In particular, $\sqrt{x}$ denotes a positive number, as in, e.g., $\sqrt{x^2} = |x|$, while positivity is not defined for complex numbers, and neither i nor $-i$ could be qualified as such.

Generalizing, we may continue with this process of halving the set of chosen points until we are left with a single one, which will be set to 1. How should the other points be selected? Depending on the degree, we tried to find a point w that is the "opposite" of 1 in a certain sense. For $n = 2, 4, 8$, such points were -1, i and $z = \frac{\sqrt{2}}{2}(1+i)$, respectively. What they have in common is that

$$w^n = 1 \qquad \text{and for} \quad 0 < j < n \qquad w^j \neq 1.$$

A complex number satisfying these two constraints will be called a *primitive n-th root of unity*. Notice that $(-1)^4 = 1$, so that -1 is also a 4th root of 1, but it is not a primitive one, because the second condition is not met, since also $(-1)^2 = 1$.

The conclusion is that in order to be able to evaluate a polynomial $P(x)$ at n points in time $O(n \log n)$ rather than $\Omega(n^2)$, it will be advantageous to choose the evaluation points as the n powers of a primitive n-th root of 1, w:

$$1 = w^0, w, w^2, \ldots, w^{n-1}.$$

This choice satisfies the conditions mentioned earlier. For example, for $0 \leq j < \frac{n}{2}$, we have

$$x_{\frac{n}{2}+j} = w^{\frac{n}{2}+j} = w^{\frac{n}{2}} w^j = -w^j = -x_j,$$

as in Eq. (8.2), since $(w^{\frac{n}{2}})^2 = w^n = 1$, so $w^{\frac{n}{2}}$ is either 1 or -1, but it cannot be 1, because w is a primitive root. The reduction is also consistent with the Divide and Conquer rules: the problem of evaluating a polynomial P of order $n-1$ at n points $1, w, \ldots, w^{n-1}$, which are powers of w, a primitive n-th root of 1, is reduced to two evaluations of polynomials PE and PO of degree $\frac{n}{2} - 1$ at $\frac{n}{2}$ points $1, w^2, w^4, \ldots, w^{n-2}$, which are powers of w^2, a primitive $\frac{n}{2}$th root of 1, see Exercise 8.2. Therefore the recurrence (8.5) applies and the complexity of the process is $O(n \log n)$.

The function f of Figure 8.3, transforming a vector of n coefficients $\mathbf{a} = [a_0, a_1, \ldots, a_{n-1}]$ into a vector of n values $\mathbf{P} = [P(1), P(w), \ldots, P(w^{n-1})]$, where w is a primitive n-th root of 1 and $P(x)$ is the polynomial defined by the coefficient vector $\mathbf{a}$, is called the *(discrete) Fourier transform*. Alternatively, the vector $\mathbf{P}$ itself is often called the discrete Fourier transform of the vector $\mathbf{a}$. The Divide and Conquer algorithm evaluating the Fourier transform in time $O(n \log n)$ is known as the *fast Fourier transform*, FFT for short.

The formal pseudo-code for FFT is given in ALGORITHM 8.1. The input consists of an integer n, which we assumed to be a power of 2, a vector of n

coefficients $[a_0, a_1, \ldots, a_{n-1}]$ defining a polynomial $P(x) = \sum_{j=0}^{n-1} a_j x^j$, and ω, a primitive n-th root of 1. The function returns a vector $P[0], P[1], \ldots, P[n-1]$, where $P[j]$ stores the value $P(\omega^j)$, for $0 \leq j < n$.

$\underline{\mathsf{FFT}(n, a_0, a_1, \ldots, a_{n-1}, \omega)}$
    if $n = 1$ then
        $\mathsf{P}[0] \longleftarrow a_0$
    else
        $\mathsf{PE} \longleftarrow \mathsf{FFT}(\frac{n}{2}, a_0, a_2, \ldots, a_{n-2}, \omega^2)$
        $\mathsf{PO} \longleftarrow \mathsf{FFT}(\frac{n}{2}, a_1, a_3, \ldots, a_{n-1}, \omega^2)$
        $x \longleftarrow 1$
        for $j \leftarrow 0$ to $\frac{n}{2} - 1$ do
            $\mathsf{P}[j] \qquad\quad \longleftarrow \mathsf{PE}[j] + x \times \mathsf{PO}[j]$
            $\mathsf{P}[j + \frac{n}{2}] \longleftarrow \mathsf{PE}[j] - x \times \mathsf{PO}[j]$
            $x \longleftarrow x \times \omega$
    return P

ALGORITHM 8.1: *Fast Fourier transform.*

The vectors PE and PO are locally defined and of size $\frac{n}{2}$, and serve to store the results of the recursive calls. According to Eqs. (8.3) and (8.4), the first recursion is with the coefficients having even indices and the second with the others, and the value of the local variable x in the j-th iteration is ω^j, for $0 \leq j < \frac{n}{2}$.

How can one find a primitive n-th root of 1? The following properties of complex numbers will be helpful.

Background concept: Polar coordinates of complex numbers

Since the complex number $z = a + ib$ can be identified with the pair of real numbers (a, b), it can also be seen as a point in a two-dimensional plane, as shown in Figure 8.4. This suggests an additional way to represent z: instead of using the abscissa a and ordinate b, the point may be defined by its distance r from the origin $(0, 0)$ and the angle θ formed by the segment $[(0, 0) : z]$ with the x axis. The representation of z as (r, θ) is called *polar coordinates* of z, and one can pass from one representation to the other by

$$a = r \ \cos(\theta) \qquad\qquad b = r \ \sin(\theta)$$

$$r = \sqrt{a^2 + b^2} \qquad\qquad \theta = \arctan\left(\frac{b}{a}\right)$$

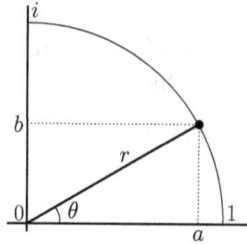

FIGURE 8.4: *Polar coordinates of a complex number.*

A major advantage of the polar representation is that multiplication is easy to perform. Indeed,

$$(r_1, \theta_1)(r_2, \theta_2) = (r_1 r_2, \theta_1 + \theta_2),$$

and thus $(r, \theta)^n = (r^n, n\theta)$. To find the n-th root of $z = 1$, whose polar coordinates are $(1, 0)$, we get the set of equations

$$r^n = 1 \qquad\qquad n\theta = 0,$$

from which one derives $r = 1$, because r is a real number, and $\theta = \frac{2\pi j}{n}$, for $0 \le j < n$. For $j = 1$, the complex numbers $(1, \frac{2\pi}{n})$ are primitive n-th roots of 1, and all of their powers are on the unit circle.

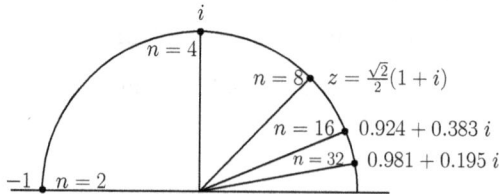

FIGURE 8.5: *Primitive n-th roots of unity.*

Figure 8.5 shows these roots for $n = 2^k$, with $1 \le k \le 5$, corresponding to the angles $\theta = \pi, \frac{\pi}{2}, \frac{\pi}{4}, \frac{\pi}{8}, \frac{\pi}{16}$, respectively.

8.4 The inverse Fourier transform

We have dealt so far with the problem of evaluating a polynomial of degree $n - 1$ at n points in time $O(n \log n)$. In terms of Figure 8.3, a function f has been presented, transforming a vector $\mathbf{a}$ of coefficients into a vector $\mathbf{P}$ of values. But this is just a part of what is needed for the polynomial multiplication, and ultimately for the pattern matching problem stated as the motivation for this chapter. There is also a need to perform the *inverse* transformation f^{-1} efficiently, that is, if the vector $\mathbf{P}$ is given, how could one derive the corresponding vector of coefficients $\mathbf{a}$?

It will be convenient to rewrite the set of equalities $P(x) = \sum_{j=0}^{n-1} a_j x^j$ for $x = 1, \omega, \omega^2, \ldots, \omega^{n-1}$ in the form of operations involving matrices:

$$
\begin{pmatrix} P[0] \\ P[\omega] \\ P[\omega^2] \\ \vdots \\ P[\omega^{n-1}] \end{pmatrix} = \begin{pmatrix} 1 & 1 & 1 & \cdots & 1 \\ 1 & \omega & \omega^2 & \cdots & \omega^{n-1} \\ 1 & \omega^2 & \omega^{2 \cdot 2} & \cdots & \omega^{2(n-1)} \\ 1 & \omega^3 & \omega^{3 \cdot 2} & \cdots & \omega^{3(n-1)} \\ \vdots & \vdots & \vdots & \ddots & \vdots \\ 1 & \omega^{n-1} & \omega^{(n-1) \cdot 2} & \cdots & \omega^{(n-1)(n-1)} \end{pmatrix} \begin{pmatrix} a_0 \\ a_1 \\ a_2 \\ \vdots \\ a_{n-1} \end{pmatrix}
$$

Remark that the $n \times n$ matrix of the right-hand side has a regular form: the element in row k and column ℓ is $\omega^{k\ell}$. It is called a *Vandermonde* matrix and we shall denote it by $\mathbf{V}_\omega$, showing that it depends on a parameter ω. The equation can thus be rewritten as

$$\mathbf{P} = \mathbf{V}_\omega \, \mathbf{a},$$

where $\mathbf{a}$ is supposed to be given and $\mathbf{P}$ has been evaluated. For the inverse problem, we assume now that $\mathbf{P}$ is given and that $\mathbf{a}$ should be derived from it. If the matrix $\mathbf{V}_\omega$ is reversible, the solution would be

$$\mathbf{a} = \mathbf{V}_\omega^{-1} \mathbf{P}. \tag{8.9}$$

Fortunately, it turns out that not only does the matrix $\mathbf{V}_\omega$ have an inverse, but that moreover, this inverse is quite similar to $\mathbf{V}_\omega$ itself, so that, ultimately, exactly the same procedure given in ALGORITHM 8.1, with minor adaptations, can be applied also to the inverse problem.

Let us evaluate the result of multiplying a Vandermonde matrix with parameter ω by another such matrix with inverse parameter.

$$
\mathbf{V}_\omega \, \mathbf{V}_{\frac{1}{\omega}} = \left(\omega^{k\ell} \right) \left(\omega^{-k\ell} \right) = \left(\sum_{m=0}^{n-1} \omega^{mk} \omega^{-\ell m} \right),
$$

the element in row k and column ℓ of the resulting matrix is therefore

$$\sum_{m=0}^{n-1} \omega^{mk} \omega^{-\ell m} = \sum_{m=0}^{n-1} \left(\omega^{(k-\ell)} \right)^m . \tag{8.10}$$

If $k = \ell$, this sum is equal to n. Otherwise, set $s = k - \ell$, then we know that $\omega^{(k-\ell)} = \omega^s \neq 1$, because ω is primitive. The elements in Eq. (8.10) are thus equal to

$$\sum_{m=0}^{n-1} \left(\omega^s \right)^m = \frac{(\omega^s)^n - 1}{\omega^s - 1} = \frac{(\omega^n)^s - 1}{\omega^s - 1} = 0,$$

because ω is an n-th root of 1. We conclude that the resulting matrix has the value n at each position of its main diagonal, and zeros elsewhere, so it suffices to divide by n to get the unity matrix. The inverse matrix is therefore

$$\mathbf{V}_\omega^{-1} = \tfrac{1}{n} \mathbf{V}_{\frac{1}{\omega}} . \tag{8.11}$$

and the solution to the interpolation problem of Eq. (8.9) is

$$\mathbf{a} = \tfrac{1}{n} \mathbf{V}_{\frac{1}{\omega}} \mathbf{P}.$$

The same algorithm can thus be used for both the interpolation and the evaluation, because if ω is a primitive n-th root of 1, then so is $\frac{1}{\omega}$, see Exercise 8.2. Of course, the inverse matrix as defined in Eq. (8.11) will never be calculated, because the division by n of all its elements alone takes already time $\Omega(n^2)$; we rather derive $\mathbf{V}_{\frac{1}{\omega}} \mathbf{P}$ using ALGORITHM 8.1 in time $O(n \log n)$, and then divide the resulting vector by n to get $\mathbf{a}$ in time $O(n)$.

8.5   Concluding remarks

8.5.1   *Summary of fast polynomial multiplication*

The algorithm we have seen is a classical example of the fact that the shortest and straightforward way is not always the fastest and most efficient one. The principle behind the idea of applying FFT to multiply polynomials is quite similar to the "beat-the-red-light" algorithm that I have learned years ago from my then ten-year-old son. We used to live close to an intersection like that in Figure 8.6, the crossing of which was regulated by a complicated system of several traffic lights, controlling left turns and straight traversals separately.

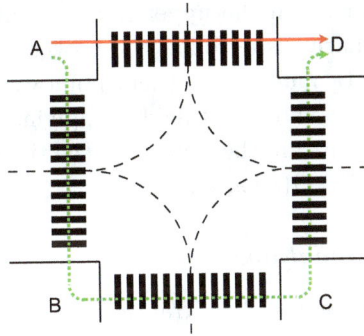

FIGURE 8.6: *Fast crossing of an intersection.*

The destination was D, but very often, just when we approached A, the light turned red. While I patiently waited at A, I noticed that my son, who obviously knew more about the timing dependencies between the different lights, immediately crossed over to B, which he reached just in time to turn left to C and then again to D, where he usually arrived while I was still waiting at A for my own green light. Though his route was longer, it was nonetheless faster.

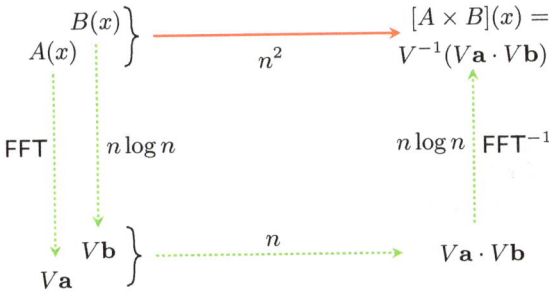

FIGURE 8.7: *Fast multiplication of polynomials.*

The analogy to fast multiplication of polynomials, as summarized in the diagram of Figure 8.7, should be obvious. Given are two polynomials $A(x)$ and $B(x)$, both of degree $n - 1$, defined, respectively, by the vectors of coefficients $\mathbf{a} = [a_0, a_1, \ldots, a_{n-1}, 0, \ldots, 0]$ and $\mathbf{b} = [b_0, b_1, \ldots, b_{n-1}, 0, \ldots, 0]$; the vectors have already been extended to length $2n - 1$ to match the length of their product. The trivial way to multiply and get $[A \times B](x)$

is given by the solid arrow in the upper line and takes time $O(n^2)$. The seemingly longer alternative is to follow the broken line arrows, using FFT to evaluate $V\mathbf{a}$ and $V\mathbf{b}$, multiplying them pointwise yielding a vector of $2n - 1$ values, and then interpolating by applying the inverse function $V^{-1}(V\mathbf{a} \cdot V\mathbf{b})$. The results are the same, but the time has been reduced to $O(n \log n + n + n \log n) = O(n \log n)$.

8.5.2   *Using approximations*

Some care is required when handling irrational and often even decimal numbers, because the rounding errors in our calculations could accumulate. The problem may be aggravated by our lack of awareness that there has been any rounding at all. Everybody knows that $\frac{1}{3}$ cannot be accurately represented in our finite precision computers, but we settle for approximations like 0.33 or 0.333333, depending on the intended application. When one evaluates a polynomial $P(x)$ with integer coefficients at integer points x, all the operations could be performed with integer arithmetic, so one does not expect any rounding errors. Yet, using FFT, say with $n = 8$, we saw that one of the primitive roots was $z = \frac{\sqrt{2}}{2} + \frac{\sqrt{2}}{2}i$, which forces us to process irrational numbers.

Here is a more subtle example of hidden approximations. Consider the following loop:

```
x ⟵ 0.2
for i ⟵ 1 to 100 do
    x ⟵ x + x
    if x ≥ 1 then
        x ⟵ x − 1
print x
```

We may convince ourselves that the value of x at the end of the iteration forms a periodic sequence $0.4, 0.8, 0.6, 0.2, 0.4, 0.8, \ldots$. Since the period is 4, we would expect the value of x printed after exiting the loop to be 0.2. This might be true mathematically, but not computationally. If you try to program it, you will probably be surprised to get a different output. So what went wrong?

The secret here is that while 0.2 is a decimal number, in the sense that it has a finite representation in base 10, this is not the case for a binary representation, to which our computers convert all the numbers. Indeed, in binary, 0.2 is the infinite periodic string $A = 0.001100110011\cdots$. This can be seen by multiplying A by 16, that is, shifting it four places to the left, to get $11.001100110011\cdots$. Subtracting 3, we get back to A, so that $16A - 3 = A$, showing that $A = 0.2$. The initial assignment of 0.2 to x is therefore already an approximation as far as the actually stored number is concerned, unless infinite precision has been used. The loop essentially shifts the input by one bit to the left and discards any 1-bit to the left of the binary point, getting ultimately to 0.

It has been shown that if the polynomials are of order n, a precision of only $\log n$ bits is needed to get the correct results for $V^{-1}(V\mathbf{a}\cdot V\mathbf{b})$. Since modern implementations use at least 32 bits for a computer word, this allows the processing of polynomials of degree in the billions, more than we will probably ever need.

8.5.3 *Summary of extended pattern matching*

We can now return to the motivation for introducing FFT, namely the extended pattern matching problem, presented in Section 8.1. We shall restrict the discussion to a binary alphabet $\Sigma = \{0, 1\}$, but there is no loss of generality, because if the characters of the text and the pattern are drawn from a larger alphabet, for example if $\Sigma = \text{ASCII}$, one can always translate all the strings into their binary equivalents and perform the search with the additional constraint that the matching positions should be aligned on byte boundaries.

Recall that to find the number of mismatches at each position, the text T and the reversed pattern P^R are "multiplied" in the sense defined in Eq. (8.1). The multiplication is then accelerated by means of the FFT algorithm. This, however, raises a technical problem. The correctness of FFT depends on the fact that the set of numbers over which the operations are performed has a structure called a *field*. The set of complex numbers $\mathbb{C}$, defined as above in the form of matrices and with the according addition and multiplication operators, is indeed a field — but this is not true for the special kind of multiplication in Eq. (8.1) we needed to solve the extended pattern matching problem.

To overcome this difficulty, note that for a binary alphabet, the operation defined in Eq. (8.1) is just the Boolean XOR, and it can thus be rewritten as

$$a \text{ XOR } b = a \cdot \bar{b} + \bar{a} \cdot b,$$

where $\bar{x}$ is the binary complement of x, and the operators $\cdot$ and $+$ stand for Boolean AND and OR. The multiplication of text and pattern can thus be replaced by

$$T \cdot \overline{P^R} + \overline{T} \cdot P^R, \tag{8.12}$$

and since in the binary case, logical AND and regular multiplication are identical, the operations in Eq. (8.12) can be performed with two applications of FFT over the complex field according to the standard definition of multiplication.

A last comment concerns the time complexity of the extended matching. One should bear in mind that the length n of the text T is usually of a larger order of magnitude than the length m of the pattern P. The problem of counting the mismatches of P with substrings of T has been transformed into a problem of multiplying polynomials, but the FFT algorithm uses quadratic matrices, which correspond to the multiplication of polynomials of the same degree. This was justified by the fact that one can always pad the shorter polynomial with zero-coefficients, but the resulting algorithm had complexity $O(n \log n)$. Since $\log n$ might be larger than m, it is not clear whether this is preferable to the naïve $O(mn)$ procedure.

Actually, the complexity can be lowered to $O(n \log m)$ by the following technique, which breaks the text into parts handled separately and then assembles the results. Partition the text $T = t_0 t_1 \cdots t_{n-1}$ into $\frac{n}{m}$ substrings T_j, each of length $2m - 1$ (padding the end of the text with blanks, if necessary), so that the first $m - 1$ bits of T_{j+1} overlap with the last $m - 1$ bits of T_j, for $0 \leq j < \frac{n}{m} - 1$. In other words, define

$$T_j = t_{jm} t_{jm+1} \cdots t_{jm+2m-2} \qquad \text{for} \qquad 0 \leq j < \tfrac{n}{m}.$$

The FFT procedure can then be applied on each $T_j \cdot P^R$, each application yielding a piece of m elements of the vector D defined in Section 8.1 counting the mismatches. The overall complexity is therefore $O(m \log m)$ to evaluate $T_j \cdot P^R$ times the number of substrings $\frac{n}{m}$, which gives $O(n \log m)$.

8.6   Exercises

8.1 What is wrong with the following "proof" that all the integers are equal?

$$i = i$$
$$\sqrt{-1} = \sqrt{-1}$$
$$\sqrt{\frac{-1}{1}} = \sqrt{\frac{1}{-1}}$$
$$\frac{\sqrt{-1}}{\sqrt{1}} = \frac{\sqrt{1}}{\sqrt{-1}}$$
$$\sqrt{-1} \times \sqrt{-1} = \sqrt{1} \times \sqrt{1}$$
$$-1 = 1$$

from which the claim follows, QED.

8.2 Show that if ω is a primitive n-th root of 1, then $\frac{1}{\omega}$ is also a primitive n-th root of 1, and ω^2 is a primitive $\frac{n}{2}$th root of 1. What about $-\omega$?

8.3 What is the Fourier Transform of the polynomial $5x^4 + 4x^3 + 3x^2 + 2x + 1$? Simulate running ALGORITHM 8.1 on the corresponding input and show the result of all the recursive calls.

8.4 The Fourier Transform of a polynomial $P(x)$ is $(3, 1-6i, -1, 1+6i)$. What is $P(x)$ if we assume that the primitive 4th root of 1 used was i?

8.5 Consider the following alternative to the FFT algorithm. Instead of splitting the polynomial $P(x) = \sum_{j=0} a_j x^j$ into two parts according to the odd or even powers, assume now that n is a power of 4 and split $P(x)$ into four parts, using

$$P_0(y) = a_0 + a_4 y + a_8 y^2 + a_{12} y^3 + \cdots + a_{n-4} y^{n/4-1}$$

$$P_1(y) = a_1 + a_5 y + a_9 y^2 + a_{13} y^3 + \cdots + a_{n-3} y^{n/4-1}$$

$$P_2(y) = a_2 + a_6 y + a_{10} y^2 + a_{14} y^3 + \cdots + a_{n-2} y^{n/4-1}$$

$$P_3(y) = a_3 + a_7 y + a_{11} y^2 + a_{15} y^3 + \cdots + a_{n-1} y^{n/4-1}$$

(a) Write $P(x)$ as a function of P_0, P_1, P_2, P_3.

(b) Show how to get $P(x)$ and $P(-x)$ efficiently.

(c) In the original FFT algorithm, $P(x)$ is evaluated at n special points $1, \omega, \omega^2, \ldots, \omega^{n-1}$, where ω is a primitive n-th root of 1. Which n points should be chosen in the current case?

(d) What is the complexity of this alternative FFT procedure?

Chapter 9

Cryptography

This chapter deals with yet another transformation of an input text, in addition to the data compression methods seen in Chapter 6. Actually, the technical goal of *Cryptography*, the subject of the current chapter, is also to remove as much *redundancy* as possible, but the motivation is different, as already mentioned. While compression strives to reduce redundancy in order to minimize the required space, the objective of encrypting is to hide information from whoever it is not intended for.

Methods for encoding a message so that it may be communicated safely only to an intended partner, but without being intelligible to a malevolent or even occasional eavesdropper, are probably as old as the civilized world. Here is an early written example, in which the command (*come* or *go*) has been "encoded" by some harmless sounding instructions of collecting arrows that are seemingly unconnected:

If I say, Behold, the arrows are on this side of thee; then come thou —
But if I say, Behold, the arrows are beyond thee; go thy way

1 Samuel 20:21–22, KJV English Bible

There is no need to insist on the usefulness and even indispensability of encryption methods to many aspects of modern life. Electronic commerce and banking and many if not most transactions we wish to perform daily from our computers and mobile devices would be jeopardized if it were not possible to keep the transmitted data confidential.

Historically, important messages have been signed and appropriately sealed and were then transmitted via trustworthy dispatch riders, which gave rise to many masterpieces of the world literature in the case of betrayal. This cumbersome setup has meanwhile been replaced by mathematical functions, and modern cryptography is the well-established field of the study of such functions and their properties.

In a general setting, two complementing encoding and decoding functions $\mathcal{E}_K$ and $\mathcal{D}_K = \mathcal{E}_K^{-1}$ are given, both depending on some secret key K that has been shared by the encoder and decoder in advance. A message M, called the *cleartext*, is encrypted by applying

$$C = \mathcal{E}_K(M),$$

thereby producing the *ciphertext* C, which, at least apparently, does not convey any information. Nonetheless, whoever is in possession of the secret key K may apply the inverse function for decoding and get

$$\mathcal{D}_K(C) = \mathcal{D}_K(\mathcal{E}_K(M)) = M,$$

that is, reproduce the original message M.

One of the oldest examples is the so-called *Caesar's cipher*: rewriting the alphabet $\{A, B, C, \ldots, Z\}$ as $\{c_1, c_2, \ldots, c_{26}\}$ so that we can identify a letter with its index; the secret key K is just one of the possible indices, $1 \leq K < 26$. We then define

$$\mathcal{E}_K(c_i) = c_j, \qquad \text{where} \qquad j = \big((i + K - 1) \bmod 26\big) + 1,$$

that is, in the encoding, each character is shifted cyclically K positions forward in the alphabet. For example, if the secret key is $K = 10$, then the encoding of the cleartext

$$\text{T O P S E C R E T M E S S A G E}$$

would be the ciphertext

$$\text{D Y Z C O M B O D W O C C K Q O.} \tag{9.1}$$

If the fact that a Caesar's cipher has been used is known, an attempt to decode the cleartext could be based on trying all the possible shifts, as there are only 25 options: for natural languages, only one of these alternatives will produce a reasonable cleartext. Moreover, not only would then this specific message be known, but also the secret key K, which enables the decoding of all subsequent texts enciphered in the same way. Guessing or taking some action to unveil the secret key is called *breaking* the code, and this is one of the objectives of the enemy.

A generalization of Caesar's cipher is known as the more general class of *substitution ciphers*, relying on some permutation π of the set $\{1, 2, \ldots, 26\}$. The secret key is the permutation π itself or its index within the ordered sequence of all 26! permutations, and the encoding function is

$$\mathcal{E}_K(c_i) = c_{\pi(i)},$$

that is, each character is replaced by the one having the same index in this permutation. Contrary to Caesar's method, the secret key cannot be found by exhaustive search, as the number of possibilities is $26! \simeq 4 \cdot 10^{26}$. Nonetheless, substitution ciphers are not secure either, since one could derive enough information from the frequencies of the occurrences of the characters in a long enough text to ultimately break the code. For example, even for the short ciphertext in (9.1), the most frequent character is O, so one may reasonably guess that it stands for the most frequent character in English, which is E.

So are there at all any secure encoding methods? The answer is yes, for example, the use of a *one-time pad*. The key is then a (generally randomly generated) bit-vector of the same length as the cleartext M which has to be created afresh for each new message, and the ciphertext C is obtained by XORing M with the key, bit per bit, that is,

$$C = \mathcal{E}_K(M) = M \text{ XOR } K.$$

Decoding is then just an additional application of the same procedure:

$$\mathcal{D}_K(C) = C \text{ XOR } K = M.$$

Figure 9.1 illustrates a one-time pad on the cleartext $M = $ TOPSECRET, shown in the second line in its ASCII form. The first bits of the binary expansion of π served as a "random" key K and are shown in the third line. The resulting ciphertext C appears in the fourth row of the figure, and would yield the string ¥@è±d+ÉqÉ if interpreted as an ASCII encoding (using code page 437 for the extended set).

	T	O	P	S	E	C	R	E	T
M	01010100	01001111	01010000	01010011	01000101	01000011	01010010	01000101	01010100
K	11001001	00001111	11011010	10100010	00100001	01101000	11000010	00110100	11000100
C	10011101	01000000	10001010	11110001	01100100	00101011	10010000	01110001	10010000
	¥	@	è	±	d	+	É	q	É

Figure 9.1: *Example of the use of a one-time pad.*

While one-time pads may provide a theoretical solution to our security problem, they are not really practical. They have to be as long as the message to be encoded itself, and need to be generated from scratch for every new instance, even when the same sender and receiver are involved.

In addition, if in a community of n parties, each one wishes to communicate safely with the others, then $\theta(n^2)$ one-time pads are needed.

In fact, it seems that there has only been a shift in the problem. Instead of securely transmitting a message M, we first need to share somehow a secret key K, but the transmission of this key seems just as difficult as the transmission of the original M itself. There is, however, a difference: the message M is a very specific one, chosen by the sender, whereas the exact form of the key K is not important; only the fact that it is shared by the two sides is.

9.1   Public key cryptography

These thoughts, as well as the increased use of electronic and wireless data transfer, which can easily be intercepted, led to the advent of a new paradigm known as *public key cryptography*, in which the secrecy for some elements could be relaxed, without compromising the overall security of the communication systems. It relies on the fact that the difficulty of applying a function and that of finding its inverse are not necessarily symmetric. The very fact of this asymmetry might come as a surprise, as it does not hold for many of the functions we use. For instance, if $y = f(x) = 1 + \frac{1}{2}e^{\frac{3}{x^4-5}}$, one can easily derive the inverse $x = f^{-1}(y) = \sqrt[4]{\frac{3}{\ln 2(y-1)} + 5}$. Actually, for this example, this is only one of the inverses, since $f(-x) = f(x)$. There are, however, functions for which the inversion is not straightforward, and the design of public key schemes is a clever exploitation of this fact.

Here is an example of such a function. Consider a prime number p, then the set $\mathbb{Z}_p = \{0, 1, \ldots, p-1\}$ of the integers modulo p, on which addition and multiplication are defined, is known to satisfy the axioms of a finite *field*. A *primitive element* of this field is a generator of its multiplicative group, that is, a number $a \in \mathbb{Z}_p$ such that each non-zero element $b \in \mathbb{Z}_p$ can be written as $b = a^i \bmod p$ for some integer i.

Table 9.1: $a^i \bmod 17$. 6 *is a primitive element for* $p = 17$, 2 *and* 4 *are not.*

$a \backslash i$	1	2	3	4	5	6	7	8	9	10	11	12	13	14	15	16
2	2	4	8	16	15	13	9	1	2	4	8	16	15	13	9	1
4	4	16	13	1	4	16	13	1	4	16	13	1	4	16	13	1
6	6	2	12	4	7	8	14	16	11	15	5	13	10	9	3	1
16	16	1	16	1	16	1	16	1	16	1	16	1	16	1	16	1

Table 9.1 shows the powers modulo $p = 17$ for several elements a. We see that neither 2 nor 4 or 16 are primitive elements, because not all the numbers $\{1, 2, \ldots, 16\}$ appear in the corresponding rows. On the other hand, 6 is a primitive element, and so are 3, 5, 7, 10, 11, 12 and 14. The fact that the last column, corresponding to 16, contains only 1s is no coincidence, since $a^{p-1} \equiv 1 \pmod{p}$ for every prime p by Fermat's Little Theorem, as we saw in Section 5.2.1.

If we do not restrict the calculations to integers, then both exponentiation and its inverse, which is a logarithmic function, are easy to calculate. For a given real constant $a > 1$, if

$$y = f(x) = a^x,$$

then

$$x = f^{-1}(y) = \log_a y.$$

Surprisingly, when turning to integer arithmetic modulo some prime number p, this symmetry is broken. Modular exponentiation

$$m = f(n) = a^n \bmod p$$

can be done in time $O(\log n)$, as we have seen in Section 5.2.3, and the inverse function $n = f^{-1}(m)$ of finding the exponent n which yields a given value m exists if a is a primitive element. This inverse is called the *discrete log*, and the best-known algorithms for its evaluation still require exponential time. For example, if one asks for which n do we get that $6^n \bmod 17 = 5$, we may scan the third line of Figure 9.2 until we find the (only) column holding the value 5; this is column 11, and indeed

$$6^{11} = 362,797,056 = 21,341,003 \times 17 + 5.$$

For this example, there is of course no problem of checking all the 16 options, but when p is a large prime such as those we dealt with in Chapter 5, such an exhaustive search for the proper exponent cannot be performed in reasonable time.

A function that is easy to compute but hard to invert is called a *one-way function*. Actually, it is still an open question whether the discrete log is one-way, because there is no *proof* for the difficulty of its evaluation, though it is widely believed to be a hard problem, and it is therefore used in several practical applications.

For one of these applications, due to Diffie and Hellman (1976), we return to the problem mentioned before of sharing some secret key K, which can then serve to securely encode some message M. The main innovation here is that there is no need for all the available information to be hidden. In particular, the encoding function $\mathcal{E}$ can be made public, and this will not compromise the security of the system, because the derivation, without some additional knowledge, of the corresponding decoding function $\mathcal{D} = \mathcal{E}^{-1}$ is not considered to be feasible.

The communicating parties are traditionally called **Alice** and **Bob**, but we shall replace them by the main characters of Molière's *Imaginary Invalid*, **A**rgan and **B**éline. We also assume that some hostile eavesdropper, called **C**léante, has access to all the messages exchanged between Argan and Béline by tapping their telephone lines or by other means.

The algorithm, known as *Diffie–Hellman key exchange*, consists of the following steps:

(a) Argan chooses randomly a large prime number p and a primitive element s, and sends them to Béline. As their communication is not secure, we suspect that Cléante also knows both p and s.

(b) Argan then chooses, again randomly, an integer $a < p$ and keeps it secret. He sends the number $A = s^a \bmod p$ to Béline. Both Béline and Cléante then know A, but not a, which they could get only if they had an efficient algorithm for the discrete log.

(c) Béline now acts symmetrically to Argan in step (2): she chooses randomly an integer $b < p$ and keeps it secret. She sends the number $B = s^b \bmod p$ to Argan. Both Argan and Cléante then know B, but not b.

(d) Argan and Béline now share a common key $D = s^{ab} \bmod p$, which is unknown to Cléante.

Indeed, both Argan and Béline can derive the value of D from what they know, because

$$A^b = (s^a)^b = s^{ab} = (s^b)^a = B^a \quad (\bmod \ p),$$

Argan knows B and a, and Béline knows A and b. On the other hand, Cléante knows A and B, but not a or b, which have both been kept secret and have never been communicated. Note that it is not useful for Cléante to multiply A by B, as this would give $s^{a+b} \bmod p$, not D. Cléante thus needs one of the numbers a and b in order to reveal D, but to get them, he needs the discrete log. Argan and Béline, on the other hand, need only modular exponentiation.

Here is an alternative solution to the same problem. Not that the first solution is not good enough, but the second solution is a nice application of another interesting mathematical fact, known as the *birthday paradox*.

Background concept: The birthday paradox

How likely would you estimate is the event that in a randomly chosen class of about 25 students, there are at least 2 students born on the same date (month and day, not necessarily year)? The intuition of most people leads them to think that this occurs with very low probability, yet it can be shown to be more likely to occur than not. For reasons that are more psychological than mathematical, this fact seems to be so surprising that it has become known as a *paradox*.

To evaluate the probability, let us consider the complementing event of all students in the class having different birthdays. In a more general setting, consider a set of n elements, from which k not necessarily different elements are drawn randomly, and we ask for the probability of the event that all k elements will nevertheless be different.

The probability of the second element being different from the first is $\frac{n-1}{n}$. The probability of the third element being different from the first two drawn elements is $\frac{n-2}{n}$. For the general case, define the event D_i as the fact that the i-th element is different from the $i-1$ preceding ones, thus $\Pr(D_i) = \frac{n-i+1}{n}$. The events $D_2, \ldots, D_k$ should occur simultaneously, and they are mutually independent, therefore the probability of all k elements being different is

$$\left(\frac{n-1}{n}\right)\left(\frac{n-2}{n}\right)\cdots\left(\frac{n-k+1}{n}\right) = \prod_{i=1}^{k-1}\left(1-\frac{i}{n}\right). \qquad (9.2)$$

All of the factors of the product are smaller than 1 and are getting smaller with increasing i. There will thus be an index for which the product will already be smaller than $\frac{1}{2}$. Formally, let us look at the last $\frac{k}{2}$ factors. As each of the left out factors is smaller than 1, the product of just the last $\frac{k}{2}$ factors will be larger.

$$\prod_{i=1}^{k-1}\left(1-\frac{i}{n}\right) < \prod_{i=k/2}^{k-1}\left(1-\frac{i}{n}\right) < \prod_{i=k/2}^{k-1}\left(1-\frac{k/2}{n}\right) = \left(1-\frac{k/2}{n}\right)^{k/2}.$$

If we choose k as $k = 2\sqrt{n}$, we get

$$\left(1-\frac{1}{\sqrt{n}}\right)^{\sqrt{n}} \xrightarrow[n\to\infty]{} \frac{1}{e} = 0.368,$$

so that it suffices to choose randomly just about $\sqrt{n}$ of the n elements to get already collisions with probability at least $\frac{1}{2}$. In particular, if n stands for the 365 days of the year and there are $k = 23$ students in the class, the exact value of the product in (9.2) is 0.493, and the probability of 2 students having the same birthday is 0.507. For a group of 50 students, it would already be 0.970.

As an illustration, among the 27 laureates of the Nobel Prize in the years 2018 and 2019, Frances H. Arnold and John B. Goodenough were both born on July 25.

Here is how to exploit this paradox to let Argan and Béline share some key K, without letting Cléante or anybody else know it. Assume that there is some publicly known set S of fixed integers, and that its size $n = |S|$ is so large that any action to be performed on all of its elements would take too long for our current technology, but that $O(\sqrt{n})$ steps can be done in reasonable time. In addition, we rely on the existence of a pair of encoding and decoding functions $\mathcal{E}$ and $\mathcal{D} = \mathcal{E}^{-1}$, the first being easy and the second hard to evaluate. As mentioned, modular exponentiation and the discrete log are believed to be such a pair of functions. The protocol allowing us to share a secret key is then:

(a) Argan chooses, at random, a subset $A \subset S$ of size $|A| = \sqrt{n}$, $A = \{a_1, a_2, \ldots, a_{\sqrt{n}}\}$, and sends the set $\mathcal{A} = \{\mathcal{E}(a_1), \mathcal{E}(a_2), \ldots, \mathcal{E}(a_{\sqrt{n}})\}$ to Béline.

(b) Béline acts symmetrically: she chooses a random subset $B \subset S$ of the same size $|B| = \sqrt{n}$, $B = \{b_1, b_2, \ldots, b_{\sqrt{n}}\}$, and sends the set $\mathcal{B} = \{\mathcal{E}(b_1), \mathcal{E}(b_2), \ldots, \mathcal{E}(b_{\sqrt{n}})\}$ to Argan.

(c) By the birthday paradox, there are good chances that $\mathcal{A} \cap \mathcal{B} \neq \emptyset$. If not, repeat the process until reaching a common value in $\mathcal{A}$ and $\mathcal{B}$. If the intersection is not a singleton, choose the largest value.

(d) Let $d \in \mathcal{A} \cap \mathcal{B}$ be the chosen value, and let k and ℓ be the indices of d in A and B, respectively, that is $d = \mathcal{E}(a_k) = \mathcal{E}(b_\ell)$. Both Argan and Béline then know the element $a_k = b_\ell$ of S that yielded d, so they can use this element as their shared secret key. Cléante, on the other hand, only knows $d = \mathcal{E}(a_k)$. To get a_k itself, he therefore must either derive the inverse function $\mathcal{E}^{-1}$ or apply $\mathcal{E}$ on up to n elements of $\mathcal{S}$, both of which we assumed to be prohibitively expensive operations.

9.2 The RSA cryptosystem

Sharing a secret key does not completely solve our problem of allowing public key encryption. In particular, it would force every pair of communicating parties to generate its own key, which is not practical when many parties are involved. In fact, we need a one-way function $\mathcal{E}_K$ depending on some secret key K in the sense that having a knowledge of the function alone does not permit the derivation of its inverse $\mathcal{D}_K = \mathcal{E}_K^{-1}$ without knowing K, but with the knowledge of K, the inversion should be easy.

An elegant solution to this problem has been suggested by Rivest, Shamir and Adleman [Rivest et al. (1978)] and is known by the initials of their names as the RSA cryptosystem. It is based on the fact that *factoring*, that is, the decomposition of a composite number into a product of primes, is (still) a difficult task for which no fast algorithm is known. This leads to the idea of not just using some large prime number as modulus in our quest for an appropriate function $\mathcal{E}_K$, but rather the product n of *two* large prime numbers p and q.

To understand the mathematical details, we need a reminder of some algebraic facts.

Background concept: Multiplicative group of integers modulo n

For a given integer n, consider the set of integers that are co-prime to n, that is, those that have no common divisor with n except the trivial divisor 1. We denote the set by $\mathbb{Z}_n^* = \{a \mid \gcd(a, n) = 1\}$. The size of this set is known as *Euler's totient* function and is generally denoted by $\varphi(n) = |\mathbb{Z}_n^*|$. If n is a prime number, then $\varphi(n) = n - 1$, but the set $\mathbb{Z}_n^*$ forms a multiplicative group modulo n also for composite n, which means that

1. it is closed under multiplication modulo n;
2. multiplication is associative, commutative and there exists an identity element;
3. every element has a multiplicative inverse.

As an example, let us choose $n = 21 = 3 \times 7$, for which

$$\mathbb{Z}_{21}^* = \{1, 2, 4, 5, 8, 10, 11, 13, 16, 17, 19, 20\}.$$

Table 9.2 shows the multiplication matrix modulo 21 for $\mathbb{Z}_{21}^*$. As can be seen, the matrix is symmetric (commutativity), the identity element 1 appears in every row and column exactly once, in the grayed cells, and if such a cell is at the intersection of row u with column v, then u and v are mutual inverses.

Table 9.2: *Multiplicative group* $\mathbb{Z}_{21}^*$.

×	1	2	4	5	8	10	11	13	16	17	19	20
1	1	2	4	5	8	10	11	13	16	17	19	20
2	2	4	8	10	16	20	1	5	11	13	17	19
4	4	8	16	20	11	19	2	10	1	5	13	17
5	5	10	20	4	19	8	13	2	17	1	11	16
8	8	16	11	19	1	17	4	20	2	10	5	13
10	10	20	19	8	17	16	5	4	13	2	1	11
11	11	1	2	13	4	5	16	17	8	19	20	10
13	13	5	10	2	20	4	17	1	19	11	16	8
16	16	11	1	17	2	13	8	19	4	20	10	5
17	17	13	5	1	10	2	19	11	20	16	8	4
19	19	17	13	11	5	1	20	16	10	8	4	2
20	20	19	17	16	13	11	10	8	5	4	2	1

There are thus four elements which are their own inverses modulo 21: 1, 8, 13 and 20, and four pairs of elements for which one is the inverse of the other:

$$2 \longleftrightarrow 11 \qquad 4 \longleftrightarrow 16 \qquad 5 \longleftrightarrow 17 \qquad 10 \longleftrightarrow 19.$$

The size of $\mathbb{Z}_n^*$ for composite $n = p \times q$, where p and q are primes, is

$$\varphi(n) = \varphi(p) \times \varphi(q) = (p-1)(q-1),$$

and for our example $\varphi(21) = 2 \times 6 = 12$.

An additional twist in the RSA algorithm is that while all the calculations are performed modulo $n = p \times q$ for two large primes p and q, the ultimate step which allows us to derive the decompression key is based on an inversion modulo $\varphi(n)$ rather than modulo n. According to the principles of public key encryption, the encoding function $\mathcal{E}_K$ will be known to all. RSA uses a family of exponential functions for this purpose, and the key K will consist of a pair of numbers (n, e) which act as parameters for the function:

$$\text{RSA-}\mathcal{E}_{(n,e)}(M) = M^e \bmod n = C,$$

where M is the message to be encoded, or rather the number represented by the binary string into which the message has been transformed. The integer e can be chosen arbitrarily by the encoder, but it must satisfy the strange constraint that it should have an inverse modulo $\varphi(n)$, that is,

$$d = e^{-1} \pmod{\varphi(n)}.$$

This is not a trivial request, because $\varphi(n) = (p-1)(q-1)$ may have quite a few divisors, and their multiples do not have such inverses. Nevertheless, for large enough numbers, an arbitrary choice of e will generally be successful in locating an element that is prime to $\varphi(n)$ and thus has an inverse modulo $\varphi(n)$, and if not, one can try another choice for e until an appropriate element is found.

A nice feature of the RSA scheme is that its encoding and decoding functions are symmetric—just the parameters are changed. Indeed, the decoding will be

$$\text{RSA-}\mathcal{D}_{(n,d)}(C) = C^d \bmod n,$$

and we shall see that $C^d \bmod n = M$ for all messages M.

The modulus n and the encoding parameter e will be published, but the two primes p and q and the decoding parameter d are kept secret. If factoring n is really difficult, as is believed, one cannot derive the primes p and q from the knowledge of n, nor will it be possible, for large p and q, to know $\varphi(n) = n - p - q + 1$, which is needed to calculate d. As an example, consider the primes $p = 2^{400} - 1205$ and $q = 2^{399} - 937$. Their product n is an 800-bit integer, which in decimal notation needs 241 digits:

3334007216 4399271370 3992589536 0628898572 3791611579
5408019812 8905882018 6189088160 3576071610 0435777145
3714649552 9671661624 7570713512 2639087648 9411272752
8634492465 5107291674 0927643145 9748628892 9174374633
7572892192 6218653878 4458049786 0081701542 1.

The corresponding value of $\varphi(n)$ is

3334007216 4399271370 3992589536 0628898572 3791611579

5408019812 8905882018 6189088160 3576071610 0435777145
3714649552 9671661624 3697338695 1335458804 1023396952
4116680970 9520399235 7400718547 0713817921 4844122015
8378445389 6298447131 2919502206 4169577750 0.

The non-overlapping parts of n and $\varphi(n)$, which are at least the right-most half of their bits and are not grayed out in this example, still provide enough variability for not enabling the calculation of p and q on the basis of their sum $p + q = n - \varphi(n) + 1$. The formal RSA algorithm is summarized in ALGORITHM 9.1.

RSA-scheme

1 choose large primes p and q and set $n \leftarrow p \times q$
2 choose randomly an integer e which is prime to $\varphi(n)$
3 evaluate $d \leftarrow e^{-1} \bmod \varphi(n)$
4 publicize (n, e) but keep (p, q, d) secret

encoding: RSA-$\mathcal{E}_{(n,e)}(M) = M^e \bmod n = C$
decoding: RSA-$\mathcal{D}_{(n,d)}(C) = C^d \bmod n = M$

ALGORITHM 9.1: *RSA public key cryptosystem.*

Since the security of the system seems to depend on the inability of the opponent to find the factors of the publicly known modulus n, the larger the prime numbers p and q are chosen, the better will the secret message be protected. Using 800 bits as in the example above is a reasonable choice, though the current recommendation (in 2020) is for a minimum of 2048 bits, which restricts the message M to be encoded also to the same length 2048 bits, enough to ASCII encode a string of 256 characters. If the message is longer, it is broken into chunks of up to 256 characters, and each chunk is encoded independently.

To work through an example, let us again consider the message $M =$ TOPSECRET, whose translation into a 72-bit ASCII string is shown in the second line of Figure 9.2. The decimal equivalent of this binary string would be $M = 6{,}075{,}162{,}742{,}045{,}689{,}172$.

For improved readability, we shall however use smaller prime numbers and consequently build chunks of at most two characters, each of which is

T	O	P	S	E	C	R	E	T
01010100	01001111	01010000	01010011	01000101	01000011	01010010	01000101	01010100
084	079	080	083	069	067	082	069	084

M	84079		80083		69067		82069	84
C	29723966		47338390		4015496		63373736	24448115

Figure 9.2: *RSA encoding example.*

first translated into its 3-digit ASCII index, as shown in the third line of Figure 9.2. The numbers for each chunk are concatenated, yielding one decimal number of up to six digits for each chunk, as can be seen in the line headed M, on each of which the RSA encoding will be applied.

The prime numbers in our example are

$$p = 9901 \qquad \text{and} \qquad q = 8803 \qquad \text{yielding} \qquad n = 87158503,$$

where numbers that are kept secret appear, here and below, in gray. Next, we have to select the exponent e of the encoding function. We saw in Section 5.2.3 that the complexity of modular exponentiation depends on the number of 1-bits in the binary representation of the exponent, so picking a number e with just a few such 1-bits will accelerate the encoding. Alternatively, we could first select d and derive e as $e = d^{-1} \pmod{\varphi(n)}$, if we prefer the decoding to be faster, rather than the encoding. We choose

$$e = 289 = 100100001.$$

Since $\varphi(n) = (p-1)(q-1) = 87139800$, we get

$$d = 289^{-1} \bmod 87139800 = 1507609,$$

which can be double-checked by

$$e \times d = 435699001 = 5 \times 87139800 + 1 = 1 \pmod{87139800}.$$

The first number we wish to encode is 84079, standing for the first two characters TO of our example string:

$$84079^{289} = 29723966 \pmod{87158503},$$

and the corresponding decoding would be

$$29723966^{1507609} = 84079 \pmod{87158503}.$$

The other numbers of the ciphertext are derived similarly and are shown in the last line, headed C, of Figure 9.2.

A point that still needs to be clarified concerns the choice of e and consequently the calculation of d. A random choice of e, as suggested in ALGORITHM 9.1 does not ensure that e will be prime to $\varphi(n)$, without which it will not have the required inverse. Moreover, finding this inverse when it exists does not seem straightforward. Fortunately, both concerns are taken care of by *Euclid's extended algorithm*, which not only finds the greatest common divisor of two integers a and b as we saw in ALGORITHM 5.3, but also shows how to rewrite the GCD as a linear combination of the arguments

$$a \cdot x + b \cdot y = \mathsf{GCD}(a, b), \tag{9.3}$$

where x and y are appropriate integers.

Background concept: Euclid's extended algorithm

The fact that for any integers a and b, there are integer coefficients x and y as given in equation (9.3) is known as *Bézout's lemma*. Euclid's extended algorithm shows how to derive these numbers x and y. It receives the pair (a, b) as input and returns the triplet $(\mathsf{GCD}(a, b), x, y)$.

$\underline{\mathsf{GCDext}(a, b)}$
    if $b = 0$ then
        return $(a, 1, 0)$
    else
        $(D,\ oldx,\ oldy) \leftarrow \mathsf{GCDext}(b,\ a \bmod b)$
        $x \leftarrow oldy$
        $y \leftarrow oldx - \lfloor b/a \rfloor \times oldy$
        return $(D,\ x,\ y)$

ALGORITHM 9.2: *Euclid's extended* GCD *algorithm.*

The complexity of the algorithm is $O\big(\log(\min(a, b))\big)$, as for the original ALGORITHM 5.3.

To check whether an arbitrarily selected e is prime to $\varphi(n)$, one may calculate their GCD. If the result is not 1, another value of e should be chosen. Once an appropriate e is found, its inverse is given by the parameter y returned by Euclid's extended algorithm. Indeed, applying the algorithm with $(a, b) = (\varphi(n), e)$, one gets

$$\varphi(n) \cdot x + e \cdot y = \mathsf{GCD}(\varphi(n), e) = 1,$$

and taking the equation modulo $\varphi(n)$ then yields

$$e \cdot y = 1 \quad (\text{mod } \varphi(n)),$$

which means that $y = e^{-1}$ (mod $\varphi(n)$) as claimed.

Of course, the RSA scheme works for every possible input M, as is shown in the following theorem.

Theorem 9.1: *For every choice of prime numbers p and q and encoding exponent e which is prime to $\varphi(p \times q)$, and for every number $M < p \times q$ serving as input,*

$$\text{RSA-}\mathcal{D}\left(\text{RSA-}\mathcal{E}\left(M\right)\right) = M.$$

Proof: Defining $n = p \times q$, $\varphi(n) = (p-1)(q-1)$ and $d = e^{-1}$ (mod $\varphi(n)$), it has to be shown that for every $M < n$

$$(M^e)^d = M \quad (\text{mod } n).$$

Since e and d are inverses modulo $\varphi(n) = (p-1)(q-1)$, there exists some integer k for which we can rewrite the exponent ed as

$$ed = 1 + k\,(p-1)(q-1).$$

We shall show that $M^{ed} = M$ holds both modulo p and modulo q; it then follows from the *Chinese remainder theorem* that it is true modulo $p \times q$.

If M is a multiple of p, then both M mod p and M^{ed} mod p are zero and thus $M^{ed} = M$ (mod p). If $M \neq 0$ (mod p), then

$$M^{ed} = M^{1+k(p-1)(q-1)} = M \left(M^{k(q-1)}\right)^{p-1} = M \quad (\text{mod } p),$$

where the last equality follows from Fermat's Little Theorem, since p is prime. Similarly, if M is a multiple of q, then $M^{ed} = M = 0$ (mod q), and if not,

$$M^{ed} = M^{1+k(p-1)(q-1)} = M \left(M^{k(p-1)}\right)^{q-1} = M \quad (\text{mod } q). \qquad \square$$

9.3 Useful applications of public key encryption

We have dealt so far with methods allowing the secure communication of a message M between interested parties, without letting an eavesdropper decipher M on its own. If Argan wants to send a message M to Béline, he uses her publicly known encryption function $\mathcal{E}_B$ and sends her the cipher-text $\mathcal{E}_B(M)$, which only Béline is able to decipher. But how can she be sure that indeed Argan did send the message? After all, a hostile Cléante has also access to $\mathcal{E}_B$ and could pretend being somebody else in his message.

This is the problem of *authentication*, which has traditionally been solved by requiring, on contracts or other important documents, a hand-written signature that can supposedly not be counterfeit, so that there can be no doubt about the authenticity of the document. It turns out that there is no need to investigate this seemingly different problem, and that the same mechanism of public key encryption can be applied also in this context. The only additional constraint is that it should be allowed to apply encoding and decoding in any order, so that not only

$$\mathcal{D}\big(\mathcal{E}(M)\big) = M \qquad \text{but also} \qquad \mathcal{E}\big(\mathcal{D}(M)\big) = M,$$

for all possible messages M.

This might seem strange at first sight. What sense does it make, and how could it be possible, to decode a message that has not been encoded at all? Can one shut an already closed window and expect to get it somehow doubly secured, and then restore it to its simply closed initial state by "opening" it? If we remember that our encoding and decoding procedures are just mathematical functions, these questions are rhetorical. The inverse of $\mathcal{E}(M) = M + 7$ is $\mathcal{D}(C) = C - 7$, and there is no problem to subtract before adding. The same is true for RSA, as

$$\mathsf{RSA}\text{-}\mathcal{D}\big(\mathsf{RSA}\text{-}\mathcal{E}\,(M)\big) = M^{ed} \bmod n = M^{de} \bmod n = \mathsf{RSA}\text{-}\mathcal{E}\big(\mathsf{RSA}\text{-}\mathcal{D}\,(M)\big).$$

If Argan wishes to disseminate some message M so that there should henceforth be no doubt that he is the author, he applies his *decoding* function $\mathcal{D}_A$ to M and can then publicize $\mathcal{D}_A(M)$ along with a statement like "Important message from Argan." Anyone can then have access to M, because the corresponding encoding function $\mathcal{E}_A$ is publicly known. More-over, nobody besides Argan could have been able to "sign" the message by applying $\mathcal{D}_A$, which is secret and known only to him.

Note that authentication can be useful also in a related, though essen-tially different scenario, in which Argan does not wish to convince the world that he is the author of the message, but quite on the contrary, tries to deny any connection with it. It will be in the interest of Béline to get a signed document from Argan, for example, if the document M is a contract or a promise from some politician. Since only Argan knows how to invert his public key, Béline may use $\mathcal{D}_A(M)$ as a proof that Argan has sent M, a fact that he cannot repudiate.

Authentication and encoding can be combined, which may be important for highly sensitive communication in military and political contexts. If Prime Minister Argan wishes to send a message M to his Chief of Staff Béline, it should be in a secure way, and Béline needs a proof that the message is authentic. Argan will thus first underwrite the message with his signature and then encode the result with Béline's key, sending

$$C = \mathcal{E}_B\big(\mathcal{D}_A(M)\big). \tag{9.4}$$

Only Béline can decode C by applying $\mathcal{D}_B$ and she will be convinced that Argan is the sender after removing his signature by applying his "encoding," which yields

$$\mathcal{E}_A\big(\mathcal{D}_B(C)\big) = M.$$

We close this chapter with an example of the usefulness of cryptographic protocols to many situations that arise in our digital world. It is generally presented as a solution to the cardinal problem of playing a card game without requiring the players to be physically at the same location, and nonetheless ensuring fairness and an impossibility to cheat. It should, however, be noted that this ludic interpretation of the algorithmic framework is just an appealing didactic way to present the ideas, which are pertinent as well to electronic commerce, trading, banking, politics and many other areas.

The problem is known as playing *mental poker*, but could be defined for many, if not most, other card games, like Rummy, Bridge, Blackjack, Canasta, Kvitlekh, etc. What they have in common is a deck of n cards, often $n = 52$, from which k cards have to be dealt arbitrarily to each of the players. Depending on the game, each of the players may request more cards, which are taken from the remaining deck. The constraints of the game are as follows:

1. Disjointness: the hands and the set of remaining cards are all disjoint;
2. Secrecy: each player can only see her or his own cards, not those of any opponent or those which remained in the deck;
3. Fairness: all the possible choices (of the hands and the additional cards) are equally likely;

4. Verifiability: at the end of the game, all cards can be exposed so that each player can verify that all the steps were according to the rules.

The following steps will allow Argan and Béline play the game, using their encryption / decryption functions, that is, $\mathcal{E}_A$ and $\mathcal{E}_B$ are public, but $\mathcal{D}_A$ and $\mathcal{D}_B$ are secret and only known to Argan and Béline themselves, respectively. We also require here the different encoding functions to be commutative, that is,

$$\mathcal{E}_A\big(\mathcal{E}_B(M)\big) = \mathcal{E}_B\big(\mathcal{E}_A(M)\big),$$

for all messages M, as is the case for RSA, since

$$\big(M^{e_A}\big)^{e_B} \bmod n \; = \; \big(M^{e_B}\big)^{e_A} \bmod n$$

for all possible exponents. We identify the cards by their sequential indices, so that the set of cards consists of $\{1, 2, \ldots, n\}$. To ensure fairness, Argan will shuffle the cards, but Béline will select the cards of the hands, so that both will not have any incentive to cheat by not really shuffling or by choosing a specific subset of the cards rather than a randomly chosen one.

Protocol for remotely playing a card game

Step 1) Argan chooses a random permutation π of the integers $\{1, 2, \ldots, n\}$, shuffles the deck by rearranging it to

$$\big(\pi(1), \pi(2), \ldots, \pi(n)\big),$$

and sends it in encoded form to Béline as

$$\mathcal{C} = \big(\mathcal{E}_A(\pi(1)), \mathcal{E}_A(\pi(2)), \ldots, \mathcal{E}_A\pi(n))\big).$$

Step 2) Béline chooses a random subset $EH_A = \{a_1, \ldots, a_k\}$ of size k from $\mathcal{C}$ and sends it to Argan, who applies his decoding function on each of the cards to get his hand

$$H_A = \{\mathcal{D}_A(a_1), \ldots, \mathcal{D}_A(a_k)\} \subset \{1, \ldots, n\}.$$

Argan now has k cards in his hand, but Béline has no knowledge about them, as she saw them only in encoded form.

Step 3) Béline chooses another random subset $EH_B = \{b_1, \ldots, b_k\}$ of size k from the remaining deck $\mathcal{C} \setminus EH_A$, but before sending it to Argan, she encodes the already encoded elements with her own function to prevent their disclosure to Argan. She thus sends the set

$$EEH_B = \{\mathcal{E}_B(b_1), \ldots, \mathcal{E}_B(b_k)\}.$$

Recall that each b_i is $\mathcal{E}_A(\pi(j))$ for some j, so that $\mathcal{E}_B(b_i)$ is $\mathcal{E}_B(\mathcal{E}_A(\pi(j)))$. Therefore Argan now holds a doubly encoded hand intended for Béline, consisting of k cards, to each of which he applies his decoding, which cancels one of the encoding layers:

$$\mathcal{D}_A\Big(\mathcal{E}_B(\mathcal{E}_A(\pi(j)))\Big) = \mathcal{D}_A\Big(\mathcal{E}_A(\mathcal{E}_B(\pi(j)))\Big) = \mathcal{E}_B(\pi(j)),$$

where we have used the commutativity of $\mathcal{E}_A$ and $\mathcal{E}_B$. Argan sends this set $\{b'_1, \ldots, b'_k\}$ of k elements to Béline, who can apply her decoding to recover the k original cards

$$H_B = \{\mathcal{D}_B(b'_1), \ldots, \mathcal{D}_B(b'_k)\} \subset \{1, \ldots, n\}.$$

This set is disjoint of Argan's hand H_A, and Argan saw only encoded versions of these cards.

Step 4) If Argan or Béline request more cards, repeat Step 2) or Step 3) with $k = 1$ as often as needed.

Step 5) At the end of the game, Argan and Béline reveal their secret decoding functions $\mathcal{D}_A$ and $\mathcal{D}_B$.

If the playing rules impose the partial divulgence of certain cards during the game, while the others should still be kept hidden, Argan and Béline could use n different coding functions, one for each of the n cards. For example for RSA, they could choose fixed prime numbers p_A, q_A, p_B and q_B and different encoding exponents $e_{A_1}, \ldots, e_{A_n}$ and $e_{B_1}, \ldots, e_{B_n}$.

9.4 Exercises

9.1 The modulus n used by RSA is the product of two primes p and q. Does it make sense to choose n itself as a prime, while leaving all the other details of RSA unchanged? How about choosing n as a product of three primes?

9.2 Using the same pair of primes as in our running example $p = 9901$ and $q = 8803$, but another encoding exponent $e = 31770653$, what will be the decoding exponent

$$d = 31770653^{-1} \bmod 87139800 = ?$$

What will be the encoding of the first number 84079,

$$\mathcal{E}(84079) = 84079^{31770653} \bmod 87158503 \ = \ ?$$

9.3 When both authentication and encryption are required, does it matter whether Argan first signs and then only encrypts the message, as shown in Eq. (9.4), or could he inverse these actions to perform

$$C = \mathcal{D}_A\big(\mathcal{E}_B(M)\big)?$$

Does Béline have to be informed about the chosen order of applying the functions?

9.4 Why is it important to generate a new one-time pad for every message? After all, once we have produced a random string K, the computer does not check whether K has been used before, so what is the problem in generating, once and for all, a "good" and long enough string K to be used repeatedly thereafter?

PART 6
Intractability

Chapter 10

$\mathcal{NP}$ Completeness

A book on algorithms is expected to present various methods and techniques on how to tackle computational problems. The contents of the present chapter will thus be rather surprising, because it deals more with challenges we are unable to solve than with those we can. And yet, there are good reasons to investigate these topics, as is also done in many other textbooks on algorithmics.

Many beginning programmers reach soon the wrong conclusion that it is just a matter of choosing a proper approach to succeed in solving any problem whatsoever. They are not aware of the fact that there are problems for which it has been *proven* that no program solving them could possibly exist! A famous example is the *Halting problem* asking whether, given a program P and an input A, the program P will halt when applied on input A, or loop forever. We shall not be concerned with such problems. Nevertheless, even when we can prove the correctness of a program we have written, this does not mean that we can always expect getting a solution in reasonable time. We have seen the trivial primality test for n in ALGORITHM 5.1 of Chapter 5, checking divisors up to $\sqrt{n}$; such an algorithm is useless for an input number n of hundreds of bits in length, such as those needed in the cryptographic applications mentioned in Chapter 9.

A common trait of all the problems we shall deal with in this chapter is that for all of them, it will be quite easy to come up with a proper solution, which, however, will not be scalable. These solutions will thus be adequate for small-sized input parameters, but will be out of reach for larger ones, not only for our current technologies, but probably forever. The interesting fact is that so far, no alternative solution working with a realistic time complexity has been found for any of these problems. This led to the evolution of an ever-growing web of interconnected problems, for which it is known that feasible solutions exist either for all of them, or for none. Most experts believe the latter to be true, that is, that these problems are really all very difficult, though nobody has been able to prove this conjecture.

The question of whether or not a new problem X one is confronted with belongs to this set is therefore not just a theoretical one, driven by our curiosity. It could have very practical implications about how to manage our time and computational resources. If one is able to show that X belongs to this exclusive club of hard problems, this may be a good enough justification for not wasting any time on trying to derive an optimal solution. A better idea could then be to invest efforts in finding possibly sub-optimal approximations or other alternatives, as we shall see in Chapter 11.

10.1 Definitions

The informal discussion above did not specify when we shall deem a complexity function to still be acceptable. To explain the consensus adopted among computer scientists, consider polynomial and exponential functions, for example n, n^3, 1.5^n and e^n, which are plotted on a logarithmic scale y-axis in Figure 10.1.

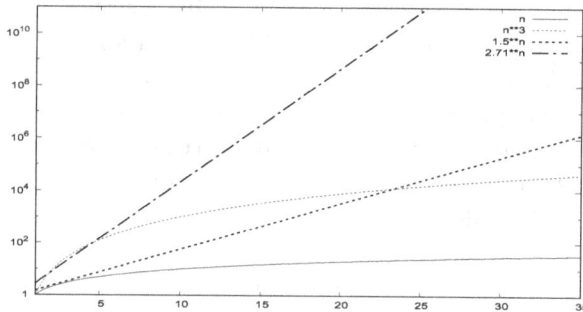

FIGURE 10.1: *Comparing complexity functions n, n^3, 1.5^n and e^n on logarithmic scale.*

Any exponential function will ultimately assume larger values for large enough arguments than does any polynomial function, and the figure shows that even when n is less than 100, an exponential function of n may be prohibitive as complexity of an algorithm. Of course, this does not mean that for specific values, especially smaller ones, and with appropriate coefficients, a polynomial complexity will always be preferable. If we compare the polynomial n^{28} with the exponential $1.1^{n/5}$, the former will be larger than the latter by many orders of magnitude up to more than 10000; only for $n > 14026$ will the exponential function take over. We are, however, mainly concerned with asymptotic values when the input parameter n grows to infinity, which leads to the following definition.

Definition 10.1: A problem will be called *tractable* if it can be solved by an algorithm whose time complexity is bounded by a polynomial. Otherwise, the problem is *intractable*.

One may wonder why such a broad definition has been chosen. Clearly, a polynomial like the example n^{28} above, whose values are already as large as 10^{56} for $n = 100$, can hardly be considered as a desirable complexity, whereas the corresponding value for the exponential $1.1^{n/5}$ will just be 6.7. Wouldn't it be more adequate to define tractability by polynomials bounded in their degrees, say, up to n^8 or so? The reason is that we shall define classes of complexity functions, and it is important that they should be closed under common operations like multiplication and composition.

A crucial tool in the study of the complexity of problems is the ability to transform one problem into another by means of a *reduction*. Both the solution of a problem and the transformation of one into another can be viewed as the application of functions, and it is desirable to be able to compose the functions without transcending the limits of the complexity class. This is not the case for polynomials of some bounded degree, but unbounded polynomials can be multiplied and composed without restriction and still yield polynomials.

We have already encountered several reductions in this book:

- compressing a set of correlated bitmaps by building a minimum spanning tree, see Exercise 3.4;
- maximizing the probability of properly functioning for an electricity network by reduction to a minimum spanning tree problem in Exercise 3.5;
- finding an optimal parsing by reduction to a shortest path problem in Figure 6.8;
- deriving an optimal merge algorithm for a set of integer sequences from Huffman coding in Exercise 6.14;
- building the suffix array by means of the Burrows–Wheeler trans-

form [Burrows and Wheeler (1994)] as shown in Figure 7.14;
- transforming pattern matching with wildcards into a problem of multiplying polynomials in Section 8.5.3.

All these algorithms are tractable. We shall see more reductions in this chapter, involving problems for which it is not known whether they are tractable or not. The purpose of a reduction from A to B will precisely be to find a connection between these problems, so that there exist polynomial algorithms either for both of them, or for neither.

To simplify the discussion, we shall restrict it to *decision problems*. These can be solved by algorithms whose output is either yes or no, such as the primality tests we have seen in Chapter 5. The loss of generality incurred by this restriction is not as bad as it may seem. Indeed, if we know how to solve, for instance, an optimization problem like finding some shortest path, the same algorithm will, in particular, answer the question of whether there exists a path whose cost is bounded by some constant B. But the converse is also true: if we have a procedure for solving the decision problem, we can apply it repeatedly with varying constants. Start with a value for which the answer is trivially yes. This could be, for the shortest path example, the sum S of the weights of all the edges in the graph. Then continue with $S/2$, then $S/4$ or $3S/4$, etc., in a binary search until the optimal bound is reached in $O(\log S)$ trials.

The conclusion is that an optimization problem and its corresponding decision problem are related in the sense that a solution for the one can be transformed in polynomial time into a solution for the other, so they are equivalent from the point of view of their tractability.

To be more precise, it is not a problem that is transformed in a reduction, but rather its *instances*. An instance of a problem is obtained by the assignment of specific values to its parameters. For example, if the problem is finding a minimum in a set A of size n, one of its instances could set $n = 8$ and $A = \{31, 41, 59, 26, 53, 58, 97, 93\}$, as shown in Figure 1.1. An instance of the Minimum Spanning Tree problem could be the weighted graph shown in Figure 3.2, and an instance of the String Matching problem is shown in Figure 7.1. Denote by $\mathcal{I}_\pi$ the set of instances of a problem π.

Definition 10.2: A problem π_1 is said to be *reducible* to a problem π_2, and we shall use the notation $\pi_1 \propto \pi_2$ for this fact, if there exists a function $f : \mathcal{I}_{\pi_1} \to \mathcal{I}_{\pi_2}$ from the set of instances of π_1 to the set of instances of π_2 such that

(a) for each $x \in \mathcal{I}_{\pi_1}$,

the answer of x is yes $\iff$ the answer of $f(x)$ is yes

(b) f is polynomial in the size of x.

There is a subtle point in the definition, because being polynomial in the size of the input depends on how we have chosen to represent the parameters. For the problem of finding the minimum of n integers, the input is of size $\theta(n)$ assuming that we use a constant number of bits, like 32 or 64, per number. However, for the primality problem, the input consists of the number m to be tested. This number is usually represented in just $b = \theta(\log m)$ bits, so that m itself is exponentially larger than the input size, $m = 2^b$. The trivial primality test of ALGORITHM 5.1, performing up to $\sqrt{m} = 2^{b/2}$ iterations, is therefore exponential in the size b of the input. Had we chosen a non-efficient representation like the unary encoding shown in Section 6.1.1, or by using Roman numerals (the largest of which is M, representing 1000), the input size would have been $\theta(m)$, so the complexity would be bounded by a function that is linear in b.

We may thus turn many problems into tractable ones by simply choosing a bad representation for their input, though this is obviously not the intention of developing this theory. The assumption will henceforth be that the input is represented as succinctly as possible.

Theorem 10.1: $\forall \pi_1, \pi_2$, if $\pi_1 \propto \pi_2$ and π_2 is tractable, then so is π_1.

Proof: Let g be a function producing the correct output for an instance $y \in \mathcal{I}_{\pi_2}$. Since π_2 is tractable, the function g is (bounded by a) polynomial. The fact that π_1 is reducible to π_2 implies the existence of a polynomial function f as in Definition 10.2. Composing the functions, we get that an algorithm for solving an instance $x \in \mathcal{I}_{\pi_1}$ is to apply $g(f(x))$, and that this solution is also polynomial, so π_1 is tractable. $\qquad \square$

The equivalent transposition of Theorem 10.1 is

Theorem 10.1': $\forall \pi_1, \pi_2$, if $\pi_1 \propto \pi_2$ and π_1 is intractable, then so is π_2.

A useful mnemotechnical aid to remember the interpretation of reducibility is by means of the similarity of the $\propto$ notation to the symbol for inequality

$$\propto \;\; \rightarrow \;\; \prec \;\; \rightarrow \;\; < \, .$$

The fact that $\pi_1 \propto \pi_2$ means that π_2 may be harder to solve than π_1. Keeping this in mind, we now seek how to identify the most difficult problems within a given class. This led to the following definitions.

Definition 10.3: For a family $\mathcal{S}$ of decision problems, a problem π will be called $\mathcal{S}$-*hard* if

$$\forall \pi' \in \mathcal{S} \quad \pi' \propto \pi. \tag{10.1}$$

If in addition we know that $\pi \in \mathcal{S}$, then π is said to be $\mathcal{S}$-*complete*.

The $\mathcal{S}$-complete problems are thus the most difficult problems within $\mathcal{S}$, while the $\mathcal{S}$-hard ones are problems at least as hard as the most difficult of $\mathcal{S}$, but they may be beyond the limits defining the family $\mathcal{S}$. Many such families have been suggested and we shall concentrate here only on the two most famous, known as $\mathcal{P}$ and $\mathcal{NP}$, for the definition of which we recall the details of *Turing machines*. More details on the theory of $\mathcal{NP}$-completeness can be found in the book by Garey and Johnson (1979).

Background concept: Turing machines

A Turing machine (TM) is a theoretical device, simulating a primitive computer with a very simple and limited set of possible instructions, but capable of solving any problem a full-fledged standard computer does. Its basic structure is that of a finite automaton, but with the additional ability of changing what is written on its working tape. It consists of

(a) a semi-infinite tape partitioned into cells, each containing a letter of some alphabet Σ, and
(b) a control unit with a read/write head, capable, in each step, of reading from and writing onto a single cell of the tape, and moving the head one step to the left, right, or not moving at all.

Formally, a TM is defined as a 5-tuple $\langle Q, \Sigma, q_0, H, \delta \rangle$, where

$Q = \{q_0, q_1, \ldots, q_r\}$ is a set of states
$\Sigma = \{\sigma_0, \sigma_1, \ldots, \sigma_s\}$ is an alphabet, including a blank character ⊔
$q_0 \in Q$ is a starting state
$H \subseteq Q$ is a set of accepting states
$\delta : Q \times \Sigma \longrightarrow \Sigma \times \{L, R, N\} \times Q$ is a transition function effectively describing the possible steps of the TM. It assigns to each pair (q, σ) of possible state–letter combinations at step t a triple (σ', Δ', q'), with $\Delta' \in \{L, R, N\}$, that describes the TM at step $t + 1$. The interpretation is that if, at step t, the machine is in state q and the

head points to a cell containing σ, then at step $t+1$, the cell containing σ will be overwritten by σ', the head will move, according to Δ', one cell to the left, right, or remain in place without moving, and the state will be switched to q'.

A TM starts working in the initial state q_0, with its head pointing to the first cell of the tape. The input is written at the beginning of the tape and the rest of the tape is filled with blanks. The answer to the instance defined by the input is yes if the final state in which the TM stops belongs to the accepting states H. If during execution, the TM reaches a pair (q, σ) with no defined transition, the TM halts. In fact, the initial state q_0, the input and the function δ already determine the entire execution path, which is why this TM is called *deterministic*. It can be shown that a TM with $k \geq 1$ tapes and k read/write heads is equivalent to the model with $k = 1$.

A *non-deterministic* TM (NDTM) is defined similarly to the deterministic one, the only difference being that more than one possible transition is allowed for each (q, σ) pair. The transition function is accordingly

$$\delta : Q \times \Sigma \longrightarrow 2^{\Sigma \times \{L,R,N\} \times Q},$$

where 2^A denotes the power set of a set A. One could have

$$\delta(q, \sigma) = \{(\sigma'_1, \Delta'_1, q'_1), (\sigma'_2, \Delta'_2, q'_2), \ldots, (\sigma'_\ell, \Delta'_\ell, q'_\ell)\},$$

meaning that if, at step t, the machine is in state q and the head points to a cell containing σ, then at step $t+1$, the status of the machine will be according to one of the triplets $(\sigma'_i, \Delta'_i, q'_i)$ for some $1 \leq i \leq \ell$. This may be interpreted as the NDTM guessing an index i that will ultimately lead to an accepting state. Alternatively, one could imagine that the machine duplicates itself and pursues all ℓ options in parallel. The acceptance for an NDTM is defined by at least one of these copies reaching an accepting state.

Definition 10.3: The class $\mathcal{P}$ is defined as the set of problems that can be solved in polynomial time on a deterministic TM.

Definition 10.4: The class $\mathcal{NP}$ is defined as the set of problems that can be solved in polynomial time on a NDTM.

One may ask why the primitive Turing machines have been chosen in these definitions, rather than models of more powerful computers provided by our modern technologies. While it will be easier to show that a certain problem can be solved by a more sophisticated device, the simplicity of the TM model is preferable for showing negative results of not being able to solve certain problems.

The following example should clarify these definitions. We consider the *subset sum* problem, which will be shown below to be $\mathcal{NP}$-complete.

Definition 10.5: Subset sum (SUS) problem: Given a multi-set (repetitions are allowed) $A = \{a_1, a_2, \ldots, a_n\}$ of strictly positive integers and a target value B, is there a subset $A' \subseteq A$ summing to B, that is, such that $\sum_{a_i \in A'} a_i = B$?

We build an NDTM with $k = 2$ tapes. The input is supposed to be written at the beginning of the first tape, and in spite of our earlier discussion requesting a succinct representation of the input numbers, we shall use here a unary notation for simplicity. The machine is defined as $\langle Q, \Sigma, \mathsf{St}, \{\mathsf{Ac}\}, \delta \rangle$, where $Q = \{\mathsf{St}, \mathsf{De}, \mathsf{Co}, \mathsf{Nc}, \mathsf{Su}, \mathsf{Ac}\}$ is the set of states, $\Sigma = \{0, 1, \mathsf{D}, \sqcup\}$ is the alphabet, and the transition function δ is defined by Table 10.1. For example, if $A = \{3, 1, 4, 1, 5\}$ and $B = 7$, the first tape could contain

$$1\,0\,0\,0\,1\,0\,1\,0\,0\,0\,0\,1\,0\,1\,0\,0\,0\,0\,0\,\mathsf{D}\,0\,0\,0\,0\,0\,0\,0\,\sqcup\sqcup\sqcup\cdots,$$

where the integer $m \in A$ is represented by a string of m zeros and preceded by 1, acting as separator. The sequence A is terminated by a delimiter D, followed by the target value B, also in unary representation.

A typical line defining δ represents the transition

$$(q, \sigma^1, \sigma^2) \longrightarrow (\sigma'^1, \Delta'^1, \quad \sigma'^2, \Delta'^2, \quad q'),$$

where the superscripts 1 and 2 denote the indices of the tapes. The algorithm begins in the St art state shown in line 1 by adding the delimiter D at the beginning of the second tape; this will be used when scanning this tape backwards in the last loop of the algorithm, and prevent falling off the beginning of the tape.

The De cision state shown in lines 2a and 2b is the only non-deterministic step and therefore shown in gray. The NDTM chooses whether the following element of A will be written to the second tape in the Co py loop of lines 3–5 or not be written, in the Nc loop of lines 6–8, which just scans the zeros without copying them. Note that the strings of 0s are copied without their separating 1s.

When the delimiter D on the first tape is encountered, we turn to the Su mmation loop of lines 9–10, proceeding to the right on the first tape and to the left on the second, thereby comparing B with the sum of the elements of A that have been non-deterministically chosen in the given instance. Line 10 checks if the first blank ⊔ on the first tape and the delimiter D on the beginning of the second tape are reached simultaneously. If so, there is equality and the machine enters the Accept state. The other two possibilities for (σ^1, σ^2) are

(a) (⊔, 0), meaning that the end of B has been reached before the zeros of the chosen a_i were exhausted; and

(b) (0, D), meaning that the sum of the chosen a_i is smaller than B.

In both cases, the machine should halt, so there is no need for a transition for them.

Table 10.1: *Algorithm for* SUS *on* NDTM *with two tapes.*

	$(q$	σ^1	$\sigma^2)$	$(\sigma'^1 \Delta'^1$		$\sigma'^2 \Delta'^2$		$q')$	Explanation
1	St	1	⊔	1	R	D	R	De	Add marker on tape 2 and goto Decision step
2a	De	0	⊔	0	R	0	R	Co	Non-deterministic choice of
2b				0	R	⊔	N	Nc	Copy or Nc (non-copy)
3	Co	0	⊔	0	R	0	R	Co	Loop and copy while seeing 0s
4		1	⊔	1	R	⊔	N	De	Return to Decision step
5		D	⊔	D	R	⊔	L	Su	Pass to Summation loop
6	Nc	0	⊔	0	R	⊔	N	Nc	Loop without copying while seeing 0s
7		1	⊔	1	R	⊔	N	De	Return to Decision step
8		D	⊔	D	R	⊔	L	Su	Pass to Summation loop
9	Su	0	0	0	R	0	L	Su	Compare sum of subset to B
10		⊔	D	⊔	N	D	N	Ac	If sum is equal to B, Accept

In general, it will not be necessary to build a NDTM to show that a given problem belongs to $\mathcal{NP}$. Most of the problems we shall deal with are of the form "*Is there a subset of . . . satisfying . . .*" and considering the non-determinism as simulating a guessing module, one has just to check in polynomial time that the conditions are satisfied.

But how do we show that a problem π is $\mathcal{NP}$-hard? The definition in Eq. (10.1) requires a proof of $\pi' \propto \pi$ for every $\pi' \in \mathcal{NP}$, and that might be infinitely many reductions! Fortunately, there exists a shortcut to this process: it will be sufficient to find a single problem π_0 that has already been shown to be $\mathcal{NP}$-complete, and to prove for it that $\pi_0 \propto \pi$. Indeed, we then have that

$$\forall \pi' \in \mathcal{NP} \qquad \pi' \propto \pi_0 \propto \pi,$$

which implies the $\mathcal{NP}$-completeness of π because the reduction operator $\propto$ is transitive.

10.2   Existence of $\mathcal{NP}$-complete problems

The class of known $\mathcal{NP}$-complete problems does already contain hundreds of known decision problems and new ones are constantly adjoined to it, so that for a given new problem to be shown $\mathcal{NP}$-hard, there is an ever-growing choice of potential problems π_0 from which the reduction can emanate. An adequate selection of π_0 is nonetheless an art for which quite some intuition may be necessary. Even if theoretically, there exists a polynomial reduction from every π_1 to every π_2 if both are $\mathcal{NP}$-complete problems, this does not mean that these reductions are easy to derive.

An additional difficulty of this strategy of picking some $\mathcal{NP}$-complete π_0 and reducing it to π is that one needs some problem to initialize the reduction chain, because it is not self-evident that the set of $\mathcal{NP}$-complete problems defined as above is not empty! This missing link of the theory has been filled by Cook (1971), who showed that the *satisfiability* problem of Boolean expressions in conjunctive normal form is $\mathcal{NP}$-complete. The problem deals with formulas of first-order logic using n variables, for example,

$$(x_2 \vee \overline{x_3} \vee x_4) \rightarrow \left(x_1 \wedge (x_3 \vee \overline{x_5} \vee x_6) \right), \tag{10.2}$$

where the Boolean variables $x_1, \ldots, x_n$ can appear in their original form x_i or may be negated $\overline{x_i}$. Any such formula can be converted into an equivalent one in *conjunctive normal form* (CNF), consisting of a conjunction (ANDing of) *clauses*, each of which is a disjunction (ORing of) several literals. A *literal* is one of the variables or its negation. Using the definition of $x \rightarrow y$ as $\overline{x} \vee y$, De Morgan's law and the distributivity of $\wedge$ over $\vee$, the above formula (10.2) is equivalent to

$$(x_1 \vee \overline{x_2}) \wedge (x_1 \vee x_3) \wedge (x_1 \vee x_4) \wedge (x_3 \vee \overline{x_5} \vee x_6). \tag{10.3}$$

Definition 10.6: Satisfiability (SAT) problem: Given a set of n Boolean variables $\{x_1, \ldots, x_n\}$ and a Boolean expression in conjunctive normal form

$$E = (\ell_{11} \vee \ell_{12} \vee \cdots \vee \ell_{1r_1}) \wedge (\ell_{21} \vee \cdots \vee \ell_{2r_2}) \wedge \cdots \wedge (\ell_{s1} \vee \cdots \vee \ell_{sr_s}),$$

where each literal ℓ_{ij} for $1 \leq i \leq s$ and $1 \leq j \leq r_i$ is either x_t or its negation $\overline{x_t}$ for some $1 \leq t \leq n$, does there exist a truth assignment T or F for each of the n variables such that the expression E is true?

Setting $x_1 = $ T, $x_5 = $ F and any value for the variables x_2, x_3, x_4, x_6 results in formula (10.3) being T, so it is satisfiable. A formula that is not satisfiable is false for all possible value assignments to its variables, so it is a contradiction, for example,

$$(x_1 \vee x_2) \wedge (x_1 \vee \overline{x_2}) \wedge (\overline{x_1} \vee x_2) \wedge (\overline{x_1} \vee \overline{x_2}), \qquad (10.4)$$

which is the CNF of $(x_1 \wedge \overline{x_1}) \vee (x_2 \wedge \overline{x_2})$.

The trivial way to check the satisfiability of an expression is to build its *truth table* and find at least one row in it yielding the value T, but this requires 2^n rows for all the possible combinations of n variables. A non-deterministic machine could guess the proper values and then verify, in a linear scan of the formula, that the result is indeed T, so SAT $\in \mathcal{NP}$.

Theorem 10.2 [Cook]: SAT is $\mathcal{NP}$-hard.

Proof sketch: The main technical difficulty is to perform infinitely many reductions simultaneously, since $\pi \propto$ SAT has to be shown for all possible problems $\pi \in \mathcal{NP}$. This is handled similarly to what is done in calculus proofs for properties of infinite sets, like showing that $\ln x \leq x - 1$ for all positive reals x. One picks an arbitrary real $x_0 > 0$ and validates that the inequality is correct for this specific value x_0. The formal proof then continues by claiming that since no special properties of x_0, besides being a positive real, have been used in the proof, it holds in fact for all possible choices of $x \in \mathbb{R}_+$.

The analogue for our proof is to choose an arbitrary problem π in $\mathcal{NP}$ and reduce it to SAT, without using any special characteristics of π. The construction must therefore be general enough to encompass all possible problems in $\mathcal{NP}$, which restricts the features we may use in the reduction merely to those implied directly by the definition of a problem belonging to $\mathcal{NP}$. This is the reason why this set has been defined on the basis of a quite primitive device such as an NDTM, rather than some more sophisticated computer.

Reducing the problem π to SAT means building a Boolean expression $E(x)$ which is derived from an instance $x \in \mathcal{I}_\pi$ of the problem π, such that $E(x)$ is satisfiable if and only if the answer to x is yes, that is, when running the NDTM corresponding to π on the input x, we reach an accepting state

in polynomial time. Cook's proof defines many variables with the help
of which one can express the various states of an NDTM as well as the
transitions between the states. For example, there are variables $H[t, m]$
describing the fact that at time t, the head points to the cell indexed m
on the tape; or $S[t, m, \sigma]$ standing for the fact that at time t, cell m of the
tape contains the letter σ.

The next step is using combinations of these variables to generate logical expressions that simulate the behavior of the NDTM. There are, for
instance, expressions stating that at each point t in time, the machine is
in a unique state, scans a single cell and that every cell contains a single
letter. Other formulas say that a cell m can change at time t only if the
head points to that cell:

$$(S[t, m, \sigma] \ \wedge \ \overline{S[t + 1, m, \sigma]}) \ \rightarrow \ H[t, m], \qquad (10.5)$$

and that all the transitions are according to the choices defined by the
function δ. All these expressions have to be converted into conjunctive
normal form and properly quantified, which yields for the example in (10.5)

$$\bigwedge_{t=0}^{p(n)} \bigwedge_{m=0}^{p(n)} \bigwedge_{\sigma \in \Sigma} \left(\overline{S[t, m, \sigma]} \ \vee \ S[t + 1, m, \sigma] \ \vee \ H[t, m] \right),$$

where n is the size of the input on the tape and $p(n)$ is a polynomial
bounding the execution time of the given NDTM. Finally, there are also
expressions for starting in the initial state q_0 at time $t = 0$ and reaching an
accepting state at the end, at time $t = p(n)$.

All these formulas are combined with AND operators and the result is
satisfiable if and only if the NDTM accepts its input. The size of the combined formula is bounded by $O(p(n)^3)$, so the construction is polynomial.
We omit many technical details.   □

10.3   Reductions

Once at least one problem has been established as being $\mathcal{NP}$-complete, we
can rely on it and on its offsprings for subsequent reductions. Figure 10.2
may serve as a roadmap, showing the problems mentioned in this chapter.
Some of these reductions appear in the pioneering paper by Karp (1972).
An interesting feature, as we shall see, is connecting problems from different
domains, identified by different colors in the figure. Logic problems, like
SAT, appear in red, graph problems in blue, set related problems are shown

in yellow and number problems in green. The corresponding number of the theorem or exercise appears next to the arrow indicating the reduction, except for the dashed arrows for which the reduction is omitted.

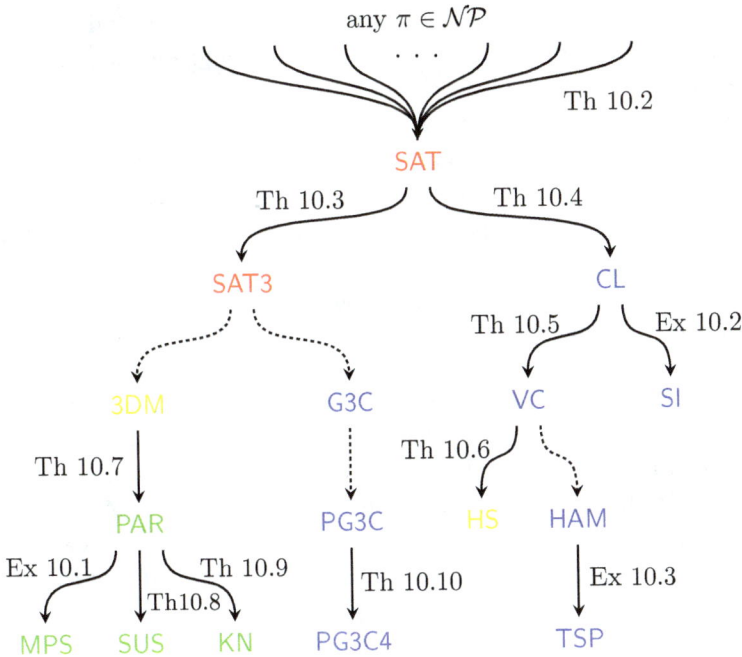

FIGURE 10.2: *Hierarchy of NP-completeness reductions.*

We now return to our challenge of showing a given problem π to be NP-hard. As mentioned, this will be done by carefully choosing a problem π_0 already known to be NP-complete, and building a reduction $\pi_0 \propto \pi$. To this end, one starts with a general instance x of π_0 with the aim to construct a special instance $f(x)$ of π which is equivalent in the sense that x and $f(x)$ are simultaneously yes-instances, or they are both no-instances.

There is, admittedly, a difficulty in understanding the structure of such a reduction, which is often conceived as being counter-intuitive. Usually, when trying to prove some fact about a problem π, one expects to deal with an instance of π as general as it could be, not one that may be restricted. Yet in the suggested framework for a reduction, the proof of π being NP-hard includes merely a very special case of an instance of π. This may raise the question of how could one possibly prove an assertion about a problem, that is, for all of its instances, if the proof refers only to a specific single case?

The answer is that there are actually two quite different forms of reductions. Those of the first type are intended to provide a solution to a problem π_1 we are working on, by means of another problem π_2 for which a solution is already known, and indeed we then show $\pi_1 \propto \pi_2$. We mentioned a list of such reductions at the beginning of this chapter, for example, the problem π_1 of optimally parsing a string of characters into codewords, which has been reduced in Figure 6.8 to the problem π_2 of finding a shortest path in a weighted directed graph; the latter can be solved, e.g., by Dijkstra's algorithm of Section 4.2.

The second type, however, relates to reductions the aim of which is to prove a rather negative result, namely, that the problem π_1 we are working on is very difficult. In this case, the reduction works in the opposite direction, that is, $\pi_2 \propto \pi_1$ for some well-chosen problem π_2 known to be $\mathcal{NP}$-complete. The idea is then that if our problem π_1 would admit a polynomial solution, we could solve *any* of its instances in polynomial time, so in particular the special instance built from the instance of π_2. But this would imply that we found a polynomial algorithm for an $\mathcal{NP}$-complete problem! As mentioned, though no proof of this fact is known so far, most experts believe that this is not possible.

10.3.1 *Logic problems*

The first example of a classical reduction deals with a *restriction* of the SAT problem. A priori, adding more constraints to a problem may turn it into one that is easier to solve, so one may could hope to find a polynomial solution to the restricted problem, even if the unrestricted one is $\mathcal{NP}$-complete. On the other hand, we shall also see examples where this is not the case, and even the more restricted form is just as hard to solve.

Definition 10.6: k-restricted satisfiability (SATk): The instances of SATk are instances of SAT with the additional constraint that each clause contains at most k literals.

Thus the formula in (10.3) is an instance of SAT3 but not of SAT2. As special cases of SAT, all the restricted variants also belong to $\mathcal{NP}$. SAT1 is clearly a polynomial problem, as every formula is in fact a conjunction of literals, so all of them have to be assigned the value T and a linear scan

can check for consistency. SAT2 is also in $\mathcal{P}$ (see Exercise 10.12). In sharp contrast, for $k \geq 3$, the problem is already hard.

Theorem 10.3: SAT3 is $\mathcal{NP}$-hard.

Proof: We show that SAT $\propto$ SAT3. Consider a general instance of SAT. If all of its clauses contain at most three literals, then this is already an instance of SAT3, we therefore concentrate on clauses containing four or more literals. Let $C = (\ell_1 \vee \ell_2 \vee \ldots \vee \ell_r)$ be such a clause, where each ℓ_i stands for a variable x or its negation $\overline{x}$, and with $r \geq 4$.

We construct a formula C' in conjunctive normal form with at most three literals per clause, which can be substituted with C in the sense that C and C' are either both satisfiable, or none of them is. We need a few new variables z_i that appear in C' but not in C. Define

$$C' = (\ell_1 \vee \ell_2 \vee z_1) \wedge (\overline{z_1} \vee \ell_3 \vee z_2) \wedge (\overline{z_2} \vee \ell_4 \vee z_3) \wedge \cdots$$
$$\wedge (\overline{z_{i-2}} \vee \ell_i \vee z_{i-1}) \cdots \wedge (\overline{z_{r-4}} \vee \ell_{r-2} \vee z_{r-3}) \wedge (\overline{z_{r-3}} \vee \ell_{r-1} \vee \ell_r).$$

In the passage from C to C', the number of literals has less than tripled, so the construction is polynomial. Suppose that C is satisfiable. Then there exists an index i_0, $1 \leq i_0 \leq r$ for which $\ell_i = \mathsf{T}$. We show how to assign values to the z_j variables, depending on i_0.

if $i_0 \leq 2$: the first clause $(\ell_1 \vee \ell_2 \vee z_1)$ is T because of ℓ_1 or ℓ_2, and each of the other clauses contains a literal of the form $\overline{z_j}$ for some j, so they are satisfied if we set all $z_j = \mathsf{F}$.

if $i_0 \geq r - 1$: the last clause $(\overline{z_{r-3}} \vee \ell_{r-1} \vee \ell_r)$ is T because of ℓ_{r-1} or ℓ_r, and each of the other clauses contains a literal of the form z_j for some j, so they are satisfied if we set all $z_j = \mathsf{T}$.

if $3 \leq i_0 \leq r - 2$: set $z_j = \mathsf{T}$ for $1 \leq j \leq i_0 - 2$ and set $z_j = \mathsf{F}$ for the other values of j: $i_0 - 1 \leq j \leq r - 3$. The first $i_0 - 2$ clauses of C' are then satisfied by the z_j, the next clause is satisfied by ℓ_{i_0} and the remaining $r - i_0 - 1$ clauses are satisfied by the $\overline{z_j}$.

Conversely, if C is not satisfiable, all the values of ℓ_i are F. Suppose then that there is an assignment of values to the z_j for which C' is satisfiable, then z_1 must be T to satisfy the first clause $(\ell_1 \vee \ell_2 \vee z_1)$. But then, the first two literals of the second clause $(\overline{z_1} \vee \ell_3 \vee z_2)$ are F, so z_2 must also be T. Repeating this argument, we reach the conclusion that all the variables z_j must be T, but then the last clause $(\overline{z_{r-3}} \vee \ell_{r-1} \vee \ell_r)$ will be F.

Summarizing, starting from a general instance x of SAT, we replace every clause of $r \geq 4$ literals by $r - 2$ clauses of three literals to obtain an instance x' of SAT3 which is satisfiable if and only if x is.   □

10.3.2   *Graph problems*

The following reduction is an example of building a connection between different domains, Boolean logic and graph theory.

Definition 10.7: Clique (CL) problem: Given is a graph $G = (V, E)$ and an integer $L \leq |V|$. Does there exist a subset $V' \subseteq V$ with $|V'| \geq L$, such that the induced sub-graph is a clique, that is, a full graph? Formally,

$$\exists V' \subseteq V \quad \big(|V'| \geq L \quad \wedge \quad \forall u, v \in V' \quad (u, v) \in E\big).$$

A trivial algorithm for solving the clique problem would be to generate all the possible subsets V' of size L and check for each of them whether there is an edge between any pair of its vertices. But there are $\binom{|V|}{L}$ possible subsets, and unless L is a (small) constant, this might be exponential, for example, if $L = \frac{|V|}{2}$, then the number is about $\frac{4^{|V|}}{\sqrt{\pi |V|}}$.

Theorem 10.4: CL is $\mathcal{NP}$-hard.

Proof: We show that SAT $\propto$ CL. Consider a general instance of SAT, that is, a set $X = \{x_1, \ldots, x_n\}$ of Boolean variables and an expression in CNF

$$\mathcal{E} = (\ell_{11} \vee \ell_{12} \vee \cdots \vee \ell_{1r_1}) \wedge (\ell_{21} \vee \cdots \vee \ell_{2r_2}) \wedge \cdots \wedge (\ell_{s1} \vee \cdots \vee \ell_{sr_s}),$$

where each literal ℓ_{ij} for $1 \leq i \leq s$ and $1 \leq j \leq r_i$ is either x_t or its negation $\overline{x_t}$ for some variable $x_t \in X$. As a function of these parameters, we have to build a special instance of CL, by defining a graph $G = (V, E)$ and an integer L.

As we saw at the beginning of Chapter 3, there are no restrictions on the set of vertices, and we shall define here a vertex for each of the literals of the given expression $\mathcal{E}$. The vertex $[i, j]$ for ℓ_{ij} will be identified by its subscripts, where i is the index of the clause the literal belongs to, and j is the index of the literal within its clause. We thus have

$$V = \{[i, j] \mid 1 \leq i \leq s, \ 1 \leq j \leq r_i\}.$$

The set of edges will be defined by connecting vertices if and only if they belong to different clause and correspond to literals that can be satisfied simultaneously:

$$E = \{([i,j],[a,b]) \mid i \neq a \wedge \ell_{ij} \neq \overline{\ell_{ab}}\}.$$

Thus there are no edges between vertices of the same clause, and if there is an edge $(u,v) \in E$, the vertices u and v may correspond to literals of different variables, like x_7 and $\overline{x_4}$, or even to the same variable x_7 and x_7, but not if one is the complement of the other, like x_4 and $\overline{x_4}$. The integer L is defined as the number s of clauses. Figure 10.3 is the graph constructed for the expression

$$(\overline{x_1} \vee x_2 \vee x_3 \vee x_4) \wedge (\overline{x_4}) \wedge (\overline{x_2} \vee \overline{x_3} \vee x_4) \wedge (x_1 \vee \overline{x_3}),$$

where the vertices of the first, second, third and fourth clauses appear near the left, upper, right and lower edges of the figure. Vertices $[3,1]$ and $[4,2]$ are connected, as they stand for $\overline{x_2}$ and $\overline{x_3}$, but there is no edge from $[1,2]$ to $[3,1]$, corresponding, respectively, to x_2 and $\overline{x_2}$.

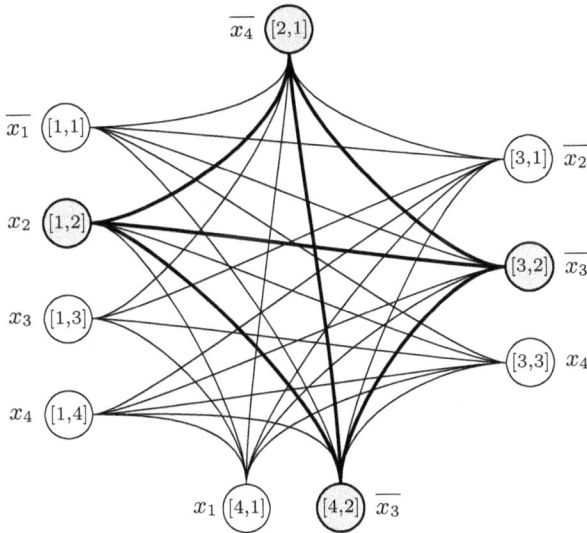

FIGURE 10.3: *Graph with clique of size 4 corresponding to*
$$(\overline{x_1} \vee x_2 \vee x_3 \vee x_4) \wedge (\overline{x_4}) \wedge (\overline{x_2} \vee \overline{x_3} \vee x_4) \wedge (x_1 \vee \overline{x_3})$$

We have to show that the expression $\mathcal{E}$ is satisfiable if and only if there is a clique of size L in G. Suppose then that $\mathcal{E}$ is satisfiable, which means that there exists an assignment of values to the variables x_i such that each clause is satisfied. More precisely, for each index i of a clause, $1 \leq i \leq s$, there is an index $g(i)$ of a literal within the clause, $1 \leq g(i) \leq r_i$, such that $\ell_{i\,g(i)} = \mathsf{T}$. For the example in Figure 10.3, the four literals are ℓ_{12}, ℓ_{21}, ℓ_{32} and ℓ_{42}. The claim is that the set of vertices $V' = \{[i, g(i)] \mid 1 \leq i \leq s\}$ forms a clique of size $s = L$ in G. In our example, the corresponding vertices [1,2], [2,1], [3,2] and [4,2] are grayed and the edges of the full subgraph of size 4 they form are emphasized.

Figure 10.4 is an example of a graph showing a no-instance of the CL problem, corresponding to the contradiction seen earlier in formula (10.4). This graph does not contain a clique of size 4, because there are only two closed paths with exactly 4 edges, shown in red and blue; however, the diagonals of these "squares" are missing.

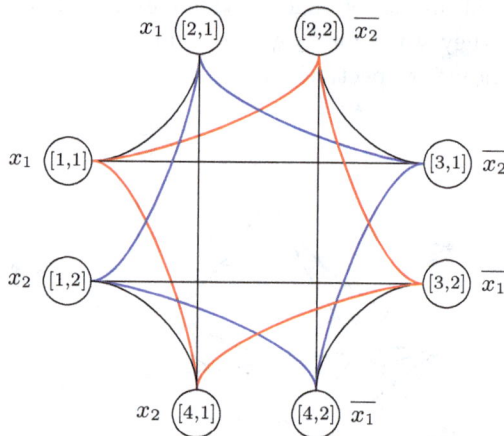

FIGURE 10.4: *Graph with no clique of size 4 corresponding to*
$$(x_1 \vee x_2) \wedge (x_1 \vee \overline{x_2}) \wedge (\overline{x_2} \vee \overline{x_1}) \wedge (x_2 \vee \overline{x_1})$$

To prove that the claim is true in general, let us pick any two elements $[i, g(i)]$ and $[j, g(j)]$ in V' and show that they are connected by an edge in E. Note that $i \neq j$, so if the vertices are not connected, it follows that $\ell_{i\,g(i)} = \overline{\ell_{j\,g(j)}}$; but this is impossible since all the literals have been chosen so that $\ell_{i\,g(i)} = \mathsf{T}$.

Conversely, suppose that there is a clique of size L in G and denote the vertices of this clique by V'. Since vertices corresponding to literals in the same clause are not connected, and the size of the clique is equal to the number of clauses, there must be exactly one element of V' for each clause in $\mathcal{E}$, so the elements of V' can be rewritten as $[i, h(i)]$, $1 \leq i \leq s$, for some function h. Let us partition the set of vertices X into three sets, $X = X_T \cup X_F \cup X_N$, defined by

$$X_T = \{x \in X \mid \exists i \in [1, s] \text{ such that } x = \ell_{[i,h(i)]}\},$$
$$X_F = \{x \in X \mid \exists i \in [1, s] \text{ such that } \bar{x} = \ell_{[i,h(i)]}\},$$
$$X_N = X \setminus (X_T \cup X_F).$$

In other words, the nodes in the clique correspond to literals in $\mathcal{E}$ and these literals are variables in X. Some of them are negated, and some are not. The variables in X_N are those that do not appear at positions in the clauses of $\mathcal{E}$ corresponding to nodes of the clique. For our example in Figure 10.3, the clique consists of the literals ℓ_{12}, ℓ_{21}, ℓ_{32} and ℓ_{42}, corresponding to x_2, $\bar{x}_4$, $\bar{x}_3$ and $\bar{x}_3$, respectively. Therefore $X_T = \{x_2\}$, $X_F = \{x_3, x_4\}$ and $X_N = \{x_1\}$.

Note that X_T and X_F must be disjoint, $X_T \cap X_F = \emptyset$, because if there is a variable y in the intersection, it must appear in one clause as y and in another as $\bar{y}$, but then there would be no edge between the corresponding nodes in the clique. On the other hand, $X_T \cup X_F$ is not necessarily the entire set X. The assignment of truth values is then as follows:

for $x \in X_T$: set the value of x to T;
for $x \in X_F$: set the value of x to F;
for $x \in X_N$: the value of x can be set arbitrarily.

Using this assignment, every clause in $\mathcal{E}$ gets the value T, that is, the expression is satisfied. □

Many other graph problems have been shown to be $\mathcal{NP}$-complete. The following reduction is from CL to the *vertex cover* problem, and it is particular because of its reversibility. Usually, one starts with a general instance of π_1 and constructs a special instance of π_2 to show that $\pi_1 \propto \pi_2$, but in our case, the two problems are equivalent, and the same reduction, with due adaptations, works in both directions.

Definition 10.8: Vertex cover (VC) problem: Given is a graph $G = (V, E)$ and an integer $K \le |V|$. Does there exist a subset $V'' \subseteq V$ with $|V''| \le K$, such that every edge in E has at least one of its end points in V''? Formally,

$$\exists V'' \subseteq V \quad (|V''| \le K \quad \wedge \quad \forall (u, v) \in E \quad \{u, v\} \cap V'' \ne \emptyset). \tag{10.6}$$

Therefore, a vertex cover is not a cover *of* vertices, but rather a cover *by* vertices. What is covered are the edges. For example, referring to the graph in Figure 10.5, the set $V'' = \{1, 3, 6\}$ is a VC of size 3. The colors

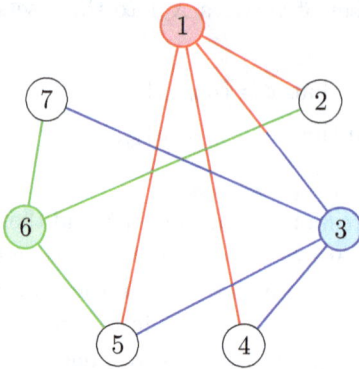

FIGURE 10.5: $G = (V, E)$. *Colored*
vertices form a vertex cover.

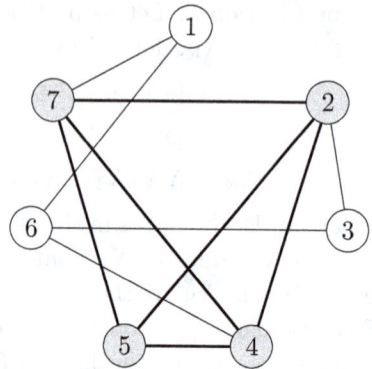

FIGURE 10.6: $\overline{G} = (V, \overline{E})$. *Grayed*
vertices form a clique.

indicate which edges are covered by which vertex; the edge $(1, 3)$ has both
of its endpoints in V''. The equivalence between the two problems is not
obvious and is only revealed by the following reduction.

Theorem 10.5: VC is $\mathcal{NP}$-hard.

Proof: We show that CL $\propto$ VC. Consider a general instance of CL, which is
a graph $G = (V, E)$ and an integer L, as given in Definition 10.7. The constructed instance of VC works with the complementing graph $\overline{G} = (V, \overline{E})$,
that is, it has the same set of vertices, but the complementing set of edges:
$(u, v) \in E$ if and only if $(u, v) \notin \overline{E}$. We also define $K = |V| - L$.

Suppose that there is a clique V' of size at least L in G. Define V'' as
the complement of V' in V, $V'' = V \setminus V''$. As to its size

$$|V''| = |V| - |V'| \le |V| - L = K.$$

It remains to be shown that V'' is a vertex cover in $\overline{G}$. To this end, we
choose an edge

$$(u, v) \in \overline{E} \tag{10.7}$$

and have to show that at least one of u or v belongs to V''. If not, then
both belong to the complement of V'', which is V', but V' is a clique in G,
so there is an edge between u and v in E, in contradiction to (10.7).

Conversely, suppose there is a vertex cover W of size at most K in $\overline{G}$.
Define $V' = V \setminus W$, then

$$|V'| = |V| - |W| \ge |V| - K = L.$$

It remains to be shown that V' is a clique in G. To this end, we choose two vertices

$$x, y \in V' \tag{10.8}$$

and have to show that they are connected by an edge in E. If not, then they are connected by an edge in $\overline{E}$, but since W is a vertex cover in $\overline{G}$, at least one of x or y must belong to W, in contradiction to (10.8), which claims that both belong to the complement of W. $\qquad\square$

Figure 10.6 shows the complementing graph $\overline{G}$ of the graph G in Figure 10.5. The colored vertices $\{1, 3, 6\}$ are a vertex cover of G of size 3, and their complementing set $\{2, 4, 5, 7\}$ is a clique of size 4 in $\overline{G}$.

10.3.3   Set-related problems

A generalization of vertex cover to a set theoretic problem is known as the *Hitting Set* problem. Reductions to such problems are especially simple, because the particular instance of π_2 to be constructed is in fact identical to the general instance of π_1 we started with. It is therefore often sufficient to restrict the new given problem by adding a few constraints, to show that its restricted version is in fact a known $\mathcal{NP}$-complete problem.

Definition 10.9: Hitting set (HS) problem: Given is a set $A = \{a_1, a_2, \ldots, a_n\}$ of n elements, a collection $\mathcal{C}$ of m subsets of A, $\mathcal{C} = \{C_1, \ldots, C_m\}$, with $C_i \subseteq A$ for $1 \leq i \leq m$, and an integer K. Does there exist a subset A' of at most K elements of A that covers already all the sets in the collection $\mathcal{C}$? Formally,

$$\exists\, A' \subseteq A \ \left(|A'| \leq K \quad \wedge \quad \forall\, C_i \in \mathcal{C} \quad C_i \cap A' \neq \emptyset \right). \tag{10.9}$$

For example, $A = \{a, b, c, d, e, f, g, h\}$ and $\mathcal{C} = \{\{a, b, d\}, \{b, e, f, g\}, \{f, g, h\}, \{d, e, h\}, \{b, f\}, \{a, c, h\}, \{a, g\}\}$. If K is chosen as 2, then this is a no-instance, as no two elements of A cover each of the sets in the collection $\mathcal{C}$. But for $K = 3$, $A' = \{a, d, f\}$ is a Hitting Set.

Theorem 10.6: HS is $\mathcal{NP}$-hard.

Proof: We show that VC $\propto$ HS. The proof is by restriction. Choose $K = 2$ and let all the subsets of the collection $\mathcal{C}$ be of size 2. We may then identify the set A with vertices of a graph, and the collection $\mathcal{C}$ with edges. The question in formula (10.9) in the definition of HS then becomes equivalent to the question in formula (10.6) in the definition of VC, thus VC is a special case of HS. $\qquad\square$

The following fundamental problem is important enough for the survival of the human race to depend on its (efficient) solution. Consider two sets of the same size n, one of girls, the other of boys, and let us denote the sets according to the differing chromosome $X = \{x_1, \ldots, x_n\}$ for the girls and $Y = \{y_1, \ldots, y_n\}$ for the boys. Consider also the existence of a subset $\mathcal{C}$ of the Cartesian product $\mathcal{C} \subseteq X \times Y$, representing pairs of potential matches (x_i, y_j), that is, x_i would be willing to marry y_j and vice versa, lacking any gender related asymmetry. We assume here an idyllic world without unrequited love, in which x wants to marry y if and only if reciprocally, y wants to marry x as well. In addition, the set of potential grooms for x, $\{y_j \mid (x, y_j) \in \mathcal{C}\}$, as well as the set of potential brides for y, $\{x_i \mid (x_i, y) \in \mathcal{C}\}$, are not considered as ordered according to any preferences, so that each of the possible mating choices is equally acceptable for all involved marriage candidates. The goal of the optimization problem is to marry off as many couples as possible without violating the additional constraint of monogamy, allowing any girl or boy to have at most one spouse. The decision problem asks whether all the girls and boys can be matched.

Fortunately enough, the solution to this *two dimensional matching* problem is polynomial, and can be derived by a reduction to a *flow* problem in weighted graphs. Interestingly, extending the problem to three dimensions by adjoining a third set, $W = \{w_1 \ldots, w_n\}$, say of dogs or other pets, and redefining a basic family to consist of a triplet (girl, boy, dog) rather than of classical pairs alone, turns the matching problem into an $\mathcal{NP}$-complete one. The formal definition of the corresponding decision problem is as follows:

Definition 10.10: Three-dimensional matching (3DM) problem: Given are three sets $X = \{x_1, \ldots, x_n\}$, $Y = \{y_1, \ldots, y_n\}$ and $W = \{w_1 \ldots, w_n\}$ of the same size n and a collection $\mathcal{C}$ of $m \geq n$ triples $\mathcal{C} \subseteq X \times Y \times W$. Does there exist a matching, that is, a sub-collection $\mathcal{C}' \subseteq \mathcal{C}$, with $|\mathcal{C}'| = m$ and such that

$$(x, y, w), (x', y', w') \in \mathcal{C}' \quad \rightarrow \quad x \neq x' \wedge y \neq y' \wedge w \neq w'?$$

It can be shown that SAT3 $\propto$ 3DM, but we shall not give the reduction here, and rather concentrate on one of the important offsprings of matching, the *partition* problem, in yet another transition between different problem domains: from set-related problems to those featuring numbers.

10.3.4 Number problems

Definition 10.11: Partition (PAR) problem: Given a multi-set of n positive integers $A = \{a_1, \ldots, a_\ell\}$, is there a possibility to partition it into two halves of equal weight, i.e., does there exist a subset $A' \subseteq A$ such that

$$\sum_{a_i \in A'} a_i = \sum_{a_i \in A \backslash A'} a_i ?$$

For example, if $A = \{\mathbf{3}, 1, \mathbf{4}, \mathbf{1}, 5, \mathbf{9}, 2, 6, \mathbf{5}\}$, the answer is yes, because

$$a_1 + a_4 + a_6 + a_9 = a_2 + a_3 + a_5 + a_7 + a_8 = 18,$$

where the elements of A' are emphasized in boldface.

Theorem 10.7: PAR is $\mathcal{NP}$-hard.

Proof: We show that 3DM $\propto$ PAR. The main difficulty is how to pass from an instance dealing with sets and their elements to an instance manipulating numbers. The missing link is provided by *bitmaps*, as mentioned already when we dealt with Karp and Rabin's algorithm in Section 5.3.1. A bitmap can, on the one hand, act as an occurrence map for the elements of a set, but it can also be considered as the binary representation of some integer. Starting with a general instance of 3DM, we shall therefore generate a set of bitmaps, each of which will be considered as a number.

The given instance of 3DM is defined by three sets X, Y and W and a collection $\mathcal{C}$ as given in Definition 10.10. Let $\mathcal{C} = \{c_1, \ldots, c_m\}$ and let us denote the triplet c_r as

$$c_r = (x_{i_r}, y_{j_r}, w_{k_r}), \quad 1 \leq i_r, j_r, k_r \leq n, \quad 1 \leq r \leq m.$$

We define an occurrence bitmap B_r for each of these triplets. Each map is partitioned into three blocks, corresponding to the sets X, Y and W, and each block consists of n regions, one for each element of the corresponding set.

In a first approach, define each region as a single bit, so that the length of each of the B_r is $3n$ bits. For the triplet $c_r = (x_{i_r}, y_{j_r}, w_{k_r})$, the bits in the regions corresponding to x_{i_r}, y_{j_r} and w_{k_r} are set to 1, and all the other bits to zero, so that all the bitmaps B_r have exactly three 1-bits. The upper part of Figure 10.7 shows the bitmap constructed for the triple

X					Y					W				
x_1	x_2	x_3	x_4	x_5	y_1	y_2	y_3	y_4	y_5	w_1	w_2	w_3	w_4	w_5
0	1	0	0	0	1	0	0	0	0	0	0	0	1	0

X					Y					W				
x_1	x_2	x_3	x_4	x_5	y_1	y_2	y_3	y_4	y_5	w_1	w_2	w_3	w_4	w_5
0000	0001	0000	0000	0000	0001	0000	0000	0000	0000	0000	0000	0000	0001	0000

FIGURE 10.7: *Example of bitmaps representing the triple* (x_2, y_1, w_4).

(x_2, y_1, w_4) assuming $n = 5$. The number represented by this bitmap is $2^1 + 2^9 + 2^{13} = 8706$.

The triplets belonging to a matching C' do not have any shared elements, so the 1-bits in the corresponding bitmaps must all be in different regions. By adding the numbers represented by the n bitmaps corresponding to C', one therefore obtains the number $\mathcal{B} = \sum_{i=0}^{3n-1} 2^i = 2^{3n} - 1$ whose binary representation is a string of $3n$ 1s.

The reverse implication, however, is not true. The fact that the numbers add up to $\mathcal{B}$ does not necessarily mean that different triplets cannot have elements in the same region. Consider, for example, the triplets in the left column of Table 10.2, the corresponding bitmaps in the middle and the integers they represent in the right column, assuming again $n = 5$. As can be seen, the sum of these integers is $\mathcal{B}$, even though w_5 appears several times, and w_4 does not appear at all in the shown triplets.

Table 10.2: *Counter-example, showing that $\sum B_i = \mathcal{B}$ does not imply a matching.*

(x_2, y_2, w_2)	0 1 0 0 0	0 1 0 0 0	0 1 0 0 0	8,456
(x_3, y_3, w_3)	0 0 1 0 0	0 0 1 0 0	0 0 1 0 0	4,228
(x_5, y_5, w_5)	0 0 0 0 1	0 0 0 0 1	0 0 0 0 1	1,057
(x_1, y_5, w_5)	1 0 0 0 0	0 0 0 0 1	0 0 0 0 1	16,417
(x_5, y_1, w_5)	0 0 0 0 1	1 0 0 0 0	0 0 0 0 1	1,537
(x_5, y_5, w_1)	0 0 0 0 1	0 0 0 0 1	1 0 0 0 0	1,072
	1 1 1 1 1	1 1 1 1 1	1 1 1 1 1	32,767

The problem is that **PAR** deals with the addition of integers, and by the rules of binary addition, there is a carry when there is more than a single 1-bit in a given position, and such a carry may affect neighboring positions. To overcome this technical difficulty, redefine each region to consist of h bits, rather than a single one. The bit to be set to 1 will be the rightmost in each region that should contain a 1-bit, and each of the new bitmaps B'_r is of length $3nh$. The lower part of Figure 10.7 shows the same example as above, for $h = 4$.

The idea of the additional zeros in each region is to serve as buffer space, so that if there are several 1-bits in a given region, the addition will not produce carries that spill over to adjacent regions. What is then a safe choice for h? There are m triplets in $\mathcal{C}$, and even if all of them are chosen to belong to $\mathcal{C}'$, and all of them include, say, the element y_3, the 1-bits in the position of y_3 add up to m, which affects at most $\lceil \log_2(m+1) \rceil$ bits. For our example of Figure 10.7, the choice of $h = 4$ is sufficient for collections $\mathcal{C}$ up to size $m = 15$.

We have to redefine also the bitmap $\mathcal{B}$ to a sparser one $\mathcal{B}'$, having 1-bits in positions that are multiples of h, and zeros anywhere else. Its value will be $\mathcal{B}' = \sum_{i=0}^{3n-1} 2^{ih}$. Using the amended definitions in which a triplet c_r is associated with the bitmap B'_r, the claim is that

$$\mathcal{C}' \text{ is a matching} \quad \longleftrightarrow \quad \sum_{c_r \in \mathcal{C}'} B'_r = \mathcal{B}'. \tag{10.10}$$

Indeed, if the triplets form a matching, all their $3n$ elements are different and appear exactly once, so there is a single 1-bit in each of the $3n$ positions and the integers representing the bitmaps B'_r sum up to $\mathcal{B}'$. If, on the other hand, the sum of the $|\mathcal{C}'|$ bitmaps is $\mathcal{B}'$, then there must be a single 1-bit in each position, since because of the choice of h, there is no influence from one region on another. Therefore the corresponding set of triples is of size n and they form a matching.

We are now ready to conclude the construction phase of the special instance of **PAR**, which is a specific set of integers $A = \{a_1, \ldots, a_\ell\}$. The first m of these numbers, $a_1, \ldots, a_m$, are the bitmaps B'_r, or more precisely the numbers they represent, for $1 \leq r \leq m$. We set $\ell = m + 2$ and still need to define a_{m+1} and a_{m+2}. Denote the sum of the first m numbers by $S = \sum_{r=1}^{m} a_r$, and define

$$a_{m+1} = S + \mathcal{B}' \qquad a_{m+2} = 2S - \mathcal{B}'. \tag{10.11}$$

Note that the sum of all the numbers in A is

$$\sum_{i=1}^{m+2} a_i = S + (S + \mathcal{B}') + (2S - \mathcal{B}') = 4S,$$

so the sum of each part of the partition, if it exists, has to be $2S$.

Assume that there is a matching $\mathcal{C}' \subseteq \mathcal{C}$. According to (10.10), this implies that the sum of the corresponding numbers is $\mathcal{B}'$. Denote the set of these numbers by A'. Then the sum of the numbers is $A' \cup \{a_{m+2}\}$ is $2S$ and is therefore equal to the sum of the numbers in its complement $A \setminus (A' \cup \{a_{m+2}\})$.

Assume that there exists a partition of A into A'' and $A \setminus A''$, the sum of the elements in each of which is $2S$. The two elements a_{m+1} and a_{m+2} cannot both belong to either A'' or its complement, since their sum alone is already $3S$. Let A'' be the set including a_{m+2} and $A \setminus A''$ the set including a_{m+1}, and consider the set $D = A'' \setminus \{a_{m+2}\}$. The sum of its elements is

$$\sum_{a_i \in D} a_i \;=\; 2S - (2S - \mathcal{B}') = \mathcal{B}'.$$

The set D is thus a subset of $\{a_1, \ldots, a_m\}$ whose elements sum to $\mathcal{B}'$, so the corresponding triplets form a matching. $\square$

A last technical detail has to be clarified. Two integers have been defined in (10.11), and a_{m+1} is certainly positive, but a_{m+2} is the difference of two integers, so how can we avoid that $a_{m+2} < 0$? This may actually happen, for instance when there is no triple in $\mathcal{C}$ with x_1 or x_2 in its first position. In this case, there are no 1-bits in the first two regions of any of the corresponding bitmaps, so $2S < 2^{h(3n-1)}$, but $\mathcal{B}' > 2^{h(3n-1)}$; hence their difference is negative!

We shall therefore restrict attention only to instances of 3DM for which each of the elements of the set X, Y and W appears in at least one of the triplets of the collection $\mathcal{C}$. Instances for which this is not the case are clearly no-instances, and they can be removed from those to be considered after detecting them in a linear scan of $\mathcal{C}$. For the other instances, $a_{m+2} \geq 0$.

One of the implications of the partition problem is the *subset sum* problem we have already seen in Definition 10.5. The NDTM algorithm shown in Table 10.1 is a proof that SUS $\in \mathcal{NP}$.

Theorem 10.8: SUS is $\mathcal{NP}$-hard.

Proof: The proof is by restriction. If an instance of SUS is defined by a set of positive integers $A = \{a_1, a_2, \ldots, a_n\}$ and a target value B, it suffices to consider the special case for which $B = \left(\sum_{i=1}^{n} a_i \right)/2$, that is, B is half the sum of all the elements in A, to get an instance of PAR. □

Definition 10.12: Knapsack (KN) problem: Given is a set of items $A = \{a_1, \ldots, a_n\}$ a subset of which we would like to pack into our knapsack. The constraint is that there are two functions, $s : A \rightarrow \mathbb{R}_+$ and $v : A \rightarrow \mathbb{R}_+$, the first expressing the *size* and the second the *value* of each item. The capacity of the knapsack is limited by K, and we would like to maximize the cumulative value of the packed items. For the decision problem, there is thus a requested lower bound L on the total value. The question is: does there exist a subset of items $A' \subseteq A$ such that

$$\sum_{a_i \in A'} s(a_i) \leq K \qquad \text{and}$$

$$\sum_{a_i \in A'} v(a_i) \geq L \text{ ?}$$

Theorem 10.9: KN is $\mathcal{NP}$-hard.

Proof: Proof by restriction. Consider the special case of KN in which the set of items A contains positive integers, the functions s and v are both the identity function, and $K = L$. The constraints then imply that $\sum_{a_i \in A'} a_i = K$, which shows that SUS is a special case of the KN problem. □

The last branch of the tree of reductions in Figure 10.2 we wish to explore in this chapter concerns again graph problems, though the historic origin is the challenge faced by cartographers to produce colored maps without assigning the same color to regions having a common border. The left side of Figure 10.8 is an example of such a map with six regions labeled A to F. The fact that the regions A, C, D and E are mutually adjacent implies that at least four colors are needed to color this map, and as can be seen, four colors also suffice for this specific example. It turns out that this is true in general: four colors are sufficient to color *every* possible map, a mathematical fact that had been an open problem for many years and was finally settled in 1977 by Appel and Haken.

To reduce the coloring problem to a graph related one, each region is identified by a node, and there is an edge between two nodes if and only if the corresponding regions have a common border (not just a single point, like Arizona and Colorado). In the left part of Figure 10.8, this graph is

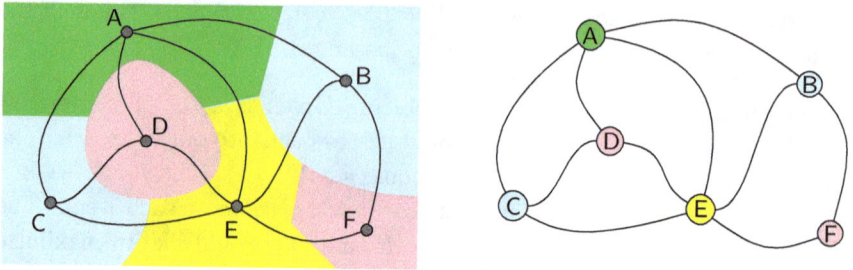

FIGURE 10.8: *Coloring a map with four colors and the corresponding graph.*

superimposed on the map, and it is reproduced without the map in the right part. The mathematical definition of the coloring problem is

Definition 10.13: Graph k-Colorability (GkC): Given a graph $G = (V, E)$ and an integer k, is it possible to color the vertices with at most k colors such that adjacent vertices have different colors? Formally, is there a function $f : V \rightarrow \{1, \ldots, k\}$, such that

$$\forall u, v \in V \qquad (u, v) \in E \quad \rightarrow \quad f(u) \neq f(v)\,?$$

A graph for which the answer is yes is said to be *k-colorable*. This definition is actually broader than is needed to describe the coloring problem above, because it mentions general graphs, whereas maps correspond to so-called *planar* graphs. These have the additional property that they can be drawn without crossing edges, such as the one in Figure 10.8. The notion of planarity is in fact quite tricky, as can be seen in Figure 10.9. This is because it is not always easy to see whether a graph *can* be redrawn into a planar form. A standard way to represent a full graph with four vertices might have crossing diagonals, as in the leftmost graph in Figure 10.9, but the same graph can be represented as the second one from the left. On the other hand, it is not possible to apply a similar technique to eliminate all crossing edges in a full graph of five vertices, and a 5-clique, like those shown in the right part of the figure, is not planar.

A graph is 1-colorable if and only if it is the empty graph, that is, it has no edges. Whether a graph is 2-colorable can be checked in polynomial time, see Exercise 10.5, but for $k \geq 3$, GkC is $\mathcal{NP}$-complete for general graphs. Note that this is the third example of problem families, after having seen SATk and kDM, which is polynomial for $k = 2$ and $\mathcal{NP}$-hard for $k = 3$. The intuitive explanation for coloring is that when only two colors are allowed,

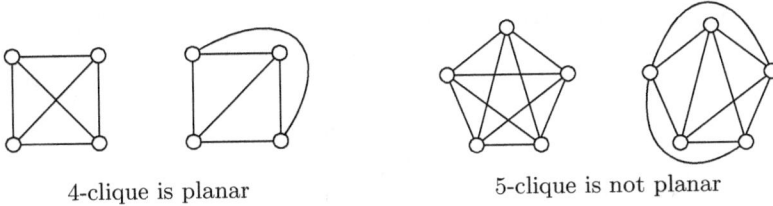

4-clique is planar   5-clique is not planar

FIGURE 10.9: *Examples of planar and non-planar graphs.*

once a node has been assigned some color, the color of the neighbors of the node are determined, and thus also that of the neighbor's neighbors, etc. However, when there are three colors, there are still two choices for each neighbor, so there are 2^2 choices for the neighbor's neighbors, etc.

The variant of the coloring problem for planar graphs is called planar graph k-colorability and will be denoted as PGkC. As special cases of the general graphs, these problems are polynomial for $k \leq 2$. For $k \geq 4$, the answer is always yes, so PG4C is not really a problem, and the only interesting case to be dealt with is for $k = 3$. It turns out that PG3C is $\mathcal{NP}$-complete. The reductions are SAT3 $\propto$ G3C and G3C $\propto$ PG3C and appear as dotted arrows in Figure 10.2 and will be omitted. Our last reduction in this chapter will be to the problem of 3-coloring a special case of a planar graph.

Definition 10.14: Planar graph 3-colorability with vertex degrees bounded by ℓ (PG3Cℓ): Given a graph $G = (V, E)$ in which the degree of every vertex is at most ℓ, is it possible to color the vertices with at most 3 colors such that adjacent vertices have different colors?

We thus add an additional constraint on the graph, bounding the degree, which is the number of neighbors, of each vertex. If the bound is too strict, the problem might be easy: the only connected graphs for which $\ell \leq 2$ consist of linear chains, and these are clearly 2-colorable. We show that for $\ell = 4$, the problem is already $\mathcal{NP}$-complete. The reduction is from planar graph 3-colorability, and uses again local substitutions, similarly to what we saw in the reduction from SAT to SAT3.

Theorem 10.10: PG3C4 is $\mathcal{NP}$-hard.

Proof: We show that PG3C $\propto$ PG3C4. Consider a general planar graph $G = (V, E)$, the aim is to construct a planar graph $G' = (V', E')$ in which the degree of every vertex in V' is bounded by 4, such that G is 3-colorable if and only if G' is.

If a vertex v has degree at most 4 in G, it will be copied to G', including all the edges having v as one of their endpoints. This is the case for example for vertex G in Figure 10.10, which appears also in the constructed instance of PG3C4 in Figure 10.11. Vertices u with higher degree, like A and E, which are emphasized by a gray ellipse in Figure 10.10, will be replaced by entire sub-graphs G_u as follows.

If the degree of u is $d \geq 5$, then G_u consists of $d-2$ copies of the triangle-shaped graph labeled $\mathcal{T}$ shown on the right side of Figure 10.11, in the insert with light blue background. These $d-2$ copies are merged into a single graph G_u by identifying the rightmost node, indexed g, of one copy of $\mathcal{T}$, with the leftmost node, indexed a, of the following copy of $\mathcal{T}$. Since A and E have degrees 5 and 6 in Figure 10.10, G_A and G_E consist of 3 and 4 copies of $\mathcal{T}$, respectively, as can be seen in Figure 10.11.

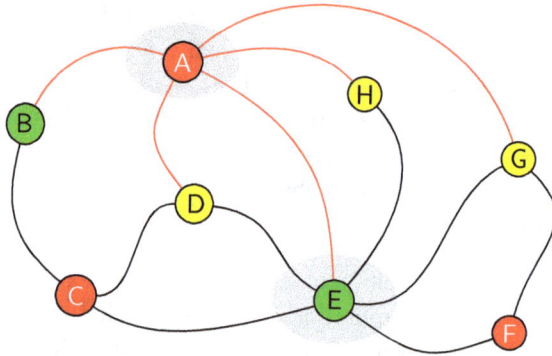

FIGURE 10.10: *A 3-colorable planar graph with no bound on vertex degrees.*

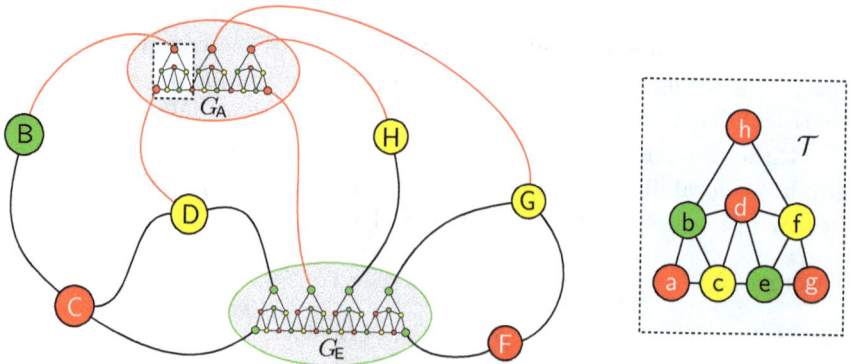

FIGURE 10.11: *Special instance of PG3C4 built for the graph of Figure 10.10.*

The number of nodes in G_u is $7(d-2)+1$, of which d nodes are identified as *extremities*: the summits of the $d-2$ large triangles, and the leftmost and rightmost nodes of G_u. The extremities are drawn in Figure 10.11 slightly larger than the other nodes of G_u. The last step of the construction of G' is to connect each of the d extremities of G_u to a different edge among the d edges emanating from u in G. For example, the five red edges in Figure 10.10 are those connected to vertex A in G, and they correspond to the five red edges connected to the five extremities of G_A in G'. The total number of nodes in G' is bounded by $7|V|^2$, hence the construction is polynomial. By construction, the degrees of all the vertices in G' are at most 4, so G' is indeed an instance of PG3C4.

It remains to be shown that G is 3-colorable if and only if G' is. If the vertices of G can be colored by three colors, use the same colors in G' for vertices that have been copied (those with degree up to 4 in G). If a vertex u of degree $d \geq 5$ has been assigned the color i, this will be the color of the extremities of G_u; the other nodes of G_u can be colored by repeating the color pattern shown for $\mathcal{T}$ in Figure 10.11.

Conversely, assume that G' is 3-colorable and consider the assignment of colors to the nodes of $\mathcal{T}$. The nodes a and b must have different colors, but once they are set, the colors of c, d, e, f, g and h, in this order, are determined, because each of these vertices closes a triangle. For our example, if a is red and b is green, then c must be yellow, so d must be red, so e must be green, etc. In particular, the three extremities of $\mathcal{T}$ must have the same color. This assignment propagates also to the adjacent copies of $\mathcal{T}$, leading to the conclusion that all extremities of G_u share the same color. By collapsing the graph G_u to a single node u, we can thus assign to u the common color of the extremities of G_u, so G is 3-colorable. $\square$

Figure 10.10 shows one of the possible assignments of three colors to the vertices of the given graph G, and the corresponding coloring of the constructed instance G' can be seen in Figure 10.11.

As a final note to conclude this chapter, we wish to address the concerns you may have about having just read a long list of formal proofs in a book about algorithms. The mathematical aspect of the presented ideas should not be misleading: the constructs used in the reductions are in fact parts of algorithms, even though the full algorithm for the solution of a given problem may not yet be available, and possibly never will be. The interconnection between all $\mathcal{NP}$-complete problems implies that if, one day, somebody will find a polynomial algorithm for even a single one of them, this would prove that $\mathcal{P} = \mathcal{NP}$, and all the proofs we have seen will turn into algorithms yielding practical solutions.

There would also be a benefit if somebody can show, on the contrary, that one of these problems is really intractable. This would imply the intractability of all the other $\mathcal{NP}$-hard problems, so if we recognize that we are faced with one of them, we shall not waste our time searching for optimal solutions, and rather try alternatives, as discussed in the next chapter.

10.4　Exercises

10.1 Multi-processor scheduling (MPS): Given is a set of tasks $A = \{a_1, \ldots, a_n\}$, a function ℓ such that $\ell(a_i)$ is the time required to solve the task a_i on any of k available processors, and a deadline D by which all the tasks need to be completed. Does there exist a partition of the set A into k mutually disjoint subsets $A_1, \ldots, A_k$, with $\bigcup_{i=1}^{k} A_i = A$, such that the deadline can be met? Formally, is there such a partition of A satisfying

$$\max_{1 \leq j \leq k} \left(\sum_{a_i \in A_j} \ell(a_i) \right) \leq D\ ?$$

Show that MPS is $\mathcal{NP}$-hard.

10.2 Sub-graph isomorphism (SI): Given are two graphs $G_1 = (V_1, E_1)$ and $G_2 = (V_2, E_2)$. Does G_1 contain a sub-graph that is isomorphic to G_2? Formally, do there exist subset $V \subseteq V_1$ of the vertices and $E \subseteq E_1$ of the edges of G_1 and a one-to-one function f from V onto V_2 that preserves neighborhood connections, i.e.,

$$\forall x, y \in V \quad (x, y) \in E \iff (f(x), f(y)) \in E_2\ ?$$

Show that SI is $\mathcal{NP}$-hard.

10.3 Hamiltonian circuit (HAM): Given a graph $G = (V, E)$, with $V = \{v_0, v_1, \ldots, v_{n-1}\}$, does it contain a *Hamiltonian circuit*, which is a closed path of n edges containing all the vertices of V? Formally, does there exist a permutation $(\sigma(0), \sigma(1), \ldots, \sigma(n-1))$

of $(0, 1, \ldots, n-1)$ such that

$$\forall i \in \{0, 1, \ldots, n-1\} \qquad (v_{\sigma(i)}, v_{\sigma(i+1 \bmod n)}) \in E \ ?$$

Traveling salesperson (TSP): Given is a directed graph $G = (V, E)$, a weight function w on the edges and a bound B. The vertices represent cities, and the weight $w(x, y)$ is the cost of traveling from x to y. Can a traveling salesperson visit all the cities exactly once and then return to the starting point of the tour without exceeding the total budget of B? Formally, is there a permutation $(\sigma(0), \sigma(1), \ldots, \sigma(n-1))$ of the vertex indices $(0, 1, \ldots, n-1)$ such that

$$\sum_{i=0}^{n-1} w(v_{\sigma(i)}, v_{\sigma(i+1 \bmod n)}) \leq B \ ?$$

The HAM problem is $\mathcal{NP}$-complete (the reduction is from VC, as indicated in Figure 10.2). Show that HAM $\propto$ TSP.

10.4 This exercise relates to the reduction from 3DM to PAR in Theorem 10.7.

(a) Is it really necessary to expand each region to consist of h bits instead of just a single one? The idea was to avoid the influence of carries due to the addition, so that if the bitmaps B_r sum up to $\mathcal{B}$, then the corresponding triples form a matching. But we may use the fact when summing bitmaps (that is, the corresponding numbers) which have more than a single 1-bit in certain positions, the number of 1-bits in the sum can only decrease. For example, in Table 10.2, the 6 triples have 18 1-bits, but the number of 1-bits in their sum $\mathcal{B}$ is only 15. Therefore if the bitmaps B_r sum up to $\mathcal{B}$, then since we know that the number of triples in a matching must be n, the corresponding triples form a matching, even for $h = 1$. What is wrong with this argument?

(b) Build an instance of PAR showing that $h = 1$ is not sufficient, that is, a yes-instance of PAR corresponding to a no-instance of 3DM.

10.5 Show that G2C $\in \mathcal{P}$ by giving an algorithm for 2-colorability.

10.6 The graph $\mathcal{T}$ shown on the right side of Figure 10.11 has been used to prove that G is 3-colorable if and only if G' is. To prove just one of the implications of this equivalence, a much simpler graph $\mathcal{T}'$ could be used. Show an example of such a graph and explain why it is not sufficient for the other implication.

10.7 Prove that G5C is $\mathcal{NP}$-hard.

10.8 Suppose an algorithm Dpar(A) will be found that solves the partition decision problem for an input array A in polynomial time. Derive from it a polynomial algorithm to find an actual partition, if it exists. You may apply Dpar(A) with various parameters, but not change the the routine Dpar itself.

10.9 The TSP of Exercise 10.3 requires the salesperson to visit each city exactly once. Consider an extended version of the traveling salesperson problem ETSP in which this constraint is removed. That is, every city has to be visited at least once, but the path does not have to be Hamiltonian.

   (a) Show that ETSP is different from TSP by building a graph which is a yes-instance for one and a no-instance for the other.
   (b) Show that ETSP is $\mathcal{NP}$-complete.

10.10 Suppose an algorithm Dtsp(G, B) will be found that solves the TSP decision problem for a weighted input graph G and bound B in polynomial time. Derive from it a polynomial algorithm to find an actual tour of the cities, if it exists. You may apply Dtsp(G, B) with various parameters, but not change the routine Dtsp itself.

10.11 The TSP of Exercise 10.3 requires the salesperson to return to the point where the tour started, closing a Hamiltonian circuit. Consider an alternative version of the traveling salesperson problem PTSP based on a Hamiltonian path rather than on a circuit. That is, all cities are visited exactly once, but the cycle is not closed. Show that PTSP is $\mathcal{NP}$-complete.

10.12 Show that SAT2 is in $\mathcal{P}$.

Chapter 11

Approximations

The previous chapter may have left us with the rather depressing impression that many, if not most, of the interesting problems we are confronted with are probably not solvable in reasonable time. Nonetheless, these and similar challenges need to be tackled as a matter of routine, and this chapter deals with possible ways to cope with such hard problems.

The algorithmic alternatives range from trying to solve a similar problem that might be strongly related to—but is not quite the same as—the hard one, to improving the running time, which remains exponential, but will be significantly faster. What they all have in common is that they try to circumvent the hardness of the given problem so as to provide a practical solution.

11.1 Trade-offs

A first attempt could be to try to prepare some auxiliary tables, with the help of which the necessary processing time may be reduced, even if it is still exponential. The rationale behind such an approach is that while from a theoretic point of view, we have defined an exponential complexity as being intractable, there are of course practical differences, depending on the size of the exponent. While executing 2^{80} operations seems, in 2021, still to be out of reach, 2^{40} operations may already be feasible.

To give an example, we return to the subset sum problem SUS defined in Definition 10.5 and shown to be $\mathcal{NP}$-complete in Theorem 10.8. A trivial solution of the problem is to generate all the 2^n subsets A' of A, and to check for each of them whether its elements sum up to the given bound B. This clearly takes time $O(2^n)$.

As an alternative, split the set A into two halves $A_1 = \{a_1, \ldots, a_{n/2}\}$ and $A_2 = \{a_{n/2+1}, \ldots, a_n\}$, where we assume that n is even to simplify the notation. If n is odd, just use $\lfloor n/2 \rfloor$ instead of $n/2$. Generate the $2^{n/2}$ subsets $A_1' \subseteq A_1$ of A_1 and for each of them, calculate the sum of its elements. Store each of these sums B_j, with $1 \le j \le 2^{n/2}$, in a table $\mathcal{B}_1$, along with a pointer to the corresponding subset of A_1. Then repeat for the $2^{n/2}$ subsets $A_2' \subseteq A_2$, creating the table $\mathcal{B}_2$. The preparation of these tables takes $O(n2^{n/2})$ operations.

The original SUS problem is thus reduced to finding whether there are two table entries $B_1 \in \mathcal{B}_1$ and $B_2 \in \mathcal{B}_2$ such that $B_1 + B_2 = B$. A simplistic approach would check all the pairs (B_1, B_2), but there are 2^n of them, so nothing would be gained. However, for the same price $O(n2^{n/2})$ as only building the table $\mathcal{B}_1$, one can as well sort its entries in $O(2^{n/2} \log 2^{n/2}) = O(n2^{n/2})$ steps. For each of the $2^{n/2}$ values of B_2, one can then use a binary search for $B - B_2$ in the sorted table $\mathcal{B}_1$ in time $O(\log 2^{n/2}) = O(n)$. The total time complexity is thus $O(n2^{n/2})$.

The last part of the algorithm, that of finding $B_1 \in \mathcal{B}_1$ and $B_2 \in \mathcal{B}_2$ such that $B_1 + B_2 = B$, can be done even faster, though this will not reduce the overall time complexity, which is then dominated by the pre-processing step of building the tables. All one needs is to sort both tables $\mathcal{B}_1$ and $\mathcal{B}_2$ and not just one of them. A linear scan of the tables will then suffice, if it is performed forwards in one table and backwards in the other, as shown formally in ALGORITHM 11.1.

Subset-Sum$(\mathcal{B}_1, \mathcal{B}_2, B)$

```
1     i ← 1      j ← 2^{n/2}
2     while i ≤ 2^{n/2} and j ≥ 1 do
3         sum ← B₁[i] + B₂[j]
4         if sum = B then
5             return subsets corresponding to B₁[i] and B₂[j]
6         else  if sum < B then
7             i ← i + 1
8         else      // sum > B
9             j ← j − 1
10    return "no subset sums to B"
```

ALGORITHM 11.1: *Using two tables to find a solution for the* SUS *problem.*

The index i points initially to the smallest element (which is 0, corresponding to the empty subset) of $\mathcal{B}_1$, whereas j points to the largest element of the other table. If the current elements sum to B, the algorithm returns the corresponding subsets and stops; otherwise, it proceeds either to the following element in $\mathcal{B}_1$ or to the preceding one in $\mathcal{B}_2$. The total number of operations is thus at most the sum of the sizes of the tables, which is $2^{1+n/2}$. Nonetheless, the algorithm is correct and covers the possibilities of all 2^n pairs (B_1, B_2).

Theorem 11.1: ALGORITHM *11.1 is correct.*

Proof: If the algorithm exits in line 5, an appropriate subset of A has been found. It remains to show correctness for no-instances of the problem. This is done by induction on the pair of indices, showing that at each step (i_0, j_0) of the algorithm,

$$\forall i, j \qquad i < i_0 \text{ and } j > j_0 \quad \longrightarrow \quad \mathcal{B}_1[i] + \mathcal{B}_2[j] \neq B. \qquad (11.1)$$

This holds at the initial step $(i_0, j_0) = (1, 2^{n/2})$, as there are no pairs (i, j) to be considered. Assume that (11.1) is true at the beginning of the iteration with (i_0, j_0). To see that it holds also at the beginning of the next iteration, consider two cases. If $\mathcal{B}_1[i_0] + \mathcal{B}_2[j_0] < B$, then also $\mathcal{B}_1[i_0] + \mathcal{B}_2[j] < B$ for $j > j_0$ since the partial sums $\mathcal{B}_2[j]$ are not larger than $\mathcal{B}_2[j_0]$. Therefore, (11.1) holds also for $(i_0 + 1, j_0)$. Similarly, when $\mathcal{B}_1[i_0] + \mathcal{B}_2[j_0] > B$, (11.1) holds also for $(i_0, j_0 - 1)$. □

This two-table variant of the SUS problem thus replaces a $O(2^n)$ time algorithm with one working, after the pre-processing, in $O(2^{n/2})$ space and $O(2^{n/2})$ time. This might still be too demanding in some cases, and can be traded for an algorithm reducing the size of the tables to $O(2^{n/3})$, at the cost of increasing the necessary time to $O(2^{2n/3})$.

To achieve the new trade-off we pass to a variant based on three tables: split the set A into three parts, $A_1 = \{a_1, \ldots, a_{n/3}\}$, $A_2 = \{a_{n/3+1}, \ldots, a_{2n/3}\}$, and $A_3 = \{a_{2n/3+1}, \ldots, a_n\}$, where we now assume that n is a multiple of 3. Then prepare three tables $\mathcal{B}_1$, $\mathcal{B}_2$ and $\mathcal{B}_3$, each of size $2^{n/3}$, of the sorted partial sums, as was done above. This pre-processing is done in time $O(n2^{n/3})$. To solve the SUS problem, process each of the elements $B_1 \in \mathcal{B}_1$ of the first table, and apply the two-table solution using $\mathcal{B}_2$ and $\mathcal{B}_3$, with parameter $B - B_1$ as target value for the sum. The resulting time complexity is thus $2^{n/3} \times 2^{n/3} = 2^{2n/3}$.

In fact, it is easy to generalize the trade-off to one working with k tables $\mathcal{B}_1, \ldots, \mathcal{B}_k$, where $\mathcal{B}_i$ is the sorted list of the partial sums of $A_i \subset A$ defined by

$$A_i = \{a_{(i-1)n/k+1}, a_{(i-1)n/k+2}, \ldots, a_{in/k}\}.$$

All these tables are of size $2^{n/k}$ and can be built in time $O(n2^{n/k})$. The solution is found by applying the $(k-1)$-table approach with parameter $B - B_1$ for each of the elements $B_1 \in \mathcal{B}_1$, yielding a total time complexity of $O(2^{(k-1)n/k})$.

All these trade-offs share a common feature, namely that the product of their time and space complexities is of the order of 2^n, but better trade-offs have been suggested, for example by Schroeppel and Shamir (1981).

11.2 Applying dynamic programming

We have seen dynamic programming in Chapter 2 as a method permitting the replacement of an expensive recursive procedure by the much cheaper filling of a table. A similar approach may be useful for certain $\mathcal{NP}$-complete problems, and we shall again deal with the subset sum problem SUS as an example.

As has been mentioned in Chapter 2, the key to the more efficient solution of the given problem lies often in the introduction of some auxiliary variable. How to choose this variable is mostly not obvious, and this is precisely the point where some intuition is required. For our current problem, we are given a multi-set $A = \{a_1, \ldots, a_n\}$ of positive integers and an integer target value B. The question is: does there exist a subset $A' \subseteq A$ the elements of which sum to the target value, that is,

$$\sum_{a_i \in A'} a_i = B.$$

To build a dynamic programming solution, we shall process the input sequence A incrementally, treating, in the i-th iteration, the multi-set $A_i = \{a_1, \ldots, a_i\}$. This will allow us to express the solution for A_i as depending on the already evaluated solutions for A_j, with $j < i$. Since SUS is a decision problem, the expected answer is yes or no, which we shall translate into T and F to build a table $\mathcal{T}$ storing Boolean values. The entries $\mathcal{T}[i, j]$ of the table will be defined for rows i with $1 \leq i \leq n$ and columns j with $0 \leq j \leq B$ and should represent the truth value of the assertion:

$$\mathcal{T}[i, j] \equiv \text{"there exists a subset } A' \subseteq A_i \text{ such that } \sum_{a_k \in A'} a_k = j \text{"}.$$

As initialization, consider the first row of the table. If the given multiset is just the singleton $A_1 = \{a_1\}$, then the only target values for which the answer is T are either $j = 0$ or $j = a_1$, and it will be F for the other columns.

For the other rows $i > 1$, inspect the value of the only additional element a_i that has been added as potential element of the sum in this i-th iteration. If a_i is larger than the current target value j, it is clearly not useful, so the answer to whether there is a subset of A_i summing to j is the same as for the set A_{i-1}. If, on the other hand, $a_i \leq j$, then we have the choice of using a_i or not. If a_i is not used, then, as before, the sum j has to be formed using elements only from A_{i-1}, but if a_i is chosen as one of the elements of A', then the other elements are still chosen from A_{i-1}, but the sum of these other elements should be $j - a_i$. This is summarized in the recursive definition of $\mathcal{T}$ given by

$$
\mathcal{T}[i,j] = \begin{cases}
\mathsf{T} & \text{if } i = 1 \wedge (j = 0 \vee j = a_1) \\
\mathsf{F} & \text{if } i = 1 \wedge (j \neq 0 \wedge j \neq a_1) \\
\mathcal{T}[i-1,j] & \text{if } i > 1 \wedge j < a_i \\
\mathcal{T}[i-1,j] \vee \mathcal{T}[i-1,j-a_i] & \text{if } i > 1 \wedge j \geq a_i.
\end{cases}
$$

As an example, consider A as the first few digits of e and the target value $B = 14$. The resulting table is shown in Table 11.1. The first row is shown in matching colors to the defining first two lines of the definition. The cells corresponding to the third line, when $i > 1$ and $j < a_i$, are shown in gray. Two examples for the condition in the fourth line are shown as we did for similar dependencies in Chapter 2: an element in row $\mathcal{T}[i,j]$ is shown in blue, and the two elements $\mathcal{T}[i-1,j]$ and $\mathcal{T}[i-1,j-a_i)]$ on which it depends are shown in different shades of pink.

Table 11.1: *Solving* SUS *using dynamic programming.*

i	a_i	0	1	2	3	4	5	6	7	8	9	10	11	12	13	14
1	2	T	F	T	F	F	F	F	F	F	F	F	F	F	F	F
2	7	T	F	T	F	F	F	F	T	F	T	F	F	F	F	F
3	1	T	T	T	T	F	F	F	T	T	T	T	F	F	F	F
4	8	T	T	T	T	F	F	F	T	T	T	T	T	F	F	F
5	2	T	T	T	T	T	T	F	T	T	T	T	T	T	T	F
6	8	T	T	T	T	T	T	F	T	T	T	T	T	T	T	F
7	1	T	T	T	T	T	T	T	T	T	T	T	T	T	T	T

Since elements in row i depend only on elements in row $i-1$, there is no need for recursion, and the table can be filled in a top-down row-by-row scan. The answer to the original question of the SUS problem is found in the rightmost cell of the last row, highlighted in green.

There is a constant number of operations per cell, and the table is of size $n \times B$, so we have found a way to solve the SUS problem in $O(nB)$, which is linear in n and B, our input parameters. Does this mean that we have just settled a long-standing open question in mathematics and shown that $\mathcal{P} = \mathcal{NP}$?

Recall the subtle point mentioned after the definition of tractability at the beginning of Chapter 10: the complexity of an algorithm is not measured as a function of its input but rather of the *size* of this input. When the input consists of a collection of n elements, like numbers or vertices, this makes no difference as the input size will be $\theta(n)$. But if the input is a number itself on which the algorithm should be applied, like in the case of primality testing, or the target value B in the SUS problem, then there are more succinct ways to represent it, for example in its standard binary representation, which requires only $b = \log B$ bits. The input number $B = 2^b$ is exponentially larger than its size b.

Therefore the complexity $O(nB)$ of the dynamic programming solution might be linear in the parameter n, but it is exponential in the parameter b—it is thus not a polynomial solution and this is not a proof of the tractability of SUS. An analogy that may help us understand this point considers the effect of multiplying all the input parameters by 100. If this is done for an array of n numbers to be sorted, it will not make any difference when using some sort based on comparisons, like heapsort. We would similarly expect that multiplying all the $a_i \in A$ and the target value B by 100 should have no impact on the algorithm solving SUS. However, in our case, there will be a 100-fold increase in the size of the dynamic programming table, and hence also of the processing time.

The previous discussion leads to the intriguing conclusion that the alleged difficulty of solving the SUS and similar problems, if indeed $\mathcal{P} \neq \mathcal{NP}$ as most experts believe, relies on the fact that very large numbers may be involved; if the numbers in A and the target value B are restricted, e.g., all smaller than some constant, then the dynamic programming approach provides a feasible solution.

The next obvious question is whether such a remark is true for all $\mathcal{NP}$-complete number problems. The answer in no, as shown by the example in Exercise 10.3. The traveling salesperson problem TSP is shown to be

$\mathcal{NP}$-hard in a reduction from the Hamiltonian circuit problem HAM. While TSP is a number problem, using a graph with weights assigned to the edges, HAM is not: a graph is given, vertices and edges, but there are no weights or other numerical parameters. Consequently, the difficulty of TSP does not stem from the possibility of using large numbers as weights, but rather from the requirement that the travel route of the salesperson must be Hamiltonian.

Indeed, the special instance of TSP built in the reduction HAM $\propto$ TSP involves only the numbers 1 and 2 (see the Appendix for a solution of Exercise 10.3). On the other hand, SUS is a special case of the partition problem PAR, and the numbers constructed in the reduction 3DM $\propto$ PAR are all large. The smallest of these numbers a_i, one corresponding to a triplet of the form (x_n, y_n, w_n), is $2^{2hn} + 2^{hn} + 1$, thus exponentially larger than the size n of the sets X, Y and W.

To distinguish between these problems, they may be partitioned into two classes:

(a) those for which only instances involving large numbers are difficult, like SUS, and
(b) those whose hardness does not depend on the size of the numbers in their instances, like TSP, or whose definition does not include any numbers at all, like HAM.

The latter are often called $\mathcal{NP}$-complete *in the strong sense*.

11.3 Optimization problems

To simplify the development of the theory of $\mathcal{NP}$-completeness, we have restricted the attention to decision problems, returning only yes or no as possible answers. However, for many of these problems, what we are really interested in are the corresponding *optimization* problems. One could, for example, ask what is the size of the smallest vertex cover, or that of the largest clique, within a given graph G. The knapsack problem KN has already been defined in Definition 10.12 also in its optimization variant, asking to maximize the cumulative value of the packed items while not exceeding the total capacity of the knapsack. For the traveling salesperson, the challenge is to minimize the overall cost of visiting all the cities.

We have seen that from the complexity point of view, the optimization variants of these decision problems are just as difficult. This suggests that we might be willing to settle for a *sub-optimal* solution of a problem, if it can be obtained in reasonable, that is, polynomial, time. Of course, not every sub-optimal solution is acceptable, and one strives to get as close as possible to the optimum.

Recall that the set of instances of a problem π has been denoted by $\mathcal{I}_\pi$. Given an optimization problem π, we denote the optimal achievable value for an instance $I \in \mathcal{I}_\pi$ by OPT(I). For example, if $\pi =$ VC, then OPT(I) would be the size of the smallest possible vertex cover for the graph in the given instance. Note that, trivial or small instances apart, the exact value of OPT(I) is generally not known.

A similar notation HEU(I) will be used to describe the outcome of some heuristic HEU to solve instances I of the optimization problem π. A *heuristic* is an algorithm for π, but without the guarantee that it will indeed reach an optimal solution. For example, for TSP, a heuristic approach NN would suggest to start the tour at an arbitrary vertex, and then move repeatedly to the *nearest neighbor* of the current vertex x, that is, a vertex y for which $w(x, y)$ is minimized; NN(I) will then be the cost of the tour built for the graph of the instance I according to the nearest neighbor strategy.

Definition 11.1: For $\alpha \geq 1$, a heuristic HEU is said to be an α-approximation for a minimization problem π if

$$\max_{I \in \mathcal{I}_\pi} \frac{\mathsf{HEU}(I)}{\mathsf{OPT}(I)} \leq \alpha.$$

For $\alpha \leq 1$, a heuristic HEU is said to be an α-approximation for a maximization problem π if

$$\min_{I \in \mathcal{I}_\pi} \frac{\mathsf{HEU}(I)}{\mathsf{OPT}(I)} \geq \alpha.$$

The aim is to find the smallest possible α for minimization and the largest for maximization problems. Getting $\alpha = 1$ would mean that the heuristic is in fact optimal, and since we only consider polynomial heuristics, this would imply that $\mathcal{P} = \mathcal{NP}$.

11.3.1   *Approximating* VC

As a first example, we revisit the vertex cover problem. A simple heuristic for its solution processes the edges in arbitrary order, includes both endpoints of the current edge (x, y) in the cover and removes all edges emanating from either x or y. We shall denote this heuristic as 2EP, referring to the two endpoints of each selected edge. Formally, we are given an undirected graph $G = (V, E)$ and build the cover $C \subseteq V$ by applying ALGORITHM 11.2.

Since edges are removed from E only if at least one of their endpoints is in C, the constructed set C is indeed a vertex cover. Moreover, at least one of the endpoints of the edges chosen in line 3 must be in the optimal vertex cover, so that the size of C at the end of the algorithm is at most twice the size of a minimal VC. It follows that 2EP is a 2-approximation for the VC optimization problem.

Figure 11.1 is an example showing that the 2EP heuristic is not optimal. The graph is a special case with $k = 6$ of a general family of graphs $G[k]$ to be used below. The graphs $G[k]$ are *bipartite*, meaning that the set of their vertices V may be partitioned into disjoint subsets U and L, $V = U \cup L$, such that each of the edges connects a vertex from U with a vertex of L, that is, there are no edges having their endpoints both in U or both in L.

$2EP(V, E)$

1     $C \leftarrow \emptyset$
2   while $E \neq \emptyset$ do
3   choose an edge $(x, y) \in E$
4        $C \leftarrow C \cup \{x, y\}$
5        $E \leftarrow E \setminus \{(a, b) \mid a = x \lor a = y\}$
6   return C

ALGORITHM 11.2: **2EP** *heuristic for* VC *including both ends of selected edges.*

In Figure 11.1, the vertices U are drawn in the Upper part on a light blue background, and those of L in the Lower part on a yellow background. There are k vertices in U, and L is partitioned into $k - 1$ subsets $L_2, L_3, \ldots, L_k$, where the size of L_i is $\lfloor k/i \rfloor$ vertices, so for our example with $k = 6$, the sizes are 3, 2, 1, 1 and 1. There are exactly i edges emanating from each of the vertices in L_i, and all the $i \times \lfloor k/i \rfloor \leq k$ edges emanating from L_i connect to different vertices in U.

The red edges show one of the possible choices of the set of edges chosen by the 2EP heuristic. In this case, the constructed vertex cover consists of the twelve colored vertices, whereas all the edges could be covered by a minimal vertex cover including only the six vertices of U. In this case, the constructed cover C comprises almost the entire set of vertices V, which itself is a trivial vertex cover in every graph.

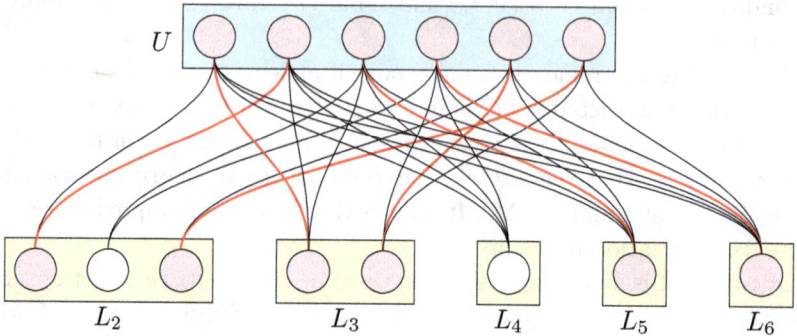

FIGURE 11.1: *Example graph $G[6]$ for vertex cover heuristics.*

Adjoining both ends of the selected edges to the cover seems to be wasteful, which leads to a second heuristic, described in ALGORITHM 11.3, processing vertices instead of edges. The chosen order will be by non-increasing degree of the vertices, in the hope to cover, with each additional vertex in C, as many edges as possible. Accordingly, we shall denote it by HDV, referring to the chosen highest degree vertex in each iteration.

HDV(V, E)

---
1    $C \leftarrow \emptyset$
2   while $E \neq \emptyset$ do
3   choose a vertex $x \in V$ of highest degree
4       $C \leftarrow C \cup \{x\}$
5       $E \leftarrow E \setminus \{(a, b) \mid a = x\}$
6   return C

ALGORITHM 11.3: *HDV heuristic processing vertices by non-increasing degree.*

Referring to the general graph $G[k]$, an example of which, for $k = 6$, is shown in Figure 11.1, the first vertex to be chosen is the sole vertex in L_k with degree k: for every vertex in U, there is at most one outgoing edge to a vertex in L_i, for $2 \leq i \leq k$, so the maximal degree of any vertex in U is $k - 1$. After erasing all the edges to L_k, the maximal degree of a vertex in U will be $k - 2$, so the next vertex chosen in line 3 will be the one in L_{k-1}, which has degree $k-1$. Repeating this argument shows that the cover constructed for $G[k]$ by the HDV heuristic consists of all the $|L| = \sum_{i=2}^{k} \lfloor k/i \rfloor$ vertices in $L = \bigcup_{i=2}^{k} L_i$. For $k = 6$, we have $|L| = 8$, which improves on the $2|U| = 12$ vertices returned by 2EP.

It is, however, not true that HDV outperforms 2EP for all possible instances $I \in \mathcal{I}_{VC}$. For $k = 16$, the size of L is

$$8 + 5 + 4 + 3 + 2 + 2 + 2 + 1 + 1 + 1 + 1 + 1 + 1 + 1 + 1 = 34,$$

which is more than twice the size of U. This already shows that HDV is not a 2-approximation of VC, so even if for many instances, it is better than 2EP, the latter has the advantage of bounding the performance for *all* instances, that is, it has a better worst case.

Actually, using the generalized $G[k]$ example of Figure 11.1, one can show that

$$\lim_{k \to \infty} \frac{\mathsf{HDV}(G[k])}{\mathsf{OPT}(G[k])} = \infty,$$

so HDV is not an α-approximation, for any constant α. More precisely, we have seen that the size of the vertex cover built by HDV for $G[k]$ is

$$|L| = \sum_{i=2}^{k} \lfloor \tfrac{k}{i} \rfloor = \theta\left(\sum_{i=1}^{k} \tfrac{k}{i} \right) = \theta\left(k \sum_{i=1}^{k} \tfrac{1}{i} \right) = \theta(k \log k),$$

where we approach the harmonic sum $\sum_{i=1}^{k}(1/i)$ by $\ln k$, as we have seen at the end of Chapter 5 in the Background concept about approximating a sum by an integral. We thus get that

$$\frac{\mathsf{HDV}(G[k])}{\mathsf{OPT}(G[k])} = \frac{|L|}{|U|} \to \frac{k \log k}{k} = \log k.$$

Summarizing this example, we have seen two different heuristics. The first, 2EP, seems wasteful in many situations, but it guarantees a bounded worst-case behavior and provides a 2-approximation. The second, HDV, is better for many instances and often reaches the optimum, but its worst case may be infinitely worse than the optimal solution.

It should also be noted that approximation ratios do not transfer automatically from one problem to a connected one, even if there is a simple direct reduction between them. For example, we saw in Theorem 10.5 of the previous chapter a reduction from clique to vertex cover and noted that it actually works in both directions. Essentially, the complement $V \setminus V'$ of a vertex cover V' in a graph G is a clique in the complementing graph $\overline{G}$, and a minimal vertex cover corresponds to a maximal clique.

Referring again to the graph $G[6]$ in Figure 11.1, since it is bipartite, there are no edges connecting vertices within U or within L. Therefore, in the complementing graph, all vertices of U are connected, as are also all those in L, yielding cliques of size 5 and 6, respectively. On the other hand, the vertex cover generated by the 2EP heuristic is of size 12 and corresponds to a clique of size 2, consisting of the white vertices, in the complementing graph. It follows that

$$\frac{2\mathsf{EPC}(G[6])}{\mathsf{OPT}_c(G[6])} = \frac{14 - 2\mathsf{EP}(G[6])}{14 - \mathsf{OPT}(G[6])} = \frac{2}{8} = \frac{1}{4},$$

where 2EPC denotes the heuristic for finding a maximal clique derived from 2EP, and $\mathsf{OPT}_c(I)$ stands for the size of the largest clique for instance I. Since 2EP is a 2-approximation for VC, we might have expected 2EPC to be a $\frac{1}{2}$-approximation of CL, but the example shows that this is not the case and that the performance ratios are not necessarily preserved by the reduction.

11.3.2 *Approximating* TSP

As a second example of approximation heuristics, we return to the traveling salesperson problem TSP defined in Exercise 10.3. Given is a graph $G = (V, E)$ with weights $w(x, y)$ assigned to its edges, and we seek, in the optimization version, a Hamiltonian circuit with minimal total weight.

The nearest neighbor heuristic NN mentioned at the beginning of Section 11.3 starts at an arbitrary vertex and then repeatedly selects an out-

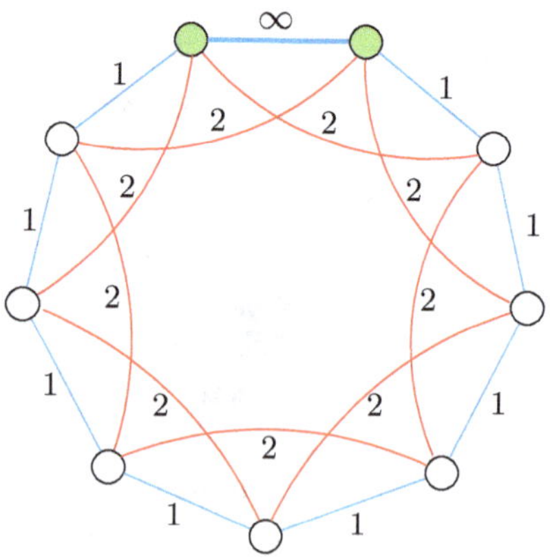

FIGURE 11.2: *Nearest neighbor heuristic for* TSP.

going edge with lowest weight. This is a *greedy* algorithm, choosing at each step the locally best option, but it does not guarantee to converge on a solution that is globally optimal. Even worse, the approximation ratio of NN can be arbitrarily bad, as can be seen in the example of Figure 11.2.

When starting at one of the green vertices, NN chooses all the blue edges with weight 1 until reaching the other green vertex; to close the Hamiltonian circuit, there is then no other choice than adding the edge with weight ∞. As an alternative, the circuit of the red edges has a total weight of 18, though none of its edges corresponds to a locally best choice, except the last one. The ratio $NN(I)/OPT(I)$ is therefore not bounded.

A noteworthy subclass of TSP instances is defined by restricting the weight functions w on the edges to those complying with one of the conditions in the definition of a metric, namely the *triangle inequality*, stating that

$$\forall\, x, y, z \in V \qquad w(x, y) \leq w(x, z) + w(z, y).$$

In many applications, this requirement is justifiable: the graph in the TSP problem represents often a map and the weights on the edges could be the Euclidian distance. Even if the weights represent travel costs or time, these are often, though not always, proportional to the distance.

The restricted problem, which we shall denote as TSP_Δ, is still $\mathcal{NP}$-complete. The graph in Figure 11.2 is not an instance of TSP_Δ, because there is a path of length 2 with weight $1 + 2 = 3$ connecting the two green vertices, while the direct edge of weight ∞ is more expensive. The NN heuristic does not yield an α-approximation of the optimal circuit of TSP_Δ for any finite value α, but the construction to show this fact is much more complicated.

The following heuristic DMST for TSP_Δ is shown in ALGORITHM 11.4 and is based on two main ideas:

(a) We are looking for a cheapest Hamiltomian circuit, but finding one is a difficult problem. An alternative to connecting somehow all the vertices is a Minimum Spanning tree MST, for which we have seen fast algorithms in Chapter 3;

(b) since the triangle inequality applies to any triple of vertices, it in fact enables shortcuts, replacing paths of any length by direct edges. This can be useful to transform a path following an MST into one without repetitions of vertices.

Contrarily to the difficult HAM problem of the existence of a tour in a graph visiting every *vertex* exactly once and closing then the cycle, the problem of deciding whether there exists a tour in a multi-graph visiting every *edge* exactly once is an easy one. Such a tour is called an *Eulerian circuit*, and it exists if and only if all the vertices have even degree. Accordingly, a multi-graph all of whose vertices have even degree is called *Eulerian*. The algorithm for finding such a circuit in an Eulerian multi-graph is straightforward and linear in the number of edges.

DMST(V, E)

1   build a MST for the graph $G = (V, E)$
2   double each of the edges of the MST
3   find an Eulerian circuit for the multi-set MST $\cup$ MST
4   perform shortcuts to avoid repeated visits at a vertex

ALGORITHM 11.4: DMST *heuristic for* TSP$_\Delta$.

An example can be found in Figure 11.3. The vertices are labeled by the characters {A, B, C, E, H, I, L, M, N, O, S, T, V} and the minimum spanning tree consists of the edges drawn by thicker lines. Doubling the edges adds the black arcs and turns the graph into an Eulerian multi-graph, since the degrees of all the vertices are now even. An Eulerian tour, starting at the vertex labeled T, would be

T A M V M A T S L H C H L S B S T N O N E N I N,

where repeated visits to a vertex are indicated by gray color and underlining. Skipping over the underlined labels results in shortcuts, which are shown as red edges in the figure, for example, the path VMATS is replaced by the direct edge VS. The resulting Hamiltonian path, after removing repetitions, is highlighted in green and consists of

T A M V S L H C B N O E I,

to which the last edge IT, shown as a dashed line, has to be added to close a circuit.

An attractive feature of this and similar heuristics is that in spite of the fact that the optimal solution is usually not known, it may still be possible to derive a useful bound on its performance. Consider an instance I and one of its optimal solutions, depicted schematically in Figure 11.4(a) as a closed circuit of cost OPT(I). If one of the edges is removed, as shown in part (b) of the figure, one is left with a Hamiltonian path whose cost OPT-(I) is obviously less than OPT(I). However, a Hamiltonian path is

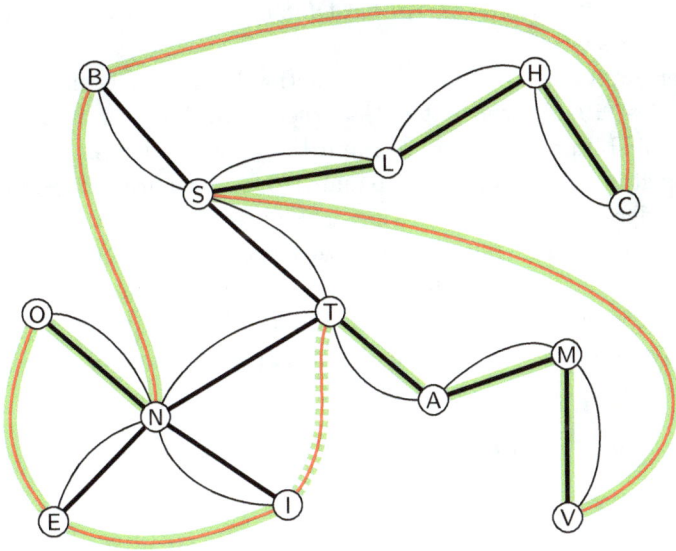

FIGURE 11.3: TSP *heuristic* DMST *based on doubling the edges of an* MST.

a very special case of a spanning tree: a tree for which all vertices have degree at most 2. The cost of such a tree must therefore be at least that of an MST of the same graph, such as the one shown in Figure 11.4(c).

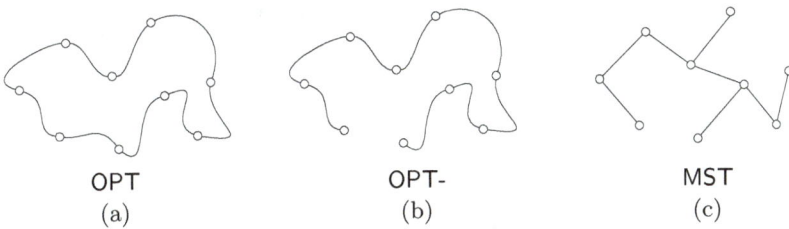

| OPT | OPT- | MST |
| (a) | (b) | (c) |

FIGURE 11.4: *Schematic proof that* DMST *is a 2-approximation for* TSP.

On the other hand, the heuristic DMST follows twice the edges of a MST, but then uses shortcuts, which cannot increase the overall cost, because of the triangle inequality. The conclusion is

$$\mathsf{DMST}(I) \le 2\,\mathsf{MST}(I) \le 2\,\mathsf{OPT\text{-}}(I) \le 2\,\mathsf{OPT}(I),$$

which holds for every instance I, so that DMST is a 2-approximation of the optimization problem.

A further improvement of the approximation bound has been suggested by Christofides (1976), who noted that there may be a cheaper way of turning the MST into an Eulerian graph than simply doubling all of its edges. For instance, the minimum spanning tree in Figure 11.5, shown as thick black edges, is the same as that in Figure 11.3, and a part of the vertices, those colored in blue, have already even degree. To extend this property to all of V, it suffices to pairwise connect the remaining vertices, those of odd degree in the MST, shown here in pink.

How do we know that the set can always be partitioned into pairs? This follows from a general property of graphs:

Lemma 11.1: *There is an even number of vertices of odd degree in every graph.*

Proof: Let $deg(v)$ denote the degree of v. Split the set of vertices V into $V_E \cup V_O$, where V_E and V_O are the vertices of even and odd degree, respectively, to get

$$2\,|E| = \sum_{v \in V} deg(v) = \sum_{v \in V_E} deg(v) + \sum_{v \in V_O} deg(v),$$

where the left equality follows from the fact that summing the degrees of all the vertices accounts twice for each edge. Considering this equality modulo 2, it follows that

$$0 = |V_E| \times 0 + |V_O| \times 1,$$

so that $|V_O| = 0 \pmod 2$; in other words, $|V_O|$ is even.   □

Given a full weighted graph with an even number of vertices $|V| = 2k$, these vertices can be partitioned into k disjoint pairs and each such partition P is called a *matching*. The cost of a matching is $\sum_{\{a,b\} \in P} w(a,b)$, that is, the sum of the weights of the edges (a,b), whose endpoints $\{a,b\}$ are the pairs of the partition P. A *minimum weight matching* MWM is one that minimizes the cost over all the possible partitions. There are simple polynomial time algorithms to construct a MWM.

Christofides' variant replaces the edge doubling of the DMST heuristic by adding instead the edges of a MWM to the edges of the MST; the result is an Eulerian graph in both cases, so one can continue as before by finding an Eulerian circuit, followed by shortcuts, as formally given in ALGORITHM 11.5.

MWMST(V, E)

1   build a MST for the graph $G = (V, E)$
2a $V_O \leftarrow$ vertices of odd degree in the MST
2b find a MWM for V_O
3   find an Eulerian circuit for MST $\cup$ MWM
4   perform shortcuts to avoid repeated visits at a vertex

ALGORITHM 11.5: *MWMST heuristic for* TSP$_\Delta$.

To see how this influences the approximation ratio, consider again an optimal TSP$_\Delta$ circuit for a given instance I, shown in Figure 11.5(a). We restrict the optimal circuit to visit only the odd-degree vertices of V_O, shown in pink in part (b) of the figure, by skipping over the blue vertices of even degree. The resulting circuit, OPT$_O$, is obtained from OPT by shortcuts, and can thus not have a larger total weight because of the triangle inequality. We know that the number of vertices in this cycle is even, and this must also be true for the number of edges. Hence, there are two possibilities of taking every second edge of the circuit, and each of these possibilities yields a matching. They are shown here in red and green, respectively, and we call them M1 and M2.

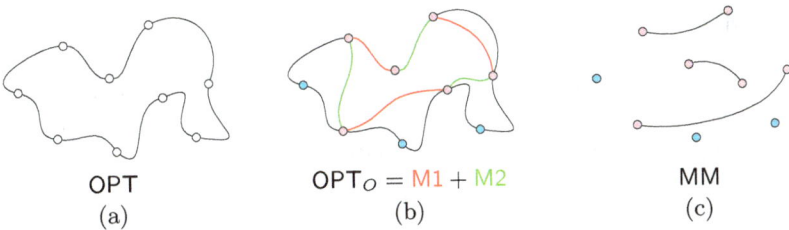

OPT   OPT$_O$ = M1 + M2   MM
(a)   (b)   (c)

FIGURE 11.5: *Schematic proof that* MWMST *is a* $\frac{3}{2}$*-approximation for* TSP.

A minimum weight matching on the set V_O may connect other pairs, as shown in Figure 11.5(c), and its cost is not larger than that of M1 or M2. Thus, for any instance I,

$$\text{OPT}(I) \geq \text{OPT}_O(I) = \text{M1}(I) + \text{M2}(I) \geq 2\,\text{MWM}(I),$$

Basic Concepts in Algorithms

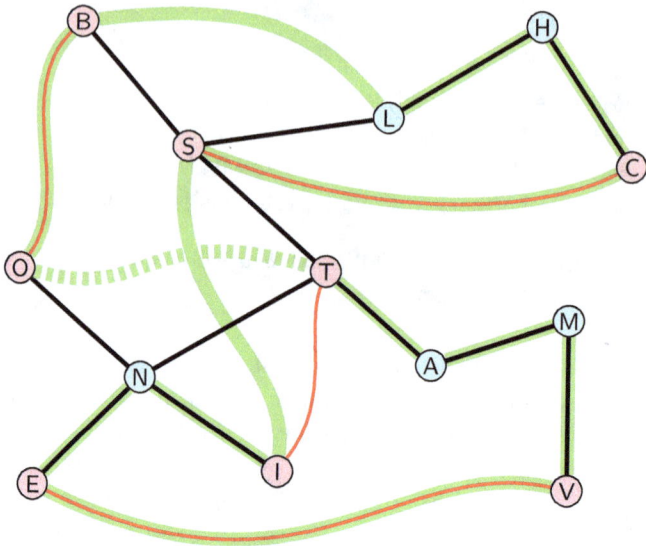

FIGURE 11.6: **TSP** *heuristic* **MMST** *based on* **MST** + **MM**.

and combining this with the fact that $\mathsf{OPT}(I) \geq \mathsf{MST}(I)$, we conclude that

$$\mathsf{MWMST}(I) = \mathsf{MST}(I) + \mathsf{MWM}(I) \leq \mathsf{OPT}(I) + \tfrac{1}{2}\mathsf{OPT}(I) = \tfrac{3}{2}\mathsf{OPT}(I),$$

improving the approximation bound from 2 to $\tfrac{3}{2}$.

Figure 11.6 revisits the same graph as that given in Figure 11.3. As above, the MST is shown as black thick edges. The 8 vertices in V_O are those colored pink and a possible MWM is shown as red edges. After adding the red edges, the graph is Eulerian. An Eulerian tour, starting at the vertex labeled T, would be

T A M V E N I T S C H L S B O,

where, as before, repeated visits to a vertex are indicated in gray and by underlining. A last edge OT, shown as a dashed green line, has to be added to close a circuit. The resulting Hamiltonian path, after removing repetitions, is highlighted in solid green and consists of

T A M V E N I S C H L B O.

11.4   Exercises

11.1 The SUS problem is solved in Section 11.1 by considering the given set of n input numbers as two disjoint sets of size $n/2$. Find a connection between this technique and related ones for different problems.

For example, to search for an element X in an array of size n, show that there is a sequence of search procedures starting with a simple $O(n)$ sequential search and reaching finally binary search in time $O(\log n)$, passing through intermediate techniques working in $O(n^{1/k})$ for $k = 2, 3, \ldots$. A similar connection can also be found to a sequence of sorting techniques starting with selection sort and ending with heapsort.

11.2 Are the following claims correct?

   (a) In the DMST heuristic, the number of edges is increased by $|V| - 1$, while the MWMST heuristic adds only at most half as many edges to the MST. Therefore, for every instance I, the approximation factor α of MWMST is at most 75% of that of DMST.

   (b) When $\alpha < \beta$, then an α-approximation of a minimization problem is always preferable to a β-approximation.

11.3 Recall that in the optimization variant of the knapsack problem (KN) we are given a set of items $A = \{a_1, \ldots, a_n\}$ a subset of which we would like to pack into our knapsack. Two functions s and v are defined, denoting the *size* and *value* of each item. The capacity of the knapsack is limited by K, and we would like to maximize the cumulative value of the packed items. Show how to build a dynamic programming table to find the optimal solution. What is its complexity? *Hint:* Define $A_i = \{a_1, \ldots, a_i\}$ and generate a recursive formula for $\mathcal{K}[i, j]$ defined as

$$\mathcal{K}[i, j] \equiv \max_{A' \subseteq A_i} \left(\sum_{a_k \in A'} v(a_k) \text{ given that } \sum_{a_k \in A'} s(a_k) = j \right),$$

that is, the maximum cumulative value of elements of A_i whose total size equals j.

11.4 Continuing with the KN problem of the previous exercise, the following heuristic DR is suggested: order the elements of A by Decreasing Ratio $\frac{v(a_i)}{s(a_i)}$, that is, consider the elements by order of their relative weight, and include as many as possible in this order in the knapsack without exceeding its capacity.

  (a) Give an example of at least 4 elements for which DR is optimal.
  (b) Give an example showing that DR is not always optimal.
  (c) Show that DR is not an α-approximation of KN for any finite α.

11.5 One of the variants of the *bin packing* problem BP is defined as follows. Given are K bins, all of equal size S, and n elements of sizes $s_1, \ldots, s_n$. We would like to store as many of the elements as possible in the K bins, without splitting any element between different bins. This problem is $\mathcal{NP}$-complete. The heuristic NDS considers the elements in order by Non-Decreasing Size $s_1 \le s_2 \le \cdots \le s_n$, and assigns as many as possible to the first bin; then to the second, etc.

  (a) Is the NDS heuristic optimal for $K = 1$?
  (b) Give an example for $K = 2$ showing that NDS is not always optimal.
  (c) Show that for all instances I and all $K \ge 1$

  $$\mathsf{NDS}(I) \ge \mathsf{OPT}(I) - K + 1.$$

Solutions to selected exercises

[**1.1**] Consider the n input elements by pairs, and for each pair, transfer the smaller element to a set S and the larger to a set L. Then find the minimum of S and the maximum of L.

[**1.2**] To evaluate F or G, there is indeed only one multiplication, but it is not a recursive call. The program multiplies two n-bit numbers, so a recursive call should multiply two $n/2$-bit numbers. The second factor, Y_1 for F and Y_2 for G, has indeed only $n/2$ bits, but the first factor $(X_1 2^{n/2} + X_2)$ is of length n bits.

If it were possible to reduce the number of recursive calls to 2, the complexity would become $O(n \log n)$, as for mergesort.

[**1.3**] Prepare an array $\mathsf{sm}[i]$ defined by

$$\mathsf{sm}[0] \leftarrow 0 \quad \text{and} \quad \mathsf{sm}[i] \leftarrow \mathsf{sm}[i-1] + A[i] \quad \text{for } 1 \leq i \leq n$$

which gives the cumulative values in the array A from its beginning. The partial sums of the elements indexed i to j can then be retrieved in time $O(1)$ as

$$PS[i,j] = \sum_{k=i}^{j} A[k] = \mathsf{sm}[j] - \mathsf{sm}[i-1].$$

Define the procedure Minsum, applying it recursively on the two halves of the input vector, and getting the output pairs (i_1, j_1) and (i_2, j_2). In addition, find, in linear time, the index i_3 that minimizes $PS[i_3, n/2]$, and the index j_3 that minimizes $PS[n/2 + 1, j_3]$. Return the pair (i_ℓ, j_ℓ) for $\ell \in \{1, 2, 3\}$ that minimizes $PS[i_\ell, j_\ell]$. The time complexity is

$$T(n) = 2T(n/2) + n = O(n \log n).$$

[1.5] (2) $n/2$

(3) $T(n) = 3T(n/2) + n = O(n^{1.58})$. It is faster to sort the array in $O(n \log n)$ and then access the median in $O(1)$.

[1.6] (1) There are $n!$ possible orderings, so the depth of the decision tree is $\log n! = \theta(n \log n)$.

(2) Denote the elements by a, b, c, d, e. Compare a with b and c with d. Without loss of generality, a $<$ b and c $<$ d. Compare a and c and assume a $<$ c. Insert e into the sorted list a $<$ c $<$ d, using binary search. That is, compare e to c, and if it is smaller, compare with a, otherwise, compare with d. This takes 2 comparisons. This yields one of 4 possible orderings: eacd, aecd, aced, or acde. We know already that b is larger than a, so the sort is completed by inserting b in the correct place to the right of a for any of the above 4 orderings. That is, b has to be inserted into cd, ecd, ced, or cde. This can again be done in 2 comparisons by binary search.

(3) $O(n \log n)$ is an asymptotic bound. The exact bound is $\log n! = \log 120 = 6.91$ in our case.

[1.7] (1) $\ell = \lceil \log_3(2 \times 12) \rceil = 3$.

(2) Compare 4 against 4 balls. If they are equal, compare 3 of the remaining against 3 of the 8 balls of the first weighing, of which we already know that they are fine. If there is again equality, the biased ball is the last one, and we can check in one comparison whether it is heavy or light. If in the second weighing there is inequality, we know whether the biased ball is heavy or light and use the last comparison to find the different one among the 3 by comparing one-to-one.

If there is inequality in the first weighing, proceed as indicated on the cover page of this book. The 4 red (blue) balls represent those that were on the heavier (lighter) side in the first weighing, the 4 white balls are those that did not participate in the first weighing, and are thus known not to be biased.

(2) $\ell' = \lceil \log_3(2 \times 13) \rceil = 3$.

(3) Compare in the first weighing r against r balls. If $r \leq 4$, then if there is equality, there are at least 5 balls left, giving 10 possibilities; this cannot be checked in 2 weighings, which have only 9 possible outcomes. If $r \geq 5$, then if there is inequality, there are at least 10 balls to be checked, though we already know whether each of them is heavy or light. Still, these are at least 10 possibilities, as in the previous case.

[2.1] Define $\mathcal{L}[i, j]$ as the longest common subsequence of $T = T_1 T_2 \cdots T_i$ and $S = S_1 S_2 \cdots S_j$. The following recurrence then holds:

$$
\mathcal{L}[i, j] = \begin{cases} 0 & i = 0 \vee j = 0 \\ \mathcal{L}[i - 1, j - 1] & i, j > 0 \wedge T_i = S_j \\ \max\left(\mathcal{L}[i, j - 1], \mathcal{L}[i - 1, j]\right) & i, j > 0 \wedge T_i \neq S_j \end{cases}
$$

The solution is $\mathcal{L}[n, m]$ and can be found filling the table top-down and left to right in time and space $O(nm)$.

[2.2] It will be convenient to extend the array M to include also $M_0 = 0$, since no time is needed at the starting point to recuperate. Define an array $T[i]$ as the minimum time needed to reach station A_i. The solution of the problem is $T[n + 1]$. The dynamic programming recurrence is based on considering, for each i, the index k of the last chalet A_k at which the alpinist rests, for $0 \leq k < i$, where $k = 0$ accounts for the possibility of taking no rest at all and proceeding directly from A_0 to A_i:

$$
T[i] = \begin{cases} 0 & i = 0 \\ \min_{0 \leq k < i}\left(T[k] + M_k + \sum_{j=0}^{n-k-1} \alpha^j\right) & 0 < i \leq n + 1 \end{cases}
$$

Time complexity is $O(n^2)$ and space complexity is $O(n)$.

[2.3] Multiplying the matrices takes $O((n - 1)p^3)$ operations. Finding the best order requires $O(n^3)$; thus the former is bounded by the latter if $p \leq O(n^{2/3})$.

[2.4] Counter-example: $(p_0, p_1, p_2, p_3) = (2, 5, 10, 5)$. The suggested algorithm would multiply $M_1 (M_2 M_3)$ in $250 + 50 = 300$ operations, while the alternative order $(M_1 M_2) M_3$ requires only $100 + 100 = 200$ operations.

[2.5] (1) An optimal path cannot contain repeated vertices, so the longest interesting path has $n - 1$ edges.

(2) The only way to get from u to v on a single road is the direct edge. Thus $H(u, v, 1) = h(u, v)$.

(3) One of the ways to get from u to v by 5 roads is to start the path with the edge (u, z). The right-hand side of the inequality gives the height constraint for this path. There is inequality because there are possibly other paths of 5 edges from u to v with better height constraints.

(4) For $k = 1$, $H(u, v, 1) = h(u, v)$.

Basic Concepts in Algorithms

For $k > 1$, $H(u, v, k)$ will be the maximum of two choices, A or B. Case A refers to the possibility of using just $k - 1$ edges, so it is $H(u, v, k - 1)$. For B, we check the possibilities of using k edges, of which the first should be (u, z), for some vertex z which is neither u nor v. This yields

$$B = \max_{z \neq u, v} \left[\min \left(H(u, z, 1), H(z, v, k - 1) \right) \right].$$

(5) Initialization $O(n^2)$. Each $H(u, v, k)$ requires $O(n)$ and there are n^2 choices for u and v and $n - 1$ choices for k, for a total of $O(n^4)$.

[3.1] We keep a linked list L of the connected components (CCs), each CC consisting of a linked list of its vertices, and being identified by its first vertex. Initially, there are $n = |V|$ CCs, each containing a single vertex. There is also an array A of length n, so that $A[i]$ gives the index of the CC the vertex v_i belongs to; initially, $A[i] = i$ for all $1 \leq i \leq n$.

For each stage, scan the list L, and seek for each CC the lowest weight outgoing edge by iterating over all the vertices in the CC. An edge (v_i, v_j) is outgoing if v_i belongs to the current CC and v_j to another one, which can be checked by inspecting $A[j]$. Add some flag to the detected cheapest edge to enable the enforcement of the rule giving it priority in case of ties. At the end of the stage, perform the following for each of the flagged edges (v_{i_0}, v_{j_0}): merge the two CCs they belong to by concatenating in $O(1)$ their lists, with the longer list in front; update the array A for the elements of the shorter list to point to the head of the longer list, which is now the head of the merged list.

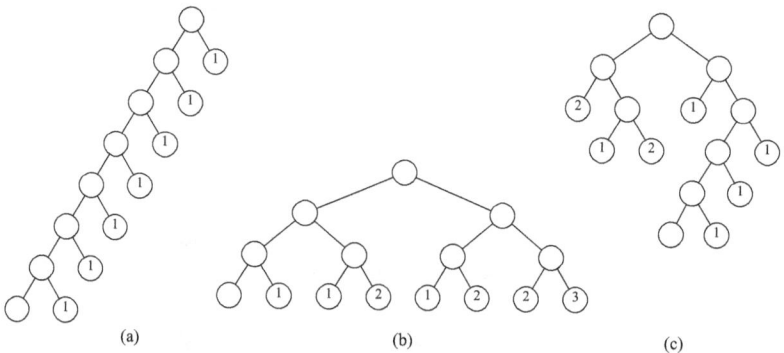

Figure Ex–3.1: Possible merging patterns.

Since at every stage, all the edges are inspected at most once, the Find operations performed to detect all the minima take $O(E)$ per stage and

$O(E \log V)$ in total. As to Union operations, we bound them globally for the entire algorithm. Exactly $n-1$ merges are performed, in $O(V)$ total time. The most expensive commands are the updates of the array A. Note that there are many ways to progressively merge the n elements of an original set into a single merged set, and each of these ways corresponds to one of the possible shapes of a complete binary tree with n leaves: the leaves are the elements and an internal node x corresponds to the merging of the elements of the subtrees of x. Refer to examples in Figure Ex–3.1, which redraws the trees of Figure 1.2.

In the leftmost tree (a) and its generalization, processing the internal node bottom-up, the smaller subtree has only one leaf at each step, so the number of updates of the array A would be $n-1$. The worst case, however, is a completely balanced tree, as in (b), because then the smaller subtree is of the same size as the larger one. In this case, the number of updates of A is

$$\frac{n}{2} + 2\frac{n}{4} + 4\frac{n}{8} + \cdots = \frac{1}{2} n \log n.$$

The leaves of the trees in Figure Ex–3.1 contain the number of times the corresponding entry in A is updated. These sum up to 7 for (a), 12 for (b) and 9 for (c), which is some arbitrary complete binary tree.

Summarizing, the complexity of the bookkeeping commands in Borůvka's algorithm is bounded by

$$O(E \log V) + O(V) + O(V \log V) = O(E \log V). \tag{S3.1}$$

[3.2] If one uses the same implementation as suggested in the solution of Exercise 3.1, one would get

$$O(E) + O(V) + O(V \log V) = O\big(\max(E, V \log V)\big),$$

instead of Eq. (S3.1). This is, however, not necessarily bounded by $E \log \log V$ when the graph is sparse. A sufficient condition in this case for the complexity of Yao's algorithm to be still $E \log \log V$ is that the graph $G = (V, E)$ satisfies $E \geq \Omega\big(\frac{V \log V}{\log \log V}\big)$.

A simpler solution would be to manage the connected components by means of the usual Union–Find structures, spending on average $\log^* V$ steps to perform a Find operation. This would imply that the cost of finding all the minimum weight edges is multiplied by a factor of $\log^* V$ to give $\frac{E}{K} \log V \log^* V$. Changing Eq. (3.3) accordingly would then yield a minimum for $K = \log V \log^* V$, and Eq. (3.4) would be replaced by

$$f(\log V \log^* V) = E \log \log V + E \log \log^* V + E = O(E \log \log V).$$

Actually, there is no need to find the optimal value of K using a derivative; just choosing $K = \log V$ yields the same bound:

$$f(\log V) = E \log \log V + E \log^* V = O(E \log \log V).$$

[3.3] The pre-processing part would then take $O(E \log E \log K)$, and since $O(\log E) = O(\log V)$, the optimum is then achieved for $K = 1$; in other words, the best choice for Yao's algorithm is to return to Borůvka's, and the complexity would remain $O(E \log V)$.

[3.4] The *Hamming distance*, denoted by $hd(A, B)$, is the number of indices at which two equal-length bitmaps A and B differ, that is,

$$hd(A, B) = popc(A \ \text{XOR} \ B),$$

and in particular,

$$popc(A) = hd(A, O),$$

where O is the bitmap consisting only of zeros.

Define an undirected weighted graph $G = (V, E)$, with

$$V = \mathcal{B} \cup O, \quad E = V \times V - \{(A, A) \mid A \in V\}, \quad w(A, B) = hd(a, b).$$

The vertices are thus the given bitmaps to which the 0-bitmap O has been adjoined. The edges are those of a full graph without loops, and the weight on edge (A, B) is the number of 1-bits in A XOR B. We are looking for a subset $E' \subseteq E$ of the edges that does not contain any cycles and touches all vertices while minimizing the sum of weights on the edges in E'. This is a *minimum spanning tree* of G.

The set $\mathcal{I}$ consists of the direct neighbors of O in E', and the best choice of $g(A)$, for any other bitmap $A \in V \setminus \mathcal{I}$, will be the successor of A in the unique directed path from A to O in E'.

[3.5] Let $G = (V, E)$ be the graph of the electricity network. We look for a subset $E' \subseteq E$ of the edges forming a spanning tree, which maximizes the probability of reaching each vertex without failure, that is, find

$$\max \prod_{(a,b) \in E'} p(a, b).$$

Change the weight of the edge (a, b) to $- \log p(a, b)$. The above maximization is then equivalent to

$$\min \sum_{(a,b) \in E'} - \log p(a, b),$$

which can be found applying any MST algorithm.

[4.1] The following figure gives a counter-example. There are two paths from s to t in the graph G: the upper path has weight 1, and the lower path weight 2. The lowest weight is $m = -2$, so adding $-m = 2$ to the weight of each edge yields the graph G' on the right side. It is now the lower path that has the smaller weight. The shortest paths are emphasized. The reason for the failure of this procedure is that not all the paths consist of the same number of edges, so adding a constant weight to each edge does not have the same effect on all the paths. For instance, the total weight of the upper path was increased by 6, and that of the lower path by only 4.

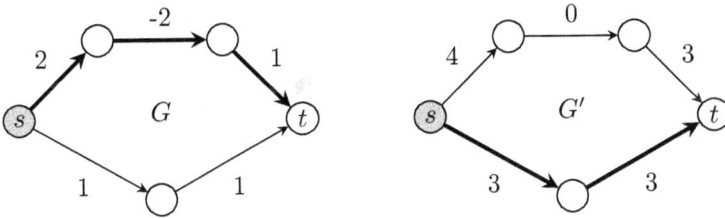

Figure Ex-4.1: *Counter-example for Exercise 4.1.*

[4.2] (a) Set the weight of all edges touching vertices A, B or C to infinity.

(b) Let ε be the smallest difference between different weights (if all weights are integers, then $\varepsilon = 1$), and decrease the weights of all the edges corresponding to airlines X or Y by $\frac{\varepsilon}{n}$, where n is the number of vertices. There are at most $n - 1$ edges in a path, so the choice of the shortest paths is not affected by this amendment, but it gives priority to the edges with reduced weights.

(c) Give the same weight to all edges.

(d) There is an edge (fr_a, to_a) with weight 0 for every flight a, where fr_a and to_a are vertices corresponding to airports. For each airport, there is an edge (to_a, fr_b) from every incoming flight a to every outgoing flight b. If a and b are operated by the same airline, the weight of the edge is 0, otherwise 1.

(e) Use the same graph as in point (d), but change the weight of the edge (fr_a, to_a) to the length of the flight a, and the weight of (to_a, fr_b) to the waiting time at the airport for the connection between a and b.

[4.3] A similar solution as for Exercise 3.5. Let $\mathcal{P}$ be the set of all the paths from s to t. We wish to maximize the probability of getting from s to t

Basic Concepts in Algorithms

without corruption, that is, find

$$\max \left\{ \prod_{(a,b)\in Q} p(a,b) \ \Big| \ Q \in \mathcal{P} \right\}.$$

Change the weight of the edge (a, b) to $-\log p(a, b)$. The above maximization is then equivalent to

$$\min \left\{ \sum_{(a,b)\in Q} -\log p(a,b) \ \Big| \ Q \in \mathcal{P} \right\},$$

which can be found by any shortest path algorithm, for example Dijkstra.

[**4.4**] Define a weighted directed graph $G = (V, E)$. The vertices $V = \{1, 2, \ldots, n, n+1\}$ correspond, in sequence, to the characters of the string, plus an additional vertex numbered $n + 1$. The edges correspond to substrings of the text that are elements of the dictionary D, setting, for all $1 \leq i < j \leq n+1$:

$$(i, j) \in E \qquad \text{if and only if} \qquad x_i x_{i+1} \cdots x_{j-1} \in D.$$

The weight on edge (i, j) is defined as $|\lambda(x_i x_{i+1} \cdots x_{j-1})|$. The sought solution is the weight of the shortest path from vertex 1 to vertex $n + 1$.

[**4.5**] (a) The algorithm would work.
 (b) The algorithm would not necessarily work.

[**5.1**] Because $2^{400} - 1 = (2^{200} - 1)(2^{200} + 1)$, so it is obviously not prime.

[**5.2**] Denote the 3×3 matrix by B, then we have

$$\begin{pmatrix} R(i) \\ R(i-1) \\ R(i-2) \end{pmatrix} = B \begin{pmatrix} R(i-1) \\ R(i-2) \\ R(i-3) \end{pmatrix} = \cdots = B^{i-2} \begin{pmatrix} R(2) \\ R(1) \\ R(0) \end{pmatrix},$$

so all we need to evaluate $R(i)$ is to raise B to the power $i - 2$. The multiplication of two 3×3 matrices takes constant time, so one can evaluate B^{i-2} just as we did for modular exponentiation in $O(\log i)$ steps, using the binary representation of i.

[**5.3**] (a) $\qquad 3^{A_5 - 1} \bmod A_5 = 3,029,026,160,$

$$3^{A_6 - 1} \bmod A_6 = 8,752,249,535,465,629,170.$$

 (b) $O(2^n)$

[5.4] Select randomly 12,000 citizens and interview the oldest among them. The probability of the chosen one not belonging to the set of the 3000 oldest is

$$\left(1 - \frac{1}{3000}\right)^{12000} = \left(1 - \frac{1}{3000}\right)^{3000 \times 4} = e^{-4}.$$

[5.5] It is $\min(\log 2^m, \log n) = b$.

[5.6] (a) Yes.
(b) No error, because in this case $\overline{Y} = Y$.
(c) Same complexity.

[5.7] The matches are 51, 17, 85 and 51, all of which are wrong.

[6.1] There are Huffman trees of depths 3, 4 or 5 for these weights. A Huffman code with maximum codeword length 4 could be $\{00, 01, 11, 101, 1000, 1001\}$.

[6.2] (a) The problem is that if $n \bmod (k-1) \neq 1$, then the number of children of every internal node cannot possibly be k. There must be at least one such node v with less than k children, and if v is not chosen carefully, optimality may be hurt.
(b) To overcome the problem, a single node with less than k children will be chosen, and these children should be on the lowest level of the tree, to minimize the waste. To achieve this, we have to change the first step of the algorithm. Instead of combining k elements, we combine $2 + (n-2) \bmod (k-1)$ elements, that is, between 2 and k. In the following steps, we continue as usual, combining k weights.
(c) A possible 4-ary Huffman code could be $\{00, 01, 02, 03, 1, 2, 30, 31, 320, 321, 33\}$.

[6.4] The code is UD. All the codewords start with 111, and this substring does not appear anywhere else, so by reversing the codewords, one gets a prefix code. Any string can thus be decoded instantaneously by scanning it backwards.

[6.5] (a) The codewords to be added are 001, 0000, 10010, 10011, 00010 and 00011.

(b) If synchronization is regained after it has been lost, there must be two codewords such that one is a proper suffix of the other. An affix code does not have such codewords.

(c) At least one such code exists as shown in (a). Take every codeword α and create two codewords $\alpha 0$ and $\alpha 1$. The new set is also an affix code, with twice as many codewords.

[6.6] There are F_r codewords of length $r+1$, where F_r is the r-th Fibonacci number. The McMillan sum thus becomes $\sum_{i=1}^{\infty} 2^{-\ell_i} = \sum_{i=1}^{\infty} 2^{-(i+1)} F_i$; denote it by S. We have

$$
S = \sum_{i=1}^{\infty} 2^{-(i+1)} F_i = \sum_{i=1}^{\infty} 2^{-(i+1)} (F_{i+1} - F_{i-1})
$$

$$
= 2 \sum_{i=2}^{\infty} 2^{-(i+1)} F_i - \frac{1}{2} \sum_{i=0}^{\infty} 2^{-(i+1)} F_i
$$

$$
= 2 \left(S - \tfrac{1}{4} \right) - \tfrac{1}{2} S = \tfrac{3}{2} S - \tfrac{1}{2},
$$

and thus $S = 1$, so the code is complete.

[6.8] Build a bitmap of size 2^{k-d}, setting the i-th bit to 1 if and only if at least one of the n elements is in the range $[2^{k-d}i, 2^{k-d}(i+1))$. It then suffices to store only the d rightmost bits of each number, plus a flag-bit indicating the last element in each range. Total size: $2^{k-d} + (d+1)n$, which can be smaller than kn.

[6.9] $a^{n/2} b^{n/2}$.

[6.10] Choosing $k = 0$ yields the Elias γ-code. If most of the runs of 0s to be encoded are long, it would be wasteful to use a low value of k.

[6.11] The string $x_1 x_2 \cdots x_{n-1} x_n$ must be such that

$$
x_2 \le x_3 \le \cdots \le x_{n-1} \le x_n \le x_1.
$$

[6.12] The strings are:
-lessessss'--ndldnnv-$-ogomaooitlllngtee-uoa-
-nnsn$-isssswBBg-eeeed-eeee
stmmasss-ivvvvvttttittEc-tlnsxnnnnnndcuuomaaaaa-aeaaeaiiiiiisaatt-$---i

[6.13] All the probabilities have to be dyadic, that is, of the form 2^{-k} with $k \geq 1$, and the order of the intervals must be the same as the order of the leaves of the Huffman tree.

[6.14] Apply Huffman's algorithm: at each stage, merge the two shortest sequences.

[7.1]

$S[j]$	N	A	N	A	H	N	A	H	M	A	N
$\Delta_2[j]$	21	20	19	18	17	16	15	14	10	6	1

$S[j]$	H	E	E	A	D	E	E	A	C	H	E	E
$\Delta_2[j]$	20	19	18	17	16	15	14	13	12	7	2	2

[7.2] Any string of 7 identical characters.

[7.3] $T = \text{CCCCCCBBA}$, $S = \text{CBACBA}$.

[7.5] $\Omega(n^2)$.

[7.6] (a) The string corresponds to an internal node with deepest level in the position tree.

(b) Find an internal node v on level ℓ of the position tree, such that there are two leaves in the subtree rooted by v with labels a and b, for which $|a - b| \geq \ell$; choose a vertex with maximal such ℓ.

(c) Define $T = S$ '#' S', that is, concatenate the strings, adding a separator # between them, and build the position tree for T. Find an internal node v on level ℓ of the position tree for T, such that there are two leaves in the subtree rooted by v with labels a and b, for which $a \leq n$ and $b \geq n + 2$; choose a vertex with maximal such ℓ.

[8.1] The problem is that the square root operator hides illegal operations. It is true that -1 can be written as both $\frac{1}{-1}$ and $\frac{-1}{1}$, but this does not extend to i, because $\frac{1}{i} = \frac{i}{-1} = -i$.

[8.2] $\left(\frac{1}{\omega}\right)^n = \frac{1}{\omega^n} = 1$. On the other hand, for $j < n$, $\left(\frac{1}{\omega}\right)^j = \frac{1}{\omega^j} \neq 1$, because ω is primitive.

$\left(\frac{1}{\omega^2}\right)^{n/2} = \frac{1}{\omega^n} = 1$. On the other hand, for $j < \frac{n}{2}$, $\left(\frac{1}{\omega^2}\right)^j = \frac{1}{\omega^{2j}} \neq 1$, because ω is primitive.

$-\omega$ is not necessarily primitive. For example, -1 is a primitive 2nd root, but $-(-1) = 1$ is not.

[8.3] We have to use $n = 8$ and as the 8th primitive root $\sqrt{i} = \frac{\sqrt{2}}{2} + \frac{\sqrt{2}}{2}i$. The Fourier transform is then a vector of 8 coordinates:

$$\left(15, \quad (-4-\sqrt{2}) + (3+3\sqrt{2})i, \quad 3-2i, \quad (-4+\sqrt{2}) + (-3+3\sqrt{2})i, \right.$$
$$\left. 3, \quad (-4+\sqrt{2}) + (3-3\sqrt{2})i, \quad 3+2i, \quad (-4-\sqrt{2}) + (-3-3\sqrt{2})i \right)$$

[8.4] $4x^3 - 2x + 1$

[8.5] (1) $P(x) = P_0(x^4) + xP_1(x^4) + x^2 P_2(x^4) + x^3 P_1(x^4)$
(2) $P(-x) = P_0(x^4) - xP_1(x^4) + x^2 P_2(x^4) - x^3 P_1(x^4)$
Define $PE(y) = P_0(y^2) + yP_2(y^2)$ and $PO(y) = P_1(y^2) + yP_3(y^2)$, then $P(x) = PE(x^2) + xPO(x^2)$ and $P(-x) = PE(x^2) - xPO(x^2)$.
(3) and (4) Same points, same complexity.

[9.1] If n is prime itself, then $\varphi(n) = n - 1$ and will be known to the decoder, since n is public. Everybody can then derive the inverse $d = e^{-1} \bmod (n-1)$ and thus decode any message.

If n is a product of three primes, at least one of them is at most $\sqrt[3]{n}$. For a given size of n, the system will thus be more vulnerable to a brute-force attack of trying to factorize n. For example, if $n \approx 2^{120}$, at most 2^{40} attempts are necessary, which might be feasible.

[9.2] $\quad d = 16517 \qquad \mathcal{E}(84079) = 76562742$

[9.3] The question is whether $\mathcal{D}_A$ and $\mathcal{E}_B$ may be commuted. This will not always be the case, but it is for RSA, for which encoding and decoding, for both parties, are just raising a number to the powers e_A, d_A, e_B and d_B. So if RSA is used, Béline does not need to be informed, as it does not matter in which order she applies $\mathcal{D}_B$ and $\mathcal{E}_A$.

[9.4] Suppose Argan uses the same key K to encode a sequence $M_1, M_2, \dots$ of messages as $M_i' = M_i$ XOR K. It then suffices for the eavesdropper Cléante to get somehow hold of a single one of these messages, say M_j, to break the system. Indeed, he could then derive the secret key K by

$$M_j \text{ XOR } M_j' = K,$$

and once in possession of the key, Cléante would have access to every encoded message.

[10.1] By restriction. Choose $k = 2$ and $D = \left(\sum_{i=1}^{n} \ell(a_i) \right)/2$. Then we have an instance of PAR.

[10.2] By restriction. Let G_2 be a clique of size L. Then we have an instance of the clique problem CL, because all cliques of the same size are isomorphic.

[10.3] Given is a general instance HAM, which is a graph $G = (V, E)$, we construct a special instance of TSP, that is, a directed graph $G' = (V', E')$ with weight function w and a bound B. Let G' be the full graph, so $E' = V' \times V'$ and define $B = |V| = n$ and the weights as

$$w(x, y) = \begin{cases} 1 & \text{if } (x, y) \in E \\ 2 & \text{if } (x, y) \notin E. \end{cases}$$

If there is a Hamiltonian circuit in G, the same path exists in G' and its total weight is n. Conversely, assume there is a circuit T of the TSP in G' the total weight of which is bounded by $B = n$. Since there are exactly n edges in T, and all the weights are either 1 or 2, it follows that all the weights in the circuit have to be 1. All the edges of T are therefore also edges in the original graph G, so G has a Hamiltonian circuit.

[10.4] (a) It is true that the definition of a matching requires the number of triples to be n, but in the proof, we start with a partition of the set of numbers A, and the size of each of the subsets is not known. The example in Table 10.2 shows that one of the subsets forming the partition is of size 6, whereas $n = 5$.

(b) One can use the example in Table 10.2, which yields $A = \{8456, 4228, 1057, 16417, 1537, 1072, 32767, 65534\}$, which can be partitioned as the last element is the sum of the 7 others, but the corresponding collection of triples is shown in the left column of Table 10.2, and does not include a matching.

[10.5] Denote the two colors by 0 and 1. Pick a node and color it with 0. Then color its neighbors with 1, continue coloring the neighbor's neighbors with color 0, etc. If during this process, one tries to color a vertex with the color i, $i \in \{0, 1\}$, but it is already colored with the opposite color $1 - i$, then a cycle of odd length has been found and the graph is not 2-colorable; stop the process and return fail. Repeat for every connected component. When all vertices are colored, return success.

[10.6] The graph $\mathcal{T}'$ of Figure Ex–10.6 would suffice to prove that if G is 3-colorable, then so is G'.
A graph G'_u consisting of $d-2$ copies of $\mathcal{T}'$ is 3-
colorable as shown (it is even 2-colorable), with all
its extremities having the same color. The converse
is not true. If G'_u is 3-colorable, this does not imply
that all its extremities must share the same color.

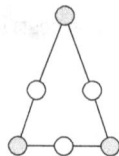

Figure Ex–10.6:
*Example of simpler
graph for $\mathcal{T}$.*

[10.7] We show that G3C $\propto$ G5C. Given is a general instance of G3C, a
graph $G = (V, E)$, we build a special instance $G' = (V', E')$ of G5C by
defining

$$V' = V \cup \{a, b\} \quad E' = E \cup \{(a,b)\} \cup \{(x,a), (x,b) \mid x \in V\},$$

that is, two new vertices are adjoined; they are connected by an edge, and
also to each of the other vertices.

If G is 3-colorable by $\{0, 1, 2\}$, then assigning the colors 3 to a and 4 to
b yields a 5-coloring of G'. Conversely, if there is a 5-coloring of G', a and
b must have different colors, and these colors must also be different from
those of any other vertex. It follows that the vertices of G are 3-colorable.

Build-Partition$(A = (a_1, \ldots, a_n))$

   if Dpar(A) is true then
      $A' \leftarrow \{a_1\}$ $K \leftarrow \sum_{i=1}^{n} a_i$ $a_1 \leftarrow a_1 + K$
      for $j \leftarrow 2$ to n do
         $a_j \leftarrow a_j + K$
         if Dpar(A) is not true then $A' \leftarrow A' \cup \{a_j\}$
         $a_j \leftarrow a_j - K$
      $a_1 \leftarrow a_1 - K$
      return $(A', A \setminus A')$

ALGORITHM Ex–10.8: *Finding a partition of a set of integers.*

[10.8] After having checked that there is a partition of the given set A,
define K as a large number and repeatedly add K to two of the elements
of A. For example, K could be the sum of all the elements of A. If after
adding K to a_i and a_j, there is still a partition, then a_i and a_j cannot

belong to the same subset of this partition, and if no partition exists, then a_i and a_j must be in the same subset; label the vertices accordingly. The formal algorithm appears in ALGORITHM Ex–10.8.

[10.9] (a) The graph in Figure Ex–10.9 is connected, but not Hamiltonian, as one cannot visit all the vertices without passing several times through the gray one in the center.

(b) Show TSP $\propto$ ETSP. The constructed graph G' will be the same as G, but with a different weight function w'. Let K be a large number, for example $K = B + 1$. Define the new bound $B' = B + nK$ and the weights as

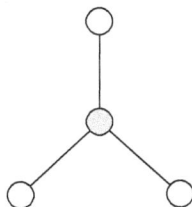

Figure Ex–10.9: *Example of graph for* ETSP.

$$w'(x, y) = w(x, y) + K \qquad \text{for } (x, y) \in E.$$

If there is a circuit in G whose cost is bounded by B, the same circuit in G' has cost $B + nK = B'$. Conversely, if there is a path with cost bounded by B' and visiting all the vertices, it contains at most n edges. For if the path has at least $n + 1$ edges, its weight must be at least

$$(n + 1)K = nK + K > nK + B = B'.$$

Therefore the path is a Hamiltonian circuit, and corresponds to a TSP circuit in G with cost at most B.

[10.10] After having checked that there is a tour for the TSP the cost of which is bounded by B, define K as a large number, for example K could be $B + 1$. Now repeatedly increase the weight of one of the edges e by K. If the answer is still **yes**, the edge e is not necessary to build the Hamiltonian circle, since its weight alone already exceeds the permitted bound; if the answer is **no**, the edge e is necessary and included in a set T. The formal algorithm appears in ALGORITHM Ex–10.10.

[10.11] We show that TSP $\propto$ PTSP. Given is a directed graph $G = (V, E)$ and a bound B.

To build $G' = (V', E')$, pick one of the vertices v and replace it by two vertices $\{v_1, v_2\}$. Copy all outgoing edges from v in G to become outgoing edges from v_2 in G', and all incoming edges to v in G to become incoming edges to v_1 in G', preserving the weights, as shown in the example in Figure Ex-10.11. Add incoming edges to v_2 and outgoing edges from v_1,

Build-TSP-Tour(G, B)

   if Dtsp(G, B) is true then
      $T \leftarrow \emptyset$ $K \leftarrow B + 1$
      for $e \in E$ do
         $w(e) \leftarrow w(e) + K$
         if Dtsp(G, B) is not true then
            $T \leftarrow T \cup \{e\}$
            $w(e) \leftarrow w(e) - K$
      return T

ALGORITHM EX–10.10: *Finding a* TSP *circuit.*

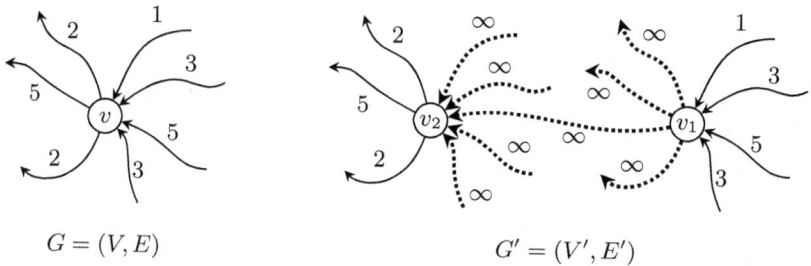

Figure Ex–10.11: *Example of a graph for* PTSP.

(x, v_2) and (v_1, x), for all $x \in V'$, which includes also the edge (v_1, v_2), and assign to all these edges a large weight, e.g., $B + 1$. These edges are shown as broken lines in the figure. Use the same bound $B' = B$.

If there is a tour of the TSP in G, then there is a Hamiltonian path of the same total weight starting at v_2 and ending at v_1 in G'. Conversely, if there is Hamiltonian path whose cost is bounded by B' in G', it must start at v_2, because v_2 has no incoming edges with lower weight than $B + 1$. A similar argument shows that the path must end at v_1. This path induces in G a circuit, with a cost bounded by B.

[11.1] For the search problem, one can partition the array into $\sqrt{n}$ sub-parts of size $\sqrt{n}$, compare X to the last elements of the sub-parts and continue sequentially just on the sub-part containing X. The search thus takes only $O(\sqrt{n})$. A further improvement partitions the array into $\sqrt[3]{n}$

sub-parts of size $\sqrt[3]{n^2}$, yielding a total search time of $O(n^{1/3})$. Ultimately, we end up with binary search.

For the sorting problem, organize the set A_1 of the n data elements into $\sqrt{n}$ subsets of $\sqrt{n}$ each, find and remove a maximal element of each of these subsets and define the set A_2 of these $\sqrt{n}$ maxima. While it takes $\theta(n)$ to locate the maximal element of the entire set, the second and subsequent elements can be found by retrieving the largest element only from the subset of A_1 from which the last element originated, and then again from the set A_2. This yields a total of $2n\sqrt{n}$. As above, the general case works in time $\theta(n^{1+1/k})$ and ultimately $\theta(n \log n)$. Details can be found in [Klein (2013)].

[**11.2**] (a) No. The number of added edges does not matter since anyway, the final tour obtained by the shortcuts is Hamiltonian and will contain exactly V edges, for both heuristics.

(b) An α-approximation heuristic is preferable to a β-approximation heuristic in the worst case only. There can well be instances for which the second heuristic is preferable to the first, as we have seen for 2EP and HDV, approximating the vertex cover optimization problem.

[**11.3**]

$$
\mathcal{K}[i,j] = \begin{cases} 0 & \text{if } i = 0 \lor j = 0 \\ \mathcal{K}[i-1,j] & \text{if } i,j > 0 \ \land \ j < s(a_i) \\ \max\left(\mathcal{K}[i-1,j], \right. & \\ \left. v(a_i) + \mathcal{K}[i-1,j-s(a_i)]\right) & \text{if } i,j > 0 \ \land \ j \geq s(a_i). \end{cases}
$$

The solution to our optimization problem is the value $\mathcal{K}[n,K]$. The complexity of building the table is $O(nK)$, which is not polynomial in the size n and $\log K$ of the input.

[**11.4**] (a) Any instance for which all elements have the same size, e.g., $s(a_i) = 1$ for all i.

(b) and (c) $A = \{a_1, a_2\}$, $v(a_1) = 1$, $s(a_1) = 1$, $v(a_2) = m - 1$, $s(a_2) = m$, capacity $K = m$. The ratios are then 1 for a_1 and $\frac{m-1}{m}$ for a_2. DR will put a_1 into the knapsack, and then there is no place to fit in a_2. The optimal solution would be to put only a_2 into the knapsack. The ratio DR$(I)/$OPT(I) for this instance is $m - 1$, which can be chosen larger than any α.

[11.5] (a) Yes.

(b) $S = 6$, $s_1, \ldots, s_4 = 1, 2, 4, 5$. NDS puts the first two elements in the first bin and the third element alone in the second bin. OPT can accommodate all four elements, as $s_1 + s_4 = s_2 + s_3 = S$.

(c) For $K = 2$, consider a single bin of size $2S$ and let r be the (optimal, since $K = 1$) number of elements that fit into the large bin. If there exists an index h such that $\sum_{i=1}^{h} s_i = S$, then this solution (h elements in the first bin, $r - h$ in the second) is also optimal for two bins. Otherwise, define h as the smallest index for which $\sum_{i=1}^{h} s_i > S$. NDS could then put elements 1 to $h-1$ into the first bin, and elements $h+1$ to r into the second. Actually, the second bin will contain elements h to $r - 1$, since $s_h \leq s_r$. In any case, at most one element is lost by passing from one bin of size $2S$ to two bins of size S each. The proof for general K is similar.

References

Appel, K., and Haken, W. (1977). Solution of the four color map problem, *Scientific American* **237**, 4, pp. 108–121, doi:10.1038/scientificamerican1077-108.

Aronovich, L., Asher, R., Bachmat, E., Bitner, H., Hirsch, M., and Klein, S. T. (2009). The design of a similarity based deduplication system, in *Proceedings of SYSTOR Conference, Haifa, Israel*, ACM International Conference Proceeding Series (ACM), pp. 1–14.

Bellman, R. (1958). On a routing problem, *Quart. App. Math.* **16**, pp. 87–90.

Borůvka, O. (1926). O jistém problem minimálním (about a certain minimum problem), *Práce morav. přírodověd. spol. v Brně (in Czech; summary in German)* **III**, 3, pp. 37–58.

Boyer, R. S., and Moore, J. S. (1977). A fast string searching algorithm, *Commun. ACM* **20**, 10, pp. 762–772.

Burrows, M., and Wheeler, D. J. (1994). A block-sorting lossless data compression algorithm, Tech. Rep. 124, Digital Equipment Corporation.

Christofides, N. (1976). Worst-case analysis of a new heuristic for the travelling salesman problem, Tech. Rep. 388, Graduate School of Industrial Administration, Carnegie-Mellon University.

Cook, S. A. (1971). The complexity of theorem-proving procedures, in *Proceedings of the 3rd Annual ACM Symposium on Theory of Computing, May 3–5, 1971*, pp. 151–158, doi:10.1145/800157.805047.

Diffie, W., and Hellman, M. E. (1976). New directions in cryptography, *IEEE Trans. Inf. Theory* **22**, 6, pp. 644–654.

Dijkstra, E. W. (1959). A note on two problems in connexion with graphs, *Numer. Math.* **1**, pp. 269–271.

Elias, P. (1975). Universal codeword sets and representations of the integers, *IEEE Trans. Inf. Theory* **21**, 2, pp. 194–203.

Floyd, R. W. (1962). Algorithm 97: Shortest path, *Communications of the ACM* **5**, 6, p. 345.

Ford, L. R. (1956). Paper p–923, *Network Flow Theory*, RAND Corporation, Santa Monica, California.

Fraenkel, A. S., and Klein, S. T. (1996). Robust universal complete codes for transmission and compression, *Discrete Appl. Math.* **64**, 1, pp. 31–55.

Garey, M. R., and Johnson, D. S. (1979). *Computers and Intractability: A Guide to the Theory of NP-Completeness* (W. H. Freeman).

Huffman, D. A. (1952). A method for the construction of minimum-redundancy codes, *Proceedings of the IRE* **40**, 9, pp. 1098–1101.

Karp, R. M. (1972). Reducibility among combinatorial problems, in R. E. Miller and J. W. Thatcher (eds.), *Proceedings Symposium on the Complexity of Computer Computations*, The IBM Research Symposia Series (Plenum Press, New York), pp. 85–103, doi:10.1007/978-1-4684-2001-2_9.

Karp, R. M., and Rabin, M. O. (1987). Efficient randomized pattern-matching algorithms, *IBM J. Res. Dev.* **31**, 2, pp. 249–260.

Klein, S. T. (2013). On the connection between Hamming codes, heapsort and other methods, *Inf. Process. Lett.* **113**, 17, pp. 617–620.

Klein, S. T. (2016). *Basic Concepts in Data Structures* (Cambridge University Press).

Klein, S. T., and Ben-Nissan, M. K. (2009). Accelerating Boyer-Moore searches on binary texts, *Theor. Comput. Sci.* **410**, 37, pp. 3563–3571.

Knuth, D. E., Morris, J. H., and Pratt, V. R. (1977). Fast pattern matching in strings, *SIAM J. Comput.* **6**, 2, pp. 323–350.

Kruskal, J. B. (1956). On the shortest spanning subtree of a graph and the traveling salesman problem, *Proceedings of the American Mathematical Society* **7**, 1, pp. 48–50.

Manber, U., and Myers, E. W. (1993). Suffix arrays: A new method for on-line string searches, *SIAM J. Comput.* **22**, 5, pp. 935–948.

McMillan, B. (1956). Two inequalities implied by unique decipherability, *IEEE Trans. Inf. Theory* **2**, 4, pp. 115–116.

Prim, R. C. (1957). Shortest connection networks and some generalizations, *Bell Syst. Tech. J.* **36**, 6, pp. 1389–1401.

Rabin, M. O. (1980). Probabilistic algorithm for testing primality, *J. Number Theory* **12**, 1, pp. 128–138.

Rivest, R. L., Shamir, A., and Adleman, L. M. (1978). A method for obtaining digital signatures and public-key cryptosystems, *Commun. ACM* **21**, 2, pp. 120–126.

Schroeppel, R., and Shamir, A. (1981). A $T = O(2^{n/2})$, $S = O(2^{n/4})$ algorithm for certain NP-complete problems, *SIAM J. Comput.* **10**, 3, pp. 456–464.

Shannon, C. E. (1948). A mathematical theory of communication, *Bell Syst. Tech. J.* **27**, 3, 4, pp. 379–423, 623–656.

Strassen, V. (1969). Gaussian elimination is not optimal, *Numer. Math.* **13**, pp. 354–356.

Teuhola, J. (1978). A compression method for clustered bit-vectors, *Inf. Process. Lett.* **7**, 6, pp. 308–311.

Ukkonen, E. (1995). On-line construction of suffix trees, *Algorithmica* **14**, 3, pp. 249–260.

Wedekind, H., and Härder, T. (1976). *Datenbanksysteme II, Reihe Informatik*, Vol. 18 (Bibliographisches Institut, Wissenschaftsverlag, Mannheim).

Weiner, P. (1973). Linear pattern matching algorithms, in *14th Annual Symposium on Switching and Automata Theory*, pp. 1–11.

Welch, T. A. (1984). A technique for high-performance data compression, *IEEE Computer* **17**, 6, pp. 8–19.

Williams, R. N. (1991). An extremely fast Ziv-Lempel data compression algorithm, in *Proceedings of the IEEE Data Compression Conference, DCC 1991, Snowbird, Utah, USA, April 8–11, 1991*, pp. 362–371.

Witten, I. H., Neal, R. M., and Cleary, J. G. (1987). Arithmetic coding for data compression, *Commun. ACM* **30**, 6, pp. 520–540.

Yao, A. C. (1975). An $O(|E| \log \log |V|)$ algorithm for finding minimum spanning trees, *Inf. Process. Lett.* **4**, 1, pp. 21–23.

Ziv, J., and Lempel, A. (1977). A universal algorithm for sequential data compression, *IEEE Trans. Inf. Theory* **23**, 3, pp. 337–343.

Ziv, J., and Lempel, A. (1978). Compression of individual sequences via variable-rate coding, *IEEE Trans. Inf. Theory* **24**, 5, pp. 530–536.

Index

www.ingramcontent.com/pod-product-compliance
Lightning Source LLC
Chambersburg PA
CBHW061620220326
41598CB00026BA/3822